A Guide to Close Binary Systems

Series in Astronomy and Astrophysics

The *Series in Astronomy and Astrophysics* includes books on all aspects of theoretical and experimental astronomy and astrophysics. Books in the series range in level from textbooks and handbooks to more advanced expositions of current research.

Series Editors:
M Birkinshaw, University of Bristol, UK
J Silk, University of Oxford, UK
G Fuller, University of Manchester, UK

Recent books in the series

Dark Sky, Dark Matter
J M Overduin and P S Wesson

Dust in the Galactic Environment, 2nd Edition
D C B Whittet

The Physics of Interstellar Dust
E Krügel

Very High Energy Gamma-Ray Astronomy
T C Weekes

Numerical Methods in Astrophysics: An Introduction
P Bodenheimer, G P Laughlin, M Różyczka, H W Yorke

An Introduction to the Physics of Interstellar Dust
Endrik Krugel

Astrobiology: An Introduction
Alan Longstaff

Fundamentals of Radio Astronomy: Observational Methods
Jonathan M Marr, Ronald L Snell, and Stanley E Kurtz

Stellar Explosions: Hydrodynamics and Nucleosynthesis
Jordi José

Cosmology for Physicists
David Lyth

Cosmology
Nicola Vittorio

Cosmology and the Early Universe
Pasquale Di Bari

Fundamentals of Radio Astronomy: Astrophysics
Ronald L. Snell, Stanley E. Kurtz, and Jonathan M. Marr

Introduction to Cosmic Inflation and Dark Energy
Konstantinos Dimopoulos

Physical Principles of Astronomical Instrumentation
Matthew Griffin, Peter A. R. Ade, Carole Tucker

A Guide to Close Binary Systems
Edwin Budding and Osman Demircan

A Guide to Close Binary Systems

Edwin Budding and Osman Demircan

CRC Press
Taylor & Francis Group
Boca Raton London New York

CRC Press is an imprint of the
Taylor & Francis Group, an **informa** business

First edition published 2022
by CRC Press
6000 Broken Sound Parkway NW, Suite 300, Boca Raton, FL 33487-2742

and by CRC Press
4 Park Square, Milton Park, Abingdon, Oxon, OX14 4RN

CRC Press is an imprint of Taylor & Francis Group, LLC

Library of Congress Cataloging-in-Publication Data

Names: Budding, E., 1943- author. | Demircan, O. (Osman), author.
Title: A guide to close binary systems / Edwin Budding, Osman Demircan.
Description: First edition. | Boca Raton : CRC Press, 2022. | Series:
Series in astronomy and astrophysics | Includes bibliographical
references and index.
Identifiers: LCCN 2021049211 | ISBN 9781138064386 (hardback) | ISBN
9781032226866 (paperback) | ISBN 9781315160429 (ebook)
Subjects: LCSH: Double stars.
Classification: LCC QB821 .B83 2022 | DDC 523.8/41--dc23/eng/20211216
LC record available at https://lccn.loc.gov/2021049211

ISBN: 978-1-138-06438-6 (hbk)
ISBN: 978-1-032-22686-6 (pbk)
ISBN: 978-1-315-16042-9 (ebk)

DOI: 10.1201/b22228

Typeset in CMR10
by KnowledgeWorks Global Ltd.

Publisher's note: This book has been prepared from camera-ready copy provided by the authors.

in Memory of Zdeněk Kopal

Contents

Foreword

It is over forty years since Miroslav Plavec, in his final remarks on IAU Symposium 88: *Close Binary Systems: Observations and Interpretation*, suggested this could be the last such conference having a programme stretching across the full field. That was already twenty years after Kopal's influential *Close Binary Systems* (*CBS*) of 1959, that purported to summarize core knowledge needed by students of this greatly developing subject to go on and produce independent research. In 1979, the confident discovery of exoplanets was still years away. Space age astronomy was yet in its infancy. Many techniques that have become routine were still being pioneered, some hardly dreamt of.

But were there not links between the *CBS* of 1959 and the wide-ranging presentations of that Toronto meeting twenty years later? In discussions of binary evolution, the fundamental role of Roche geometry was widely recognized, as was the classification of regular close binaries into detached, semi-detached and contact modes; ideas that were new in *CBS*. These binary categories connected with the significant information content of photometric light curves, the analysis of which *CBS* gave particular attention to. Clear reminders of that background were still there in the international meeting at Çanakkale, Turkey, in 2002, and its sequel, the 2014 conference in Litomysl, Czech Republic, celebrating the 100 years since Kopal's birth.* The lively discussions in those eclectic gatherings may well have sparked the concept of a sequel to *CBS*.

The stated aim of building a framework of understanding for students involves some explanation of where the procedures came from: an exposition of their connection to preceding evidence and theory. Chapter 6 in *CBS*, as well as its predecessor – the smaller *Introduction* of 1947 – clearly show the author's links to Harvard College Observatory, the pioneering role of Henry Norris Russell and the *Royal Road* that Russell, the Shapleys, their colleagues, contemporaries and students, explored in order to get a clearer view of the properties of stars. The point gains force with Kopal's note in HOC No. 451, (1947), regarding the propitious role that the *Circulars* would play in supporting the activities of the then-new IAU Commission 42 on close binary stars.

CBS, having a broad perspective for its time, could not cover everything. Naturally, a finite work implies a selection that derives from the proponent's own limited experience. Different experience will produce different emphases. Indeed sometimes, with appropriate apology, there are different viewpoints, or, more regrettably, viewpoints that are completely missed, or misunderstood. Although DQ Her had been observed to produce a nova explosion some 25 years before *CBS*, and the remnant system then modelled as an unusually close binary showing also short period (71-sec) oscillations, the subject of collapsed stars in binary systems, that figures so prominently in contemporary research, was barely touched on in *CBS*. No doubt, parallels will be found with our present endeavour. But by tracing back through the expansion of knowledge that came from the various contributions into a field, a fuller overview becomes possible.

The present work, by two former students of Zdeněk Kopal, reflects that general plan of *CBS*; its four extra chapters recognizing, to some extent, that vast development referred

---

* *New Directions for Close Binary Studies: The Royal Road to the Stars* was held at Çanakkale Onsekiz Mart University, June 24-28, 2002, ÇOMÜ Publ. Vol. 3, ISBN 975-8100-33-5, eds. O. Demircan & E. Budding. *Living Together: Planets, Host Stars and Binaries*, appeared as *Astron. Soc. Pacific Conf. Ser.*, Vol. 496, 2015, eds. S. M. Rucinski, G. Torres, & M. Zejda.

to above. Much of the text is, however, still rooted in those classical topics that featured in *CBS* and interested so many of its readers. About half of *CBS* concerned light curves: either in the theoretical exposition of their forms, or with practical methods of finding their parameters together with corresponding physical interpretation. This subject is summarized in Chapter 7 of the present book: still relatively large, though now in a more proportionate ~20% of the whole.

Part of the rationale behind reviewing such topics as modelling the light curves of close binary systems is the essential similarity of this subject to the interpretation of photometric data on transiting exoplanets. The one subject forms a continuum with the other and gives enhanced significance to topics like the nature and scale of tidal interactions between primary star and companion. Indeed, originally, there was a suggestion that this book might include a part concentrating on such fields of exoplanet research. The still growing subject areas within the range referred to by Mirek Plavec gave enough practical reasons to limit our present scope, however.

It is a pleasure to recall the helpful remarks and encouragements from friends and colleagues as this endeavour progressed, but difficult to single out names on account of knowing where to stop. The initial decision to write was made while the authors were working together in the Troad, whose invigorating properties, testified to by the classicists, are widely spoken of. How many of those inspirational philosophers of antiquity must have paused, in passing between Asia and Thrace, to pay their respects at the tomb of Dardanos, today within the precincts of the Çanakkale Onsekiz Mart Üniversitesi (ÇOMÜ)? The Rector and Staff of that university, as well as its students and all those who support it, deserve a special thanks. Zeki Eker was working with the authors at ÇOMÜ not so many years ago and his useful advice and assistance in putting together parts of this book, particularly Chapter 9, are warmly recognized. We also especially thank Ahmet Erdem, our genial and helpful host in the Physics Department at ÇOMÜ in more recent years as well as his colleagues Caner Çiçek, and Faruk and Esin Soydugan, the benevolent influence of whose former supervisor, Cafer Ibanoglu, is likewise recognized. In the western hemisphere, thanks are also due to Murray Alexander and Tim Banks, who kindly read earlier versions of parts of this book and offered useful suggestions for improvement. The ongoing encouragement and assistance of the publishing staff, particularly Kirsten Barr and Shashi Kumar, deserves mention. Finally, the steadfast support of close family and friends throughout our endeavour is appreciated much more than words alone can say.

1

Introduction

1.1 General Historical Background

The prevailing idea about stars in classical times probably derived from their apparent constancy. Apart from the seeming steady rotation about the observer, their relative positions and brightness looked permanent and invariable. They could be regarded as fixed forever into a clear 'firmament'; some rigid celestial arrangement to hold them in place. But as to the exact nature of this arrangement, or of the stars themselves, no direct evidence could be found. Speculation might then suggest, as with Aristotle, some highly special nature of the starry material, whose constancy was essentially unknown in the 'corruptible' matter familiar on Earth.

With the Renaissance, the insightful view of Giordano Bruno that stars might be something like the Sun, but at very much greater distances, gradually gathered support.* Newton could then reasonably postulate that the Sun and planets were apparently at the centre of a more or less uniform distribution of remote suns, whose small and separate gravitational attractions would have negligible short-term effect as far as any net action on known planetary motions. Such motions had been accurately summarized by Johannes Kepler more than twenty years before the birth of Isaac Newton, but in comprehensively providing the explanation of these motions Newton could have deduced that essentially similar behaviour would be exhibited by a pair of stars sufficiently close to each other.

A few years before Newton died in 1727, John Michell was born, son of a Nottinghamshire clergyman. In due course, Michell became an Anglican priest himself, taking on the rectorship of St Michael's and All Angels parish in Thornhill, near Wakefield. This was in the ancestral lands of the influential Savile family of West Yorkshire, and not very far from his place of origin on the same Savilian estate. But in his prior period of study, particularly during the twenty one years at Queen's College Cambridge before his marriage and incumbency at St Michael's, Michell was making very significant progress with basic scientific problems, including the nature of stars.

In a remarkable paper presented to the Royal Society of London in 1767, Michell, already a Fellow for six years, raised far-reaching questions on the distances of stars and their arrangement in the sky. Adopting first the working hypothesis of remote suns, he could

---

*This view is echoed, for example, in J. H. Lambert's essay in *Cosmologische Briefe über die Einrichtung des Weltbaues*, Augsberg, 1761.

DOI: 10.1201/b22228-1

1

Figure 1.1 View of the Praesepe cluster. Note the brightest stars of the cluster are frequently found with close companion stars: an appearance that caused Michell to ponder on mutual gravitational attraction between close pairs of stars. (With permission from S. Reilly, Dogwood Ridge Observatory, Va.)

show, from their apparent brightness in comparison to planets of the Solar System, that even the nearest of the stars must have a parallax less than could be measured with any optical equipment then available, still less the unaided eye. In this way, the long-standing argument for a geocentric Universe regarding the absence of any detectable parallactic effect for stars, as presented by Ptolemy at the beginning of the Almagest, was effectively countered.

Considering the distribution of angular separations between stars, Michell established, from statistical reasoning, that this was far from random. Ptolemy had indeed noted certain stars to be 'double' or have 'nebular' appearance, such as ζ Sco, and $\lambda + \upsilon$ Sco, or $\nu_1 + \nu_2$ Sgr. While he also called attention to the Pleiades and Hyades in the stellar catalogue in Book 7 of the Almagest, Ptolemy did not discuss any physical significance to be attached to stellar grouping.* Michell, however, concluded with the highest probability (the odds against being many million millions to one) that at least some stars really are collected together in self-attracting systems. The natural inference regarding very close double stars is that, in general, they are actually situated in relative proximity to each other.

Michell's paper of 1767 makes a cogent scientific case for the construction of telescopes of larger aperture. Based on his statistical analysis of the distribution of stellar magnitudes and inferred separations, Michell felt reason to believe that the Sun was within a large-scale self-gravitating system, perhaps similar to the clear groupings of the Pleiades and Hyades, but containing of the order of 1000 stars. Noting that Robert Hooke had counted an additional 78 visible stars in the Pleiades with his twelve foot telescope of 2 inch aperture, Michell argued that with a two foot mirror a number of order 1000 could become visible to confirm (or deny) his hypothesis. In his own words, "Telescopes ... of very large sizes might serve to inform us of many circumstances, both with regard to this (distribution of inherent stellar magnitudes), and ... other things, that would enable us to judge with more probability concerning the distances, magnitudes etc. of the stars

---

*In the early telescopic era Galileo recognized the star Mizar (ζ UMa) to be a close double (a finding often accredited to Riccioli), but without offering further discussion.

of our own system." Michell diverted hundreds of pounds from his own resources to this endeavour.

In the 1760s, the young musical maestro William Herschel, a refugee from disputed lands in Europe, was making a name for himself in West Yorkshire just a few miles from Thornhill. Herschel, then in his mid-twenties, had already published several symphonies through the same company as Handel. He had been recently appointed first organist of Halifax Minster. Michell was himself known as an accomplished violinist, while Herschel had been previously engaged as leader and soloist with the Newcastle orchestra. The Saviles were benevolent patrons and no doubt intermediated in the arrangement of sparkling soirées at the large rectory, the original old hall having been lost in the Civil War.* So was it perhaps the violin that first brought these two great figures of astronomy together? Although this scenario has charm and fits in with a letter published by Michell's great-grandson, based on family recollections, it requires that Michell should have been in Thornhill prior to his appointment as rector, since Herschel accepted a prominent musical position in Bath at the beginning of 1767. That is not improbable for occasional visits, given the long-standing connection with the Saviles. But the letter-writer's grandmother Mary, John Michell's daughter, must have been an infant at the time in question, and it seems unlikely that she would have retained clear, direct and extensive memories of it herself. Written correspondence involving Herschel and Michell seems sparse, but it is known that Michell was engaged in the making of his large reflecting telescope in the mid-1760s. However, it came about that Herschel's preoccupations were diverted from his highly esteemed professional activities in music to astronomy, it is likely that the musician's practical knowledge of instrument-making and soft metal engineering would have been relevant to the task before Michell. It is at least established that Herschel visited Michell at Thornhill not long before the latter's death (1793), and purchased a large telescope from him.

The reappearance of Halley's comet at the end of 1758 had a general stimulatory effect on astronomy and related sciences, with the theoretical investigations of Clairaut, Lalande and Lepaute confirming that comets too were members of the solar system with orbital details that could be computed on the basis of Newtonian physics. These events occurred just in time for the keen young observer Charles Messier to try to find out more about these newly identified members of the solar family. The first of his discoveries, the comet of 1760, was confirmed by Michell. By 1766, Messier (now also a Fellow of the Royal Society of London) and his assistant, Pierre Méchain, had discovered four new comets. Their aim was, however, frustrated by the frequent appearance of other deep-sky objects having a superficial resemblance to comets but of a completely different nature. By 1774, Messier had published a catalogue of 45 such objects, mainly to allow comet hunters to avoid being delayed. At least thirty of the Messier objects were found to be clusters or stellar systems of the type that interested Michell, so the scientific rationale for large telescope construction would have intensified by the late 1760s and 70s.

In that same year of Messier's catalogue, Herschel, having perused suitable texts on optics and astronomy, set to and completed his own reflecting telescope. This instrument was about 7 feet in length and contained a mirror of a little over six inches diameter. The mirror was made of speculum — a shiny alloy of two parts copper to one part tin. Similar alloys of copper are used for making brass musical instruments, cymbals and bells, so Herschel, whose career started in a military band, would have been familiar with relevant technology. In fact, the careful manufacturing of the mirror allowed this telescope to become probably

---

*Sir George Savile, the eighth baronet, MP for Yorkshire and Michell's friend from childhood, as well as being a Fellow of the Royal Society, was an associate of members of Dr Johnson's famous Literary Club.

Figure 1.2   There does not appear to be a surviving, fully-authenticated portrait of Rev. John Michell (1724-1793). However, the memorial plaque on the wall of the parish church of St Michael and All Angels, Thornhill, near Wakefield, Yorkshire, testifies to his scientific activities during his incumbency there.

the most effective one in the world of its time. After a year or two of general familiarization, Herschel, now around 40 and having received further encouragement from Nevil Maskelyne, the Astronomer Royal, embarked on a definite programme of double star observations. This was not necessarily to confirm Michell's probabalistic argument. In fact, an alternative view was that stars in close proximity could be of inherently different brightness and at greatly different distances. In that case, very precise measurements of the relative separations of the two stars could reveal a differential parallactic effect. This might well lead to useful estimates on the distance of at least the nearer star in the pairs. By 1782, Herschel had produced a catalogue of some 269 double or multiple stars, which he presented to the Royal Society of London.

About thirty miles north of Thornhill was another West Yorkshire country estate that would have no doubt been familiar to the Saviles: that of the Goodrickes of Hunsingore, near Wetherby. The grandson of the fifth baronet Sir John Goodricke, also called John, had acquired an active interest in astronomy. On the evening of Nov 12, in the same year as Herschel's first catalogue, young John, then 18 years old, made the startling observation that the bright star β Persei, also known as Algol (the 'demon star'), had suddenly dropped in brightness from second to fourth magnitude. Later in the year and through the following one, Goodricke continued to witness the unexpected changes of light. The star's commonly used, Arabic-derived, name is itself suggestive of something peculiar, given the long-standing idea that the stars are fixed both in their relative positions and brightness. Upon looking further into the background Goodricke found out that variations had been recorded by two separate Italian observers (Montanari and Maraldi) in the seventeenth century. After several weeks of repeated monitoring, Goodricke discovered the key property that the light variations recur regularly, with a period of two days and twenty and three-quarter hours. The periodic variability had, by this time, been independently confirmed by John's friend Edward Pigott. The paper's concluding paragraph is worth recalling:

"If it were not perhaps too early to hazard even a conjecture on the cause of this variation, I should imagine it could hardly be accounted for otherwise than either by the interposition of a large body revolving round Algol, or some kind of motion of its own, whereby part of its body, covered with spots or such like matter, is periodically turned towards the earth. But the intention of this paper is to communicate facts, not conjectures: and I flatter myself that the former are remarkable enough to deserve the attention and farther investigation of astronomers."

Again, there is no direct evidence that Michell had any part in these events, though it is tempting to think of some guiding influence on young Goodricke from the Rector of Thornhill, particularly given the former's interest in double stars. Goodricke's first speculation on the cause of the light changes did turn out to be correct and would have been scientifically plausible to Michell, although that explanation was doubted by Herschel, who was called in to check on β Per later in 1783.* Herschel, after repeated observations with his 7 foot telescope, found no visible sign of binarity in the image. But with a simple application of Kepler's third law for motion under gravity, and assuming stars of the Algol system to be basically sunlike objects, Michell would have calculated their distance apart to be of order 10 times their radii, a separation far too small for Herschel to have been able to resolve.

In fact, such calculations were presented in a paper Michell sent to the Royal Society two weeks after Goodricke sent his report on the variability of Algol on May 12, 1783. Michell referred back to his 1767 paper, but went on to develop an idea for which he has become much more well-known in recent years. This was his deduction, on the basis that light particles (as, following Newton, he took them to be) would be affected by gravitation, that stars of a certain size and density would not be seen because light, emitted with a high velocity well-known from more typical observational circumstances, would be confined to closed orbits around the source: a concept essentially similar to that of the 'black holes' of modern astrophysics. The paper does refer, somewhat obliquely, to the possibility of a normal star in orbit around such an object, but the author had separate aims in mind relating to how the diminution of light velocity from a dense star might be determined and did not pursue the question of the light variation itself.

Michell had raised the technical issue of just what is the limiting angular size that can be made out by the eye and how that is affected by the use of a telescope or other optical device. That there exists some limit, and that this was related to what may happen to light rays as they pass by the edges of apertures, had been discussed since at least the work of Grimaldi in the mid-seventeenth century. Newton's book on optics had referred to the effects of light 'inflexion' making, for example, the shadows of obstacles larger than they should be if the rays followed strictly linear paths, but the subject was not really clearly spelled out until after general adoption of the wave model for light propagation in the nineteenth century. Michell's 1767 paper argued that one might reasonably expect to see more detail in proportion to the aperture size of the telescope, though the atmosphere would also interfere with and deny this expectation in practice. But stellar parallaxes of order one arcsecond for the nearest stars might well become measurable within then foreseeable times.* On the

---

*Then President of the Royal Society, Sir Joseph Banks, wrote to Herschel as follows: (May 3, 1783) "I learnt at the Royal Society that the periodical occultation of the light of Algol happened last night at about 12 o'clock: the period is said to be 2 days 21 hours, and the discovery is now said to have been made by a deaf and dumb man, the grandson of Sir John Goodricke, who has for some years amused himself with astronomy. This is all I have yet made out."

*This actually occurred a couple of generations later, with Bessel's 1838 determination of the parallax of 61 Cyg.

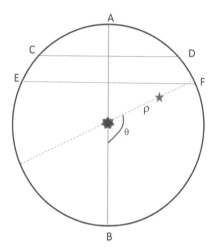

Figure 1.3   Schematic bifilar micrometer in a circular frame. This was used in the measurement of double stars since the time of William Herschel.

other hand, Michell's figure for the angular diameter of Sirius of about 1/200 of an arcsecond (surprisingly close to modern estimates) imply that it should remain out of reach for direct measurement indefinitely.

William Herschel, in the course of his ongoing programme of double star observations, had noticed the definitely disk-like appearance of Uranus in March of 1781, leading to its identification as the seventh major planet of the solar system. This event caught public imagination and led to substantial recognition and officially committed funding for Herschel's astronomical programme. In due course, Herschel collected some of his most definite findings from a quarter-century of double star observations in his 1803 paper on the "Situation of Double-stars and ... the Cause to which they are owing". At that time, Herschel announced that the bright components (A and B) of Castor form a gravitating pair, i.e. held in the same kind of orbital relationship as applies between the Sun and planets. This was taken to be a direct demonstration of the general nature of Newton's Law of Gravity, at least to distances very much greater than that of the Sun. He alluded, somewhat indirectly, to Michell's probabilistic argument in the introductory remarks on 'binary sidereal systems, or real double stars' as not being a definite proof of their existence. In discussing the apparent change in position angle of Castor A and B of about 22 deg between 1779 and 1802, Herschel referred to a different orientation of the system as observed by James Bradley in 1759 – a result communicated by Maskelyne. But Bradley's finding of a previous shift of about 30 degrees between 1719 and 1759 seems overlooked (nor was it mentioned by Michell). The great wealth of observational data contributed to this field by William Herschel, as well as his sister Caroline, can, however, be regarded as setting the subject of double star astronomy on a firm footing as a real branch of astronomical science. Fig 1.3 schematizes the type of measuring instrument at the hub of this science.

In the arrangement sketched in Fig 1.3, three very fine filaments, etch-markings or wires are set up on a frame within a graduated circle located in the focal plane of the telescope objective. The orientation of this frame with respect to the north-south reference direction can be read from the circular scale. One of the filaments (AB) is fixed along a diameter of the circle and the other two (CD, EF) are moveable and perpendicular to this first diameter. The reference direction can be checked by observing the E-W drift with the telescope stopped. The fixed filament is initially perpendicular to that direction. The two moveable filaments on this 'bifilar micrometer' are used to measure the angular separation on the sky ρ of the visual binary. The primary is centred in the field and the filaments EF and CD set to cross AB through the star centre. The circular frame is then rotated so that AB joins the two stellar images. The angular rotation of AB from its initial position θ – the position angle – is measured, conventionally from north through east. In Fig 1.3 the field has a typical arrangement of north at the bottom and east to the right. One of the two aligned filaments perpendicular to AB can be moved to measure the separation of the images, using the micrometer's linear scale. This measured separation, divided by the telescope's focal length in the same units, determines the value of ρ, in angular measure. The measurement process should ensure that the primary remains where the two diametrical filaments cross during the reading.

This manual procedure recalls the historic background to the observations of θ and ρ, but, in more recent times, the whole process has become automated, utilizing digitized high-sensitivity detectors, even from space-borne observatories. Even so, a collection of observations of θ and ρ over given dates t, that may spread over centuries, still constitute the essential data-set for a visual binary.

For many visual binaries, the change of θ and ρ with t is very slow and most historically well-known examples have not yet completed one orbit. Also, what is observed in this way is not the actual relative orbit of the secondary about the primary but its projection onto the local tangent plane of the sky. The natural assumption, since at least Michell, has been that the true orbit corresponds to classical two-mass-point motion under gravity, and so (to a high accuracy) of elliptical form and satisfying the predictions of Newtonian theory.

Let us write

$$f(\rho, \theta; \{a_1, a_2, a_3, ...a_n\}; t) = 0 \ ,$$

where ρ and θ are regarded as dependent on a set of parameters $a_1, a_2, a_3, ...a_n$; fixed for the circumstances of an individual orbit. The function f corresponds to a physical *model*: the n parameters a_i may take a range of different numbers corresponding to a range of different orbits. The orbit is *solved*, or the model deemed correct, when the parameter set reproduces what would be observed by an error-free observer. Given the known form of an elliptic orbit, we would expect this solution to conform to $\rho - r(\theta; \{a_i\}; P, t) = 0$, where r is an explicit function of the parameters $\{a_i\}$, often referred to as the orbital elements, that repeats itself with period P, ρ being observable at times t.[*]

The details of this do not concern us for the present. We seek to express the issue as typical of a class of scientific modelling problems, in which a collection of points drawn from a *data space* is provided. We have a fitting function $r(\theta\{a_i\}; t)$, through which we refer to a corresponding *parameter space*. In a *well-posed* problem, there is one clear set of parameters that allows the fitting function to match the data to an accepted level of accuracy. This process carries issues of determinacy and adequacy: in a well-posed problem all the sought parameters are uniquely determined from real observations to a specified level of uncertainty, and the model is adequate (sufficient) to explain all the observed effects. More about this is given in Section 7.1.

If Herschel had assumed sunlike masses for the visual binary in Castor, and used his period estimate of 342 y in an application of Kepler's Law, he would have found the orbit to have a semi-major axis of about 60 astronomical units: about twice the separation of the Sun from Neptune. The measured separation on the sky (ρ) of about 4 arcsec would have then

---

[*]The six conventional orbital elements derive from a background of solar system studies, whereby the periods P of the planetary orbits are related to their distances from the Sun a according to Kepler's third law. The situation for binary stars is different, because P would be regarded as another orbital parameter or element, still subject to Kepler's law, but dependent on the initially unknown masses of the component stars.

placed Castor A at a minimum distance of about 15 pc (fairly close to modern estimates). With Michell's estimate of the distance of Sirius as 1 parsec, the difference in magnitude between Sirius and the average of the two stars in Castor would have produced a distance of about half this value, though still of the right order of magnitude. Given the various uncertainties, it would have been reasonable to suppose that gravity applies to the stars in Castor A according to the same formula as in the Solar System. Conversely, if the parallax of a visual binary can be reliably determined separately, then the possibility to use Kepler's Laws to determine stellar masses opens up. At the present time, several hundred visual binary component stars have had their masses determined in this way, though the example of Castor shows that this procedure is likely to be confined to stars that are relatively near to the Sun, and applies to data that require very long collection times. The independent measurement of stellar parallaxes came into its own after Friedrich Wilhelm Bessel's use of the Fraunhofer heliometer at the Königsberg Observatory, still three decades later in the nineteenth century than the ground-breaking paper of Herschel.

Twenty years after Herschel's discovery of Uranus, on the first night of the nineteenth century, Giuseppe Piazzi at the Palermo Observatory found another new member of the Solar System. The newcomer was named Ceres after the traditional deity of Sicily. Though nearer by a factor of about 7 than Uranus, which can be seen with the unaided eye in good conditions, Ceres seemed disappointingly faint and small, at around a quarter of the Moon's diameter. Today, Ceres is cast as the prototype *minor planet*: a body with physical properties somewhat distinct from mainstream planets. Piazzi made his discovery as part of a large programme of monitoring stellar positions, during the course of which he noticed that some stars change their apparent positions by relatively large amounts that trend in a particular direction. This systematic effect is associated with the Sun's own movement relative to the average of positions of surrounding stars. In technical terms, this is a *secular parallax* arising from the solar motion relative to the *local standard of rest*. This intrinsic motion of the Sun produces a baseline of about 4 astronomical units (AU)* per year. In particular, Piazzi found the visual binary system 61 Cyg to shift by more than 5 arcsec per year, suggesting a regular parallax of order 1 arcsec.

Bessel used this information in selecting 61 Cyg as a candidate object for an attempt at direct measurement of its parallax in the late 1830s. For this purpose, he applied the *heliometer* telescope (Fig 1.4) whose objective lens consists of two adjustable halves. This arrangement had been designed by the brilliant Bavarian optician Joseph von Fraunhofer for the purpose of measuring very small differential angular displacements, originally to determine the Sun's diameter to high accuracy. 61 Cyg is among the closest stars, but the system's own relatively high speed through space produces a deceptive impression of proximity. Nevertheless, Bessel succeeded to measure a parallax of 0.29 arcsec, putting the distance at about 3.5 pc. The angular separation of the binary stars, which was 15 arcsec in Bessel's time, would thus correspond to a spatial separation of at least 52 AU. Values of the separation ρ and position angle θ were known for 61 Cyg since 1650, pointing to an orbital period of around 600-700 years. A lower limit on the masses could then be estimated as about 0.4 that of the Sun. The fuller range of data accumulating on this binary over the years has led to recent estimates of 0.6 and 0.5 solar masses for the two main components of 61 Cyg. By the 1830s, Bessel's ideas on stellar duplicity also echo something from the writings of Michell, notably in his letter to the Royal Astronomical Society of 1844, where he pointed out, on the basis of historic meridian circle measurements by himself and others, changes in the proper motions of Sirius and Procyon. These, he argued with great prescience,

---

*The mean distance between Sun and Earth.

Figure 1.4   J. Fraunhofer designed the heliometer with a specially arranged split objective that could measure the small angular differences. A heliometer was later used by F. W. Bessel at the Königsberg Observatory to measure the parallax of 61 Cyg.

were very probably accounted for by the existence of massive but unseen companions at separations in the order of a score of AUs.

A little over a year before Herschel's paper on the "situation of double stars" was read to the Royal Society, another paper of potentially far-reaching consequence to astronomy had been presented by William Hyde Wollaston. After examining carefully the dispersion of a narrow pencil of solar light by a glass prism, Wollaston noted the presence of several dark lines traversing the spectrum. The lines appeared in the same relative positions whether from daylight or the Moon, but they did not show up in the light of candles or other locally incandescent sources. Around this time, the polymath Thomas Young was presenting to members of the Royal Society clear experimental evidence to support the wave theory of light. Young's understanding of the underlying properties of light radiation led to his use of a finely ruled diffraction grating as a light dispersing agent, with which he could determine the wavelengths of the available spectral range and particular colours of the spectrum. These findings, in due course, led the young Fraunhofer, having introduced a special instrument for the purpose – the spectroscope – to make more detailed observations of the remarkable features of the solar spectrum. Fraunhofer eventually listed several hundred of the spectral lines. He also studied the spectra of Sirius and other bright stars, noting differences in the spectral features of different stars: work of inestimable significance to the subsequent development of our physical knowledge of the Universe. Even so, realization of this profound significance was rather slow.

Although intermittent attempts to follow up on Fraunhofer's stellar spectral studies occurred in the second quarter of the nineteenth century, stellar spectroscopy had to wait until the early 1860s for a concerted burst of systematic effort from several active scientists. Relevant to this development was the growth of general physical ideas on light: its wave character and its relationship to matter. Notable, for the present context, was the introduction of a principle by Christian Doppler in 1842, that the frequency with which light waves from a radiating source will pass an observer depends on the relative speed of the source with respect to the observer, now well-known as the *Doppler effect*. By 1860, Gustav Kirchhoff had spelled out the formal connection between bodies emitting and absorbing radiation while maintaining thermal equilibrium within a given enclosure, through the equation that now bears his name (though, as not infrequently happens, others had been thinking along similar lines). A key point emerging from stellar spectroscopy was that the chemical composition of stars involved the same atomic species as the matter making up the Earth and planets, and, with a few notable exceptions (particularly hydrogen and helium), in approximately comparable proportions. In 1865, James Clerk Maxwell reached deeper levels of understanding with his mathematical formulation of the wave theory of light.

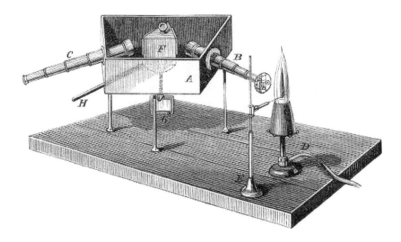

Figure 1.5 The diagram shows the classical arrangement of a laboratory spectroscope, as discussed by Bunsen and Kirchhoff in the *Philosophical Magazine*, Vol. 22, 1861. The flame source (D), coloured by the sample under study (E), illuminates a small entrance pupil whose light is directed into a parallel beam by the collimator (B) before being dispersed by a prism (F). This is kept in an isolated and controlled enclosure (A). A selected region (H) of the dispersed beam can be inspected by the telescope (C), the prism's orientation being adjustable (G). The principles of this laboratory instrument were followed in the basic design of the astronomical spectrograph in subsequent decades.

George Airy at Greenwich, Angelo Secchi at the Collegio Romano and William Huggins at his private observatory in London were among those then starting to produce important results in the newly identified field of stellar spectroscopy. By 1868, Huggins published that Fraunhofer's 'F' line, which he had identified with a line produced in a low-pressure hydrogen discharge tube (the line now commonly known as H_β), did not exactly coincide in its position with the corresponding feature in the spectrum of Sirius. Huggins, using the Doppler principle, went on to calculate a relative radial velocity for this star. By 1872 Huggins was ready to present to the Royal Society a catalogue of radial velocities for thirty stars.

The previously mentioned solar motion, if typical among the relatively nearby stars, can be understood to imply that they have velocities relative to a local 'standard of rest'[*] of a few tens of km s^{-1}, which means that the corresponding Doppler shifts of spectral lines should be on the order of 1 part in 10000. Such a relatively minute shift measured

---

[*]This is determined by averaging the radial velocities of a sufficiently large number of relatively near stars.

in situ at the telescope would require the utmost instrumental precision and control. It is therefore not surprising that Huggins' radial velocities, though of the right order of magnitude, contained errors of that same order. However, by the mid-1870s, dry photographic plates were becoming available, entailing that permanent records of observations could be measured later in stable laboratory conditions. The first photograph of a stellar spectrum, that of Vega, was produced by Henry Draper at his private observatory about 20 miles from central New York. Draper placed a quartz prism in the light path to his large reflector and was able to record previously unknown ultra-violet absorption lines.

The time was then ready for important new developments in stellar spectroscopy with the application of a larger class of telescope and photographic techniques for recording and preserving observations. Prominent in these developments were Herman C. Vogel at the newly equipped Potsdam Observatory and Edward C. Pickering at Harvard College Observatory, with its great 15-inch refractor. Within close succession they published evidence that certain stars, Mizar in the case of Pickering and Algol for Vogel, exhibited periodic line shifts: clear evidence of the close double nature of these stars. Vogel thus established the original speculation of Goodricke a century before, that Algol must be an *eclipsing binary*, as such systems became known. Vogel went on to estimate the size of the orbit and the sum of the stellar masses, thereby preparing the ground for the factual knowledge of stars.

A purely photometric analysis of the light curve of Algol, on the basis of the close binary model, had been published nine years previously by Pickering himself, initiating a field of quantitative analysis with an enormous legacy in stellar astronomy that we will encounter frequently in subsequent chapters. Data on the variation of Algol's light had been systematically collected since the time of Friedrich Argelander, a former student of Bessel. Argelander developed a procedure for magnitude determination that involved setting up a sequence of nearby comparison stars with slightly varying steps of brightness. This technique for eye-estimation, which could attain an accuracy of $\sim$0.1 magnitude after sufficient practice, was followed by many other visual observers and is still in use for patrolling variable star behaviour. It was, however, Argelander's student E. Schönfeld, at the Bonn Observatory, who came to prominence in connection with the light curves of Algol. Schönfeld's data were published in 1870 and had been cross-checked for internal consistency by others before being selected by Pickering for detailed analysis.

For this purpose, Pickering tabulated values of the 'eclipse function': numbers reflecting the amount of darkening that would affect the observed light flux from a star undergoing eclipse from a relatively close companion of comparable size. Adopting the assumptions, for initial simplicity, that the companion moved in a circular orbit, was essentially dark, and the eclipse briefly became annular at the light minimum, where the light loss could be estimated, Pickering could calculate a minimum ratio of radii of the 'secondary' to the bright 'primary' star (0.764 – usually denoted by k).

Let us denote the radius of the primary star as a fraction of the separation of the two stars by r_1, and that of the other star r_2. With Pickering's imposed condition, the separation of the centres at minimum light as a fraction of the orbital radius, i.e. $r_1(1-k)$, is equal to the cosine of the orbital inclination: the angle i between the axis of the orbit and the line of sight. The total duration of the eclipse as a fraction of the orbital period provides another equation of condition involving r_1. At the measurable orbital phase of first (or last) contact of the two disks, ϕ_1, say, it is not difficult to find that $r_1(1+k) = \sqrt{(\cos^2 i + \sin^2 i \sin^2 \phi_1)}$, where i is the inclination angle between the line of sight and the axis of the orbit.* We thus

---

*The details of this calculation are given in Section 7.3.1.

have 3 equations to solve for the 3 main geometric parameters r_1, k and i of this simplified model, noting that r_2 is simply kr_1. Pickering found, in this way, $r_1 = 0.22$, $r_2 = 0.17$ and $i = 87$ deg.

Having arrived at some 'feel' for the kind of parameter values that might apply, Pickering went on to add a few important qualifying considerations. He realized, of course, that although his solution was feasible, it was not unique. The central region of the minimum was indeed fleeting, but that was no guarantee that it was annular (i.e. the entire outline of the eclipsing star being in front of the projected surface of the primary). He experimented with different ratio of radii and checked on the differences between corresponding observed and calculated light levels. Moreover, the data that Pickering used indicated that the shape of the minimum was asymmetric, which allowed him to entertain several eccentric orbit possibilities, presenting a range of feasible models. He then discussed the important potential roles of spectroscopy, if the still unknown parallax turned out to be small; or astrometry, if it were large. Pickering's eclipse function did not take into account the darkening of the stellar surface towards its perimeter. This and other refinements have been developed over the intervening years, but the essence of the "interesting work" that Pickering envisaged for students of close binary stars still motivates.

The 1880 paper also looked into the regularity of the orbital period of Algol and found reasonable evidence for its apparent variation over the century after Goodricke's discovery: a topic that links studies of close pairs to the multiple star configurations they are often found in. The 'O – C curve' (O being the observed times of minimum light and C those calculated from predictions based on previous data) can thus play a parallel role to the much shorter-term light curve in developing physical knowledge of the stars. As deduced from the O – C variations, the third star – Algol C* – revolves around the system's centre of gravity in a period of about 1.86 y and at a mean distance of ~280 million km from the system's centre of gravity.

During the four decades that Pickering directed the Harvard College Observatory, the institution became renowned as a leading centre of practical research in astrophysics, particularly in its task of stellar classification. A number of women played an important part in bringing Pickering's ambitious plans into fruition (see Fig 1.6). Among the first of these should probably be placed Anna Palmer Draper, whose husband had died just a few years after publishing his pioneering observations of stellar spectra. Mrs Draper wished her husband's discoveries to be suitably recognized by an ongoing professional programme, to which she later made substantial endowments. The Draper bequest brought attention to the work of other women at Harvard, including Williamina Fleming, Antonia C. Maury and Annie J. Cannon. The very large amount of data that these scientists successfully organized laid the foundations for stellar spectral analysis through to the present time. An important element in this programme was the system of assigning a *spectral type* to a given star. The process of classification of stellar spectra developed over several decades, but it eventually reached the detailed form of the present day, whereby a star's spectral type can directly give important clues about its physical properties. Miss Maury, niece of Mrs Draper, discovered the second spectroscopic binary in 1889; that of the bright eclipsing binary β Aur. Twenty years later, when Mrs Fleming compiled a catalogue of known spectroscopic binaries, the number had grown to over 300. By the early 1930s, thanks to the participation of an ever-widening group of international contributors, the number reached to over 1000.

Although photography had also been employed as a more objective basis for photometry after its introduction in the 1870s, a detailed quantitative evaluation of the relationship between the influx of starlight and the corresponding image on the plate proved elusive and more prone to inaccuracy than originally expected, internal consistency of photographic

---

*This star is more recently named as β Per Ab.

Figure 1.6   The computing room at Harvard College Observatory in 1891. Antonia Maury is seen standing near the centre, with her radial velocity curve of β Aur on the wall behind. To her immediate left is Mrs Fleming, engaged with her workbook. Miss Annie J. Cannon is at the viewing desk in the foreground. The picture derives from the E. C. Pickering's computers file at Harvard College Observatory.

magnitudes being not so much better than the kind of eye estimates discussed before. This entailed typical errors for the differential light changes of a variable star of several hundredths of a magnitude. Astronomers pursuing photometric programmes then sought a more linear and controllable responder from which to take measurements. In the early years of the twentieth century, Joel Stebbins, working at the University of Illinois Observatory at Urbana, pioneered photoelectric methods in photometry by measuring the change of resistance of a cold selenium cell exposed to starlight (see Fig 1.7). In this way, he produced the first photoelectric light curve of the eclipsing variable Algol. The light curve had sufficient precision to show the shallow secondary minimum and thus definitely reveal some property of the bright star's companion that had been shrouded in mystery since Goodricke's initial postulate.

Another prominent name in American astronomy in the early 1900s was Frank Schlesinger, director of the Allegheny Observatory at Pittsburgh, Pennsylvania. Although Schlesinger's main interests were somewhat separate from binary and multiple stars, he made a notable contribution to this field in 1910 on the basis of his collection of sixty useful spectrograms of the bright Algol-like variable δ Lib. Schlesinger found systematic departures from the essentially sinusoidal (single lined) radial velocity curve concentrating towards the ingress and egress phases of the eclipse. He rightly interpreted this 'Schlesinger effect' as due to the shifting of the light centre on the source towards a greater recessional effect at (primary) ingress followed by a greater speed of approach at egress. These effects are not abnormalities of the radial projection of the orbital velocity, but result from the effect of eclipse of part of the rotating surface of the bright star, together with a net Doppler shift from the remaining uneclipsed part. Schlesinger's work with the photographic medium also bore fruit in the development of the field of *astrometry* of close pairs.

Stellar spectroscopy was being pioneered at this time, with images recorded permanently via photographic plates. Using the excellent mountain-top observatory and 36-inch Warner & Swasey telescope, with its Alvan Clark lens, at the Lick Observatory near San Jose in

Figure 1.7   Joel Stebbins' photometer on the 12-inch refractor at Urbana, Illinois. (From J. Stebbins, *Astrophys. J.*, Vol. 32, p185, 1910; ©AAS.)

California, William Wallace Campbell went on to publish a catalogue containing more than a thousand stellar spectra in 1924.

Pickering died in 1919, and one of his possible successors apparently considered by the appointments board at Harvard was Henry Norris Russell, then director of Princeton University Observatory, who had relevant scientific interests. Russell's doctoral thesis of 1900 had relevance to the orbits of binary systems, but his major opus in this field is generally regarded as two papers of 1912, and two published in the same year jointly with his student Harlow Shapley. Two years later, Shapley's own thesis appeared, containing an application of the new analytical approach developed with Russell to no less than 90 light curves. When supported by spectroscopic information the combined procedures furnished a *royal road* to the direct factual knowledge of stars, sought for so long by previous generations of astronomers. As things turned out, it was Shapley who succeeded Pickering at Harvard College Observatory, thus maintaining the Observatory's prominent role in binary star research.

By this time, Russell had become very much engaged with the work for which he is perhaps generally most well-known, that of the graphical summary of stellar properties and behaviour that has become known as the *Herzsprung Russell* diagram. Ejnar Herzsprung, then a relatively unknown astronomer from Copenhagen, had studied the relationship between the colours and luminosities of stars, publishing his findings as early as 1906. He found opportunity to discuss the subject with one of Europe's most highly respected physicists – Karl Schwarzschild at Göttingen – who understood the implications of Herzsprung's results and gave him full support. By 1913, Russell was able to call on the great wealth of observational material and definitive results available to him from the Harvard school and elsewhere, and could comprehensively distinguish between the trends of the 'dwarfs' on the *Main Sequence* and the 'giants' on their own sequence. Russell also produced an account of a possible *stellar evolution* which might connect the two trends. This subject has since become a cornerstone of stellar astrophysics, Russell thereby acquiring something of the reputation of a 'dean of American astronomy' through the first half of twentieth century.

Although Stebbins had shown the potential for accurate photoelectric photometry already by 1910, the appearance of good-quality light curves for fainter close binary stars had to wait for technical advances relating to the measurement of tiny electronic effects. Low-noise amplification was one such development, pioneered by A.E. Whitford in the 1930s, and made use of by Gerald Kron at the Lick Observatory with the definitive light curves of the eclipsing system YZ Cas towards the end of that decade. The use of the avalanche effect at the primary detection stage, whereby one initial electron movement is converted into an electron pulse of useful magnitude, was another important step, leading to the application of photomultiplier tubes to the measurement of starlight.

This type of device was used by H.L. Johnson and W.W. Morgan in the early 1950s when they set up their near-ultra-violet, blue and visual (UBV) photoelectric system, which, despite certain well-known limitations, has been in widespread use up to present times. The UBV light filters received early support from the International Astronomical Union and by now a very large number of stars have been thus characterized. Other physically advantageous schemes have appeared from time to time and continue to be presented, and the UBV magnitude system may evolve into something else. It may be noticed that the system's own V filter was intended to blend in with the photovisual magnitudes of previous schemes. Since the photoelectric UBV magnitudes were of essentially improved accuracy, however, the newer system ultimately defined the magnitudes of stars, though basically in agreement with historic representations.

A major advance occurred in 1970 with the introduction of the charge coupled device (CCD) at Bell Laboratories, New Jersey. By 1975, the CCD camera had been placed at the focal plane of some large telescopes and their diffusion throughout observational astronomy has continued ever since. CCDs, or comparable solid-state detectors, are effective for both precision and multiplicity of target coverage. If necessary, individual chips can be arrayed to cover larger areas of sky. The lower work functions of the semiconductors used in CCD light detection entail greater response in longer wavelength spectral regions (i.e. infra-red) than the earlier photoemissive detectors. In turn, this has impinged on optical filter designs so as to allow continuity in light curve studies.

The general growth of science in the second half of the 20th century, particularly after the introduction and widespread use of electronic computers in the 1960s, brings us towards contemporary times, when numerous particular specializations have become distinct within schools of astronomy and astrophysics. It was mentioned in the Foreword that, by 1979, M. Plavec foresaw increasing practical difficulties in bringing together the various fields of research on close binary stars to report within a single meeting. Our present overview thus tries only to point out some of the milestones passed since around the time of that 88th IAU Symposium. As with many of the developments noted in the previous pages, the bright prototype Algol has often played a key role. The 'dog-eat-dog' scenario for the evolution of stars in close pairs sketched by Fred Hoyle in 1955, now generally known as the Roche Lobe Overflow (RLOF) hypothesis, was an early example of theoretical study prompted by the seemingly odd combination of stars in Algol.

Data from the Orbiting Astronomical Observatory (OAO2), launched already in 1968, allowed J. A. Eaton to publish an analysis of Algol's ultra-violet light curves by 1975. From the vantage point of space, the OAO2 was one of the first astronomical satellites to yield results from hitherto inaccessible spectral regions, although in 1966 K-Y Chen and E.G. Reuning had extended the spectral coverage of Algol's light curve to about 1.6 μm in the infra-red, using the 72 cm reflector of the Flower and Cook observatory in Pennsylvania. Algol also featured in the early detections of stellar sources at radio wavelengths with C.M. Wade and R.M. Hjellming's discoveries in 1972 of microwave emission from Algol and β Lyrae.

Radioastronomy opened a new era for binary star studies with the discovery, in 1974, of the binary nature of the pulsar PSR B1913+16 by R. A. Hulse and J. H. Taylor, through use of the massive 305 m dish at Arecibo in Puerto Rico. The $\sim$17 Hz pulsation frequency was found to exhibit a sinusoidal modulation with a $\sim$7.75 h period, signalling a small orbit binarity of the source. Analysis of the data resulted in an interesting set of absolute

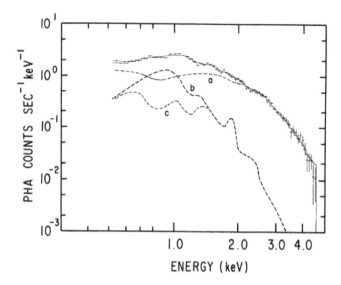

Figure 1.8 The X-ray spectrum of a presumed coronal flaring event from Algol, observed with the *Einstein* Observatory's Solid State Spectrometer in 1979. A common feature in such sources is the simultaneous appearance of high (a) and low (b) energy modelled components. Track (c) corresponds to a low temperature model with zero iron abundance. (The diagram is from the paper of N. E. White et al. *Astrophys. J.*, (Letters), Vol. 239, L69, 1980, ©AAS.)

parameters for the two neutron stars in the system, but this analysis required the application of Einstein's general theory of relativity (GTR). The special circumstances of the highly condensed stars in neutron star systems renders relativistic effects to have a predominance, which contrasts with what obtains for classical binaries, where they are usually almost negligible.

Although the early satellites capable of observing in the low nm range gave indications of intermittent X-ray emission from stars, it was not until after the Einstein Observatory was launched in 1978 that sources could be identified with high directional precision. N. E. White *et al.* later presented Algol's X-ray spectrum from the Einstein Observatory indicating flaring emission at energies higher than 6 keV (see Fig 1.8). Also in 1978, J. Tomkin and D. L. Lambert, by means of the 2.7 m telescope of the McDonald Observatory, identified spectral lines of Algol's peculiar secondary star for the first time.

Another chapter in the history of binary and multiple star systems, both regarding photometry and astrometry, opened with the launch of the HIPPARCOS satellite in 1989. This ESA supported mission included, as well as accurate positional data, near-V magnitudes measured for each of its ~120000 programme stars around 100–150 times throughout the four-year mission. Subsequent systematic analysis enabled existing light curves of known variable stars to be checked, and also brought to light several thousand previously unknown variables. As well, a new wealth of precise astrometric information was available, and could be brought to bear on the point that close binaries are often found in wider systems of higher multiplicity: a fact with wide implications for stellar cosmogony. The satellite's on-board Tycho experiment aimed at providing more precise two-colour (B and V) data for at least 400000 stars. Useful information on Hipparcos photometry has been publicly available from the 'Hipparcos Epoch Photometry Annex' of the programme, accessed from the SIMBAD database. At the time of writing, astronomers are eagerly adapting to the next generation of high-quality space-based information from Hipparcos' successor Gaia.

A major development occurred in October 1995 with the positive identification by Michel Mayor and Didier Queloz of a close orbiting companion to the sunlike star 51 Pegasi. This companion turned out to be a body of planetary mass: one of the first examples of a class of object now called exoplanets.* This discovery was made using highly refined spectrographic techniques capable of detecting changes of radial velocity of a star in the order of tens of metres per second. Further application of the technique started to reveal many more such objects, and in 1995 one example – HD 209458 – was found to show small diminutions of light as the planet crossed in front of the host star's disk in the line of sight. A remarkable light curve of this exoplanet transit was produced using the Hubble Space Telescope in mid-2000. In turn, this led to the launch of the prodigious Kepler Mission in March 2009. This NASA-supported programme has identified several thousand exoplanet systems in a relatively small area of sky in a little over four years of dedicated monitoring (see Fig 1.9). These 'tip of the iceberg' results contained many initial surprises — the follow-up burgeoning area of exoplanet research in the last and coming decades may parallel the growth of stellar astrophysics in the late nineteenth and twentieth centuries.

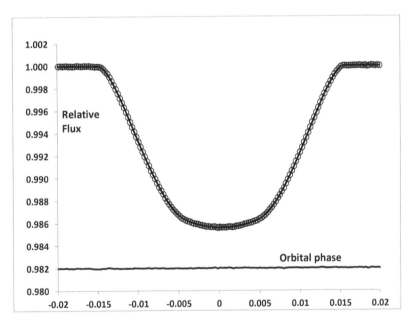

Figure 1.9 Planetary transit region of the light curve of Kepler-1, as produced by the Kepler space telescope. The model's light variation is shown as a continuous line with the observations as open circles. The residuals are shown below, displaced from their zero mean level to the position of relative flux = 0.982. The remarkable precision of data and modelling is evident. The light curve fitting implies that the mean light flux from this host star can be specified to an accuracy of better than 1 part in a million.

Algol has continued to make an interesting target in various new research areas of recent years. In Doppler tomography, for example, the computer-aided image reconstruction technique illustrates how gas can move, or be transferred, around the component stars of a close interacting system, enabling physical aspects of RLOF to be quantified.

---

*In 1992, Aleksander Wolszczan and Dale Frail, working at the Arecibo radio observatory, deduced the presence of planetary mass bodies in orbit around the pulsar PSR 1257+12 from detailed analysis of pulse timings from the unusual system's central object.

The methodology of Doppler tomography has some points of contact with ultra-high-precision astrometric source mapping using very long baseline interferometric (VLBI) techniques at radio wavelengths. J-F. Lestrade and co-workers published VLBI observations in 1988 of a source in the Algol system that they related to the magnetic field of the subgiant secondary component. The source, which shows giant arches stretching between the two main stars, was mapped in more detail with a combination of long baseline arrays stretching from New Mexico to the Effelsberg radio telescope in Germany in 2010.

Algol has also figured in the observational programme of the Center for High Angular Resolution Astronomy (CHARA) infra-red interferometer of Georgia State University located at the Mount Wilson Observatory near Los Angeles. The complex instrumentation required for CHARA became fully operational in 2004. The six 1 m telescopes used in the array could, in principle, resolve features of order 1/10 the size of the disk of Sirius. The results of extensive CHARA observations of Algol between 2006 and 2010 were published by F. Baron and co-workers in 2012. All three main stars were resolved. Previous parameter estimates could be revised with some confidence. The orbital plane of Algol C about the system centre was then found to be almost perpendicular to that of the eclipsing pair. The movements of the inner pair could, for the first time, be witnessed going through the orbital cycle, just as imagined in the original account of John Goodricke.

1.2   Basic Terminology

The preceding historical background gave rise to relevant concepts and terms now in widespread use that reappear in the subject matter of subsequent chapters. It is expected that the reader will already be familiar with most, if not all, such terms, but for completeness we present here a general preliminary review.

The expression *binary star* or *system* has come to have the more restricted meaning of a pair of stars that perform orbital motion about a common centre of gravity or *barycentre*. This is distinguished from an *optical double star*, which would be understood to refer to a pair of physically unrelated stars that just happen to lie in close angular proximity in the sky.

Angular measures on the sky often refer back to the *celestial sphere*, and associated coordinates, involving *altitude-azimuith*, *equatorial*, *ecliptic*, *galactic* or perhaps other systems. Operations using these coordinates invoke the methods and formulae of *spherical trigonometry*.

Binary systems have been subdivided into three main classes: *visual*, *spectroscopic* and *close*, the latter often used interchangeably with *eclipsing*. The classification is partly related to physical properties, but also reflects different historical backgrounds of observational methods and analysis. Eclipsing binaries are further divided into a number of types, often in connection with the form of the (periodic) *light curve** that they show. Sometimes, *ellipsoidal* close binaries are found, whose proximity is manifest by periodic light variations due mainly to tidal interactions (*ellipticity effect*), but with an orbital plane distant enough from the line of sight to prevent eclipses being seen. Certain binaries show a relatively strong variation in the light reflected towards the observer (*reflection effect*).

The study of close binaries has been thus strongly related to their *photometry*. The time-honoured *stellar magnitude* system is still in general use in this connection, where the difference in magnitude between two stars m_1, m_2 is expressed by N.R. Pogson's logarithmic

---

*The term 'light curve' usually applies to a set of discrete observations of a given source, whose trend with time derives from a continuously varying source of light.

formula

$$m_1 - m_2 = -2.5 \log(f_1/f_2) \ , \tag{1.1}$$

where f_1/f_2 represents the ratio of the received light fluxes (radiative power per unit area) from stars 1 and 2. But standard light curves are not the only function of binary star photometry: significant further information may come from the repeated timings of particular phases of data, for example *times of minimum light*. Also, the *polarization* state of the received light can tell us facts that are not evident in its total intensity.

As mentioned in the historical notes, an indication of the distances of stars, or, equivalently their *parallax*, can often be ascertained from photometry, albeit indirectly and involving parameters that have to be assumed unless supplied by separate evidence. Thus, if the distance ρ, say, of a given star were to be doubled, the flux received at the Earth would fall to a quarter of its original value in consequence of the inverse square law, itself a consequence of the principle of energy conservation. This leads to the usage of *absolute magnitude (M)*, which is the magnitude a star would have (neglecting here any intervening light absorption) if at the reference distance of 10 parsec (pc).* Elementary deductions allow us to write the extinction-free *distance modulus* formula as

$$m - M = 5 \log(\rho) - 5 \ . \tag{1.2}$$

Stars can often be assigned an absolute magnitude, based on their spectral properties, or perhaps other evidence about them, allowing (1.2) to yield a distance estimate. This would be a preliminary estimate, because interstellar material, over great distances, diminishes the light received from a star. This is usually dealt with by adding the *interstellar extinction (A)* to the right of (1.2). The value of A is derived from separate information, including the star's location, line of sight and environment.

Distances of particular stars can also be estimated by other means, especially in the case of close binary systems. Relevant photometric data would be often in the form of a table of the difference in magnitude between the object of interest and a constant reference source or known stable comparison star at given dates and times. The magnitudes are given in a *standard photometric system* involving a measuring photometer with carefully specified properties. The series of time values t_i may be converted to phases ϕ_i, according to a given *ephemeris*, which includes a reference *epoch* t_0 and orbital *period P*, so that

$$\phi_i = (t_i - t_0)/P \ , \tag{1.3}$$

where for a periodic variable the phase would normally refer only to the fractional part of (1.3), so that the phase would be in the range 0 to 1. Modelling this light curve can lead to specification of the system's physical properties, particularly if appropriate spectroscopic data are also available.

Regular photometric variations are not an essential feature for *visual binaries*, however. For pairs of stars with orbital periods of the order of a year or greater, observational methodologies may involve direct positional measurements. A distinction arises between the long-established measurements of separation and position angle, possible for pairs that are optically separated in the telescope, and the *astrometric* analysis of stellar images characterized by an irregular or perturbed appearance, whose form shows regular variations over the inferred orbital period.

The wider visual binaries are typically characterized by long-term programmes, orbital periods being on the order of at least hundreds of years. They are often relatively close, the more well-studied examples lying within tens of parsecs from the Sun. Typical data-sets list the *angular separation* of the two components ρ_i, originally in arcseconds, nowadays

---

*The parsec is defined as the distance at which a star would have a parallax of 1 arcsecond.

frequently milliarcsec (mas), together with a *position angle* θ_i, measured in an eastwardly sense from the direction of local north on the sky at given dates of observation t_i. The stellar *micrometer* (Fig 1.3) has been the instrument used, in principle, to provide these data, though modern approaches employ a considerable level of sophistication to the measurement of stellar positions.

For closer pairs, clearly the *resolution*, or ability to have a detailed image afforded by the observing system, is of major significance. With moderate-sized telescopes located in good atmospheric conditions, close pairs can be resolved down to a fraction of an arcsecond. But separations down to an order of magnitude below this have been studied by specialized astrometric techniques, involving either *interferometry*, *speckle imaging*, or other methods using dedicated ground or space-based facilities.

Remarkable progress thus came with *space-based astrometry* after the application of the HIPPARCOS and its successor, Gaia, satellites. These programmes have done much to clarify procedures in double star astronomy arising from their uniform and well-defined observing properties, particularly for systems with separations less than 1 arcsec. The special technique of using *lunar occultations*, when applicable, can push the separation limit down to $\sim$2 mas. Nowadays, astrometric methods are encroaching into the mas separation range that were once associated only with spectroscopic and photometric binaries.

The term *cluster* is in common use. Stellar astronomy distinguishes between rather large and amorphous *associations*, and *open clusters*, both tending to be concentrated towards the disk-like central plane of a galaxy; and the relatively dense *globular clusters* that are normally older, larger, and not showing much concentration to the galactic disk. A cluster could be regarded as an exaggerated case of a *multiple star*, the latter term tending to be reserved for a more compact system of up to a dozen stars that are found to be have identifiable gravitational interactions, frequently in an hierarchical orbital arrangement. Such an arrangement may resemble that of a *planetary system*, although planets would, of course, normally contribute a negligible proportion of the system's overall mass and light.

The *spectroscopic binaries* can be regarded as generally including the close binaries, but forming a larger subset with many members at wider separations of around 50–100 times their stellar radii and photometrically unremarkable. They are subdivided into *double and single-lined* binaries, depending on whether or not the spectral signatures of both components can be distinguished in the combined light. The results derived from observations with a *spectrograph*, more recently often referred to as a *spectrometer*, would usually aim at including *radial velocity* (*rv*) determinations at given times or orbital phases. The phasing may be done along the lines of Eqn (1.3), although the time chosen for the orbital reference point t_0 is not necessarily the same for spectroscopic and photometric data-sets. For close binaries, t_0 tends to be associated with the time of primary mid-eclipse, whereas for the wider spectroscopic pairs, the reference position is often that of fastest approach: the minimum *rv* value.

Very frequently, one of the stars – the *primary* – dominates the spectral appearance. It would often be the more massive component, as determined by the amplitude of its radial velocity shifts that are less than those of the *secondary*. However, with close binary systems, particularly after interactive binary evolution, it can happen that the component showing a brighter spectrum is not the more massive one. Or else, with eclipsing systems, the deeper (primary) minimum would be that of the hotter star; but that is not necessarily the more luminous one overall. It is thus sometimes necessary to distinguish between a more massive, or luminous, spectroscopic primary and a photometric, hotter one. *Tertiary*, or higher-order, members are also not infrequently found: terms such as *trinary* or *quaternary* may be used for such configurations. These topics are related to the somewhat separate class of *spectral binaries*, where an apparently single object contains two identifiably separate sets of spectral

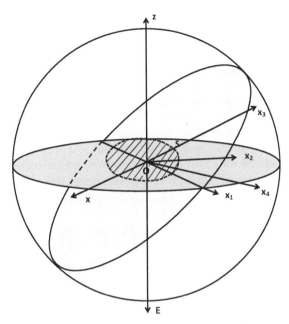

Figure 1.10   The binary system orbit (small ellipse, shaded with lines) as seen by a terrestrial observer along the direction OE, would be projected onto the local sky plane containing x, x_1 and x_4.

features, even without clear signs of radial velocity variation. This has been a productive field in studies of low mass objects.

Multiplicity is also sometimes indicated, even at great distances, by an excessive apparent luminosity of an object. Methods have thus been introduced to account for the vertical spread in a colour-magnitude diagram in terms of binarity incidence and mass-ratio distribution.

1.3   Binary System Parameters

The binary star may be regarded as the simplest example of a multiple stellar system: in examining its geometry we can find various terms that are repeated separately for the separate components in multiple star arrangements. Kepler's laws underlie much of this discussion. More details are given in what follows. Before recalling these basic discoveries, however, it is appropriate to mention that, while efforts are made to use conventional formalism, sometimes ambiguities arise. This particularly concerns frequently used symbols like a and r. It will usually be explained, or clear from the context, what a given symbol refers to; but it may also be advisable to consult the relevant sources cited in the bibliography sections.

Originally, Kepler's laws applied to the planets of the solar system, they were essentially confirmed for binary stars by the time of Herschel. In the original context, the first law states that the spatially fixed and periodically repeated orbit of the moving mass point is of elliptical form with an attracting mass centre at one of the ellipse's foci. The second is that the radius vector, joining this mass centre to a moving mass point, sweeps out equal areas in equal times; and the third law gives the connection between mean separation of the mass centre and mass point, the orbital period and the total mass involved. In its simplest form, using the astronomical unit for the ellipse's semi-major axis a, the total mass M in solar masses, and the time period P in years, Kepler's third law becomes simply $a^3 = MP^2$.

In Fig 1.10, we illustrate the geometry of a binary system, where the centre of the primary star is placed at the origin, and the downward vertical OE points in the direction to the observer; alternatively, the direction Oz is the line of sight. The plane perpendicular to OE is that of the local sky, on which the apparent orbit of the binary is observed. This orbit, referred to as the relative orbit, is simply a scaled up version of the orbits of either star about their common centre of gravity. Its mean radius is $(M_i + M_{3-i})/M_{3-i}$ times that of the ith star ($i = 1, 2$) about that centre.

The direction of local increasing declination at O is indicated by the line Ox. The true orbital plane is indicated by the great circle containing x_1, x_2 and x_3, where Ox_1 is the *line of nodes*; Ox_2 is in the direction of the *periastron*, which corresponds to the position when the stars are nearest each other, the opposite being the *apastron*. Ox_3 is the radius vector from the primary to the centre of its companion at the time of observation. On the celestial sphere at O, the angle xOx_1 is called the nodal angle Ω; x_1Ox_2 is the longitude of periastron measured from the node ω, and x_2Ox_3 is the *true anomaly* ν. Ox_3 projects on the sky as Ox_4, so xOx_4 is the measured position angle θ. For positive angles in the x, y, z system, as shown, the angular motion, seen from E, will be in the clockwise sense. The apparent separation ρ is given by

$$\rho = r \cos(x_3 O x_4) \ . \tag{1.4}$$

where the radius vector r connecting the central mass to its companion corresponds to to the line segment OS in the shaded orbit region of Fig 1.10.

Using spherical trigonometry, we have, in the triangle $x_3 x_1 x_4$,

$$\cos(\omega + \nu) = \cos(\theta - \Omega) \cos x_3 O x_4 \ , \tag{1.5}$$

or

$$\rho = r \cos(\omega + \nu) \sec(\theta - \Omega) \ . \tag{1.6}$$

The radius vector r for the relative orbit of a star moving as a mass points in accordance with Kepler's laws can be written as

$$r = r(a, e, \nu) \ , \tag{1.7}$$

where a is the semi-major axis of the orbit and e is its eccentricity. The true anomaly ν is usually expressed in terms of the independent variable (time) through a separate procedure involving Kepler's equation that connects an intermediate *eccentric anomaly E* and *mean anomaly M*. The final formulation becomes awkward to spell out explicitly, but an implicit iterative procedure can be used to enable a given accuracy of result. The mean anomaly M increases linearly with time t and involves only the additional constants P the period of the orbit, and t_0 the reference epoch of periastron passage, i.e. $M = 2\pi(t - t_0)/P$ in radians. Initially, data showing a periodic variation would be phased according to some clearly identifiable empirical feature, i.e. the time of minimum light, or radial velocity zero, occurring at HJD_0, say, and the time of periastron passage t_0 regarded as one of the parameters* to be determined. This means that M is shifted from the empirical zero phase by an amount M_0, i.e. the recorded phase $\phi = M - M_0$, or $t_0 = \mathrm{HJD}_0 - PM_0/2\pi$, M_0 then being a simple linear transformation of t_0. In principle, we can then write

$$\nu = \nu(e, t_0, P; t) \ . \tag{1.8}$$

---

*Sometimes called elements.

Since we now have r and ν in terms of time t and $a, e, \omega, \Omega, t_0$ and P, the required connections to formulate the orbit are almost in place, but there is another relationship between θ and ν from the spherical triangle zx_1x_3, i.e.

$$\tan(\theta - \Omega) = \tan(\nu + \omega)\cos i \ , \tag{1.9}$$

where i is the orbital inclination. This can be alternatively regarded as the angle between the line of sight and the direction from O to the pole of the great circle x_1x_2, or the angle between the x_1x_2 and the local plane of the sky x_1x_4. Hence, we can write, symbolically,

$$\theta = \theta(e, t_0, P, \Omega, \omega, i; t) \ , \tag{1.10}$$

and then

$$\rho = r(a, e, t_0, P, \Omega, \omega, i; t) \ . \tag{1.11}$$

Having adopted that the underlying path of the secondary star with respect to the primary is a Keplerian ellipse, its projection in the plane of the sky remains elliptical, so a given set of seven *parameters*: $a, e, t_0, P, \Omega, \omega, i$; produces a projection of the relative orbit on the sky plane $\rho = \rho(\theta)$, for varying t.

This formulation represents the *direct problem*: that is, given the set of elements — plot the apparent orbit. The issue generally faced in practice, however, is the opposite of this, i.e. the *inverse problem*: given data for an apparent orbit — produce the corresponding set of elements. Before the computer age, solving this inverse orbit problem was a rather laborious mathematical exercise. Computing machines are well adapted to iterative procedures, however, and setting up a program to solve the inverse problem is nowadays relatively straightforward. This topic provides a recurrent theme in this book.

Orbital data may be given in the more symmetrical Cartesian arrangement as Δx and Δy: changes in projected equatorial coordinates say, where $\Delta x = \rho \sin\theta$ and $\Delta y = \rho \cos\theta$ at times t. The relevant inverse problem becomes that of finding an optimal set of the seven parameters to match both data sets to the highest probability on the basis of a Keplerian model, for a given measurement error.

Observe that $r\sin(\nu + \omega) = z$, which is the displacement of the secondary in the direction of the line of sight, does not involve the nodal angle Ω. So, the nodal angle becomes ignorable for data that relates only to z or its derivative, i.e. the radial velocity. This concerns either spectroscopic binaries, or the light travel time effect that affects timing of the minima of close binaries that have a wide companion or companions.*

For normal light-curve photometry too, data usually only refers to the Stokes total intensity I, which is invariant against a rotation of the co-ordinate system about the line of sight, i.e. Ω is again redundant for the photometric intensity analysis of close binaries. The full characterization of incoming wavefronts involves all four Stokes parameters, though, and linear and circular polarization data, if available, can yield more geometrical information about the source. This approach has been developed in recent decades, though it requires technological sophistication in detection, given the generally low degree of net polarization (typically $\sim 1\%$) of the wavefronts and disturbances from unrelated agents, as well as a theory about the source of the polarization that should be independently supported.

An ambiguity occurs in defining the sense of the motion, because the apparent motion on the sky for a value of i less than 90 deg can be achieved by setting the inclination to the supplement of i. This would project the orbit in the same way to the observer but

---

*And, similarly, the timing of pulses from pulsars with companions.

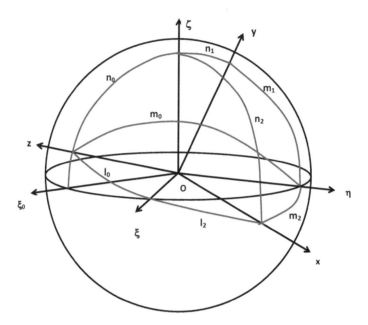

Figure 1.11   The 'natural' (ξ, η, ζ) and 'observer' (x, y, z) frames of reference, and the various angles linking them. The labels $(n_0, n_1, n_2,$ etc.) refer to the cosines of the arcs alongside which they are placed. ξ_0 marks the conjunction phase.

reverse its direction of motion, so with a reversal of that sense the original apparent motion is recovered. The ambiguity is removed if radial velocity data are available, so that, in Fig 1.10, the motion from identifiable points x_1 to x_2 could be confirmed to be a recession. Otherwise, it is conventional to set $i < 90$ deg, with the implication that the rotation sense is taken to be direct.

For a visual binary, the size of the orbit (relating to the parameter a) is measured in seconds of arc. If the distance can be estimated, this projected angular scale of the orbit can be converted into an absolute value. If radial-velocity data are also available, the projected orbit size $a \sin i$ can be independently reckoned, since, period and velocity being given, the distance travelled in one orbit follows directly. A common practice is to infer the masses of the components from their spectral types. With a given period, Kepler's third law then allows the orbit to be sized, and this may be reasonably accurate given that the semi-major axis depends on only the 1/3 power on the masses. Comparison of these two sizes will allow the inclination to be estimated.

1.3.1   Coordinates and Transformations

In Fig 1.11, the origin O is at the centre of the primary star. The direction $O\zeta$ is along the polar axis of the orbit so that the plane containing $O\xi_0$, $O\xi$ and $O\eta$ is that of the orbit: ξ, η, ζ thus making up the *natural co-ordinate system*. We can set the direction Oz to be the line of sight, so that $O\xi_0$ indicates the direction of (inferior) conjunction of the orbiting secondary object.

An *observer co-ordinate system*, used in later analysis, with its z-axis aligned to the line of sight, then has its $x - y$ plane in the local tangent plane of the sky. The secondary is instantaneously in the direction $O\xi$, and the projection of that direction from the line of sight onto the sky fixes the location of the axis Ox in this scheme. The third axis Oy then forms a right-handed system with Ox and Oz.[*]

The direction cosines (dcs) frequently encountered relate $O\xi$, for a tidal distortion, to Ox, Oy, Oz. In the notation of Kopal (1959), these are l_2, l_1 and l_0. Rotational distortion involves the dcs linking the axis $O\zeta$ to Ox, Oy, Oz: corresponding to n_2, n_1 and n_0. Easy to envisage is the constant dc n_0, which is the cosine of the angle between $O\zeta$ and Oz, so that $n_0 = \cos i$, where i is the orbital inclination. The y and ξ axes are always perpendicular in the given arrangement making $l_1 = 0$. Also fairly directly found is $l_0 = \cos\phi\sin i$. The other dcs in the complete set follow from the imposed orthogonality constraints, so that:

$$l_0 = \cos\phi\sin i \quad m_0 = -\sin\phi\sin i \quad n_0 = \cos i$$
$$l_1 = 0 \quad m_1 = n_0/l_2 \quad n_1 = -m_0/l_2$$
$$l_2 = \sqrt{1 - l_0^2} \quad m_2 = -l_0 m_0/l_2 \quad n_2 = -l_0 n_0/l_2 \ . \tag{1.12}$$

Note that n_0 is usually a small quantity for eclipsing binary systems. The other two dcs, n_1 and n_2, involve the orbital phase of $O\xi$ and will vary with time.

It may be helpful to have a sense of these arrangements for the binary system and the corresponding dcs. Thus, at the approaching (negative phase) elongation, $O\xi$ will coincide with the x-axis over to the left of the central ('vertical') ζz plane as shown in Fig 1.11. At that time, the y axis will be in this same plane, close to the direction vertically below $O\zeta$. The direction cosine n_1 relating Oy to $O\zeta$, in the highly inclined system shown, will then be close to –1. During the phases about conjunction, n_1 moves from this initial value, through zero at the conjunction, then on to near +1 at the positive elongation. The Ox axis remains 'below' the orbital plane in this transition, except at the elongation phases when Ox and $O\xi$ coincide. The dc n_2 is therefore negative through this transition, except at the elongations where it becomes zero.

In the configuration often adopted in studies of close binary systems $O\zeta$ is also the rotation axis of the primary. However, it is of interest to generalize the geometry to cover situations where the rotation and orbital axes are unaligned. If the bodily rotation axis is displaced, we can set it at $O\zeta'$, say. This is accomplished by an axis rotation about the $O\zeta$ axis from the central meridian by a *precession angle* ψ, shown in a heavier arc in Fig 1.12. A shift by *obliquity* ϵ, shown also in a heavier arc, about the new η axis completes the displacement.

In order to change the dcs from n_0, n_1 and n_2 to the displaced ones n_0', n_1' and n_2', we apply spherical trigonometry in the relevant triangles. The first of these is $\zeta\zeta'z$, where the cosine rule for the angle $zO\zeta'$ gives:

$$n_0' = n_0\cos\epsilon + \sin i\sin\epsilon\cos\psi \ . \tag{1.13}$$

The other two modified dcs involve auxiliary angles at the pole $O\zeta$. For the first of these (α) we again use the cosine rule in $\zeta z y$, where the cosine of the right angle yOz gives: $0 = n_0 n_1 + \sin i\sin\nu_1\cos\alpha$, so that

$$\cos\alpha = -n_0 n_1/\sin i\sin\nu_1 \ , \tag{1.14}$$

---

[*]Alternative arrangements for labelling axes are possible, and this may lead to ambiguities in the assignment of system parameters. An investigator should therefore initially clarify the adopted frame of reference for a given parameter set.

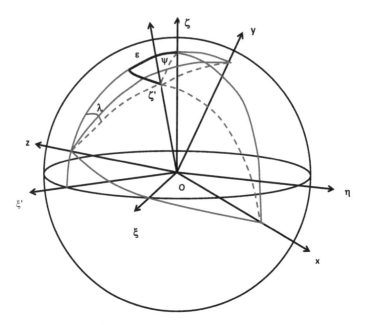

Figure 1.12   The rotation axis $O\zeta'$ is displaced from that of the orbit $O\zeta$ through the precessional angle ψ and obliquity angle ϵ (indicated by slightly heavier arcs).

where we have written ν_1 for the angle whose cosine is n_1 and α is the angle $z\zeta y$. In spherical triangle $\zeta\zeta'y$, the cosine rule now yields

$$n_1' = n_1 \cos\epsilon + \sin\epsilon \sin\nu_1 \cos(\alpha - \psi) \ . \tag{1.15}$$

Similarly, in the spherical triangle $\zeta\zeta'x$, we can write

$$n_2' = n_2 \cos\epsilon + \sin\epsilon \sin\nu_2 \cos(\beta - \psi) \ , \tag{1.16}$$

where the angle β is given from the triangle ζzx, as above for α, so that

$$\cos\beta = -n_0 n_2 / \sin i \sin\nu_2, \tag{1.17}$$

where ν_2 is the angle whose cosine is n_2.

In this way, we determine the modified dcs for the rotation axis transformation in the configuration shown, where $O\xi$ is in the positive quadrant with respect to the central meridian. For negative phases, $O\xi$ is in the fourth quadrant and the foregoing auxiliary angles become their circular complements, allowing sign ambiguities to occur. If we leave the formulae as given above we would obtain an effect, when accounting for a displaced rotation axis, that is symmetrical about the central meridian. This point is taken care of by setting $\psi = -\psi$ for the negative phases, and then the subtractions involving ψ in (1.15–16) have the appropriate values.

The *tilt angle* λ, sometimes arising in this context, is the angle $\zeta z\zeta'$ in Fig 1.12. It can be derived from the cosine rule in the triangle $\zeta z\zeta'$, thus:

$$\cos\lambda = (n_0 n_0' \cos\epsilon)/(\sin i \sin i'). \tag{1.18}$$

It may be useful sometimes to take λ and ϵ as the basic angular variables, rather than ψ (and ϵ) since λ may be determined more readily from observations. In that case, we can reverse some of the foregoing equations, noting first that

$$n_1' = \sin i \cos\lambda. \tag{1.19}$$

But, from the foregoing, we have

$$\cos(\alpha - \psi) = (n_1' - n_1 \cos\epsilon)/\sin\epsilon \sin\nu_1, \tag{1.20}$$

where α was specified above. We therefore obtain

$$\psi = \arccos\{(n_1' - n_1 \cos\epsilon)/\sin\epsilon \sin\nu_1\} - \alpha \tag{1.21}$$

i.e. ψ is derived from given values of the obliquity ϵ and tilt λ, together with the original values of the direction cosines n_0, n_1, and n_2.

1.3.2 Keplerian Elliptic Motion

Two of the great milestones in the development of science were the establishing, early in the seventeenth century, on the basis of much carefully collected and analysed observational data by Johannes Kepler, that planetary movements in the solar system satisfied the three laws set out at the beginning of this section, and that these laws could be reconciled with the single theory of gravitation proposed by Isaac Newton later in the same century. This subject has been studied in great detail and in many ways since the time of Newton and reviews abound. In the following few paragraphs, the main points relevant to later chapters are summarized.

Relative orbit and reduced mass

If we have two mass points M_1 and M_2, say, situated with radial coordinates r_1, r_2 from the origin, projecting as x_1 and x_2 along the x-axis, and similarly in the y and z directions, then the Newtonian law of force between them is the symmetrical $-GM_1M_2/r^2$, where r is the distance between M_1 and M_2. The term mass-point is an idealization, but it can be shown that, in the context of Newtonian mechanics, this is equivalent to a finite-sized body whose material is uniformly distributed in concentric spheres about its central point. In such a case, the force on M_2 accelerates it so that we can write for the x-component of that force

$$M_2\ddot{x}_2 = -\frac{GM_1M_2}{r^2}\frac{(x_2 - x_1)}{r} , \tag{1.22}$$

and similarly for the y and z components. But at M_1 we also have

$$M_1\ddot{x}_1 = \frac{GM_1M_2}{r^2}\frac{(x_2 - x_1)}{r} .$$

If we divide the first of these equations by M_2 and the second by M_1 and subtract the results we find the formula for the *relative* motion

$$\ddot{x}_2 - \ddot{x}_1 = -\frac{G(M_1 + M_2)}{r^2}\frac{(x_2 - x_1)}{r} ,$$

or

$$\ddot{x} = -\frac{\mu}{r^2}\frac{x}{r} , \tag{1.23}$$

with $\mu = G(M_1 + M_2)$. So the same orbital form holds for an individual components, M_2 say, about the system centre, as in the *relative* orbit about M_1, regarded as fixed, but with M_2 scaled down by the factor $1/(1 + q)$, where q is the mass ratio M_2/M_1. The quantity $M_2/(1 + q)$ is known as the *reduced mass* of M_2. In the solar system context, where q is very small, the mass reduction factors for the planets are very close to unity.

Angular momentum: the areal law

The forces in the two mass-point problem are always directed in the line connecting the points. There is thus no torque leading to a change of angular momentum. The motion will also remain in the plane containing the line of centres and the instantaneous velocity vector, since an increment of motion out of that plane would imply some component of acceleration, or torque, about the line of centres.

The relative orbit should thus be expressible as a function $r(\theta)$, say, in plane polar coordinates. Kepler's first law, in fact, states that the form of this function should be an ellipse, or, in polar co-ordinates,

$$\frac{1}{r} = \frac{1 + e\cos\theta}{a(1 - e^2)} \ , \tag{1.24}$$

where r is measured from the focus of the ellipse, whose semi-major axis is a and eccentricity e, and θ is the angular argument measured from the initial line joining the focus to the closest point of the ellipse to the focus: the *periastron* position for binary stars.

Kepler's second law refers to the constancy of the orbital angular momentum, usually worded as that the orbiting body sweeps out equal areas at the focus in equal times. Constancy of the orbital angular momentum entails that $\mathrm{d}(r^2\dot\theta)/\mathrm{d}t = 0$, or, integrating with respect to time t,

$$r^2\dot\theta = h \ , \tag{1.25}$$

where the constant h is twice the rate at which areas are swept out by the radial line to the orbiting point.

Kepler's third law

The average rate of angular motion $\dot\theta$ is usually denoted by n. Kepler's third law, originally for just the planets of the solar system, is

$$n^2 a^3 = \text{const.} \tag{1.26}$$

The constant can be determined using the orbital period P, since $n = 2\pi/P$, while the area of the complete ellipse is $\pi a^2\sqrt{1 - e^2}$, so that $h = na^2\sqrt{1 - e^2}$, and so, squaring the latter,

$$n^2 a^3 = \frac{h^2}{p} \ , \tag{1.27}$$

where $p = a(1 - e^2)$ is the radial separation r given by (1.24) when $\theta = 90°$. The constant h^2/p can be shown to equal the factor μ in (1.23), as will be seen below.

Newtonian motion of two mass points

The Lagrangian formulation of the equations of motion for this two mass-point problem shows that the effective force on the orbiting body is affected not just by that associated with the potential energy gradient but also by the corresponding gradient of the body's kinetic energy, so that

$$\ddot{r} = -\frac{\mu}{r^2} + r\dot\theta^2 \ . \tag{1.28}$$

Similarly, in the direction perpendicular to the line of centres

$$0 = r\ddot\theta + 2\dot{r}\dot\theta \ . \tag{1.29}$$

This latter equation is tantamount to the result that we already found in (1.25), and it allows us to eliminate time derivatives in terms of the polar coordinates. Thus, writing u

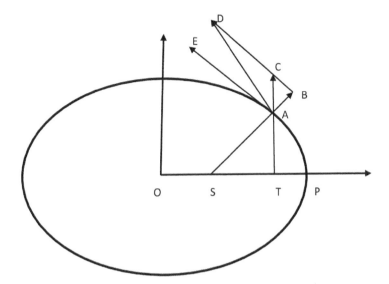

Figure 1.13 Keplerian motion in an ellipse. The focus of the relative orbit is at S. The moving particle is at A with velocity vector in the direction AD.

for $1/r$, we have $\dot{\theta} = hu^2$ and $d/dt = hu^2 d/d\theta$. Hence, $\dot{r}$ becomes $-h\,du/d\theta$, and $r\dot{\theta}$ becomes hu. In this way, after dividing by u^2 and a little rearrangement, (1.28) can be recast in the standard form

$$\frac{d^2u}{d\theta^2} + u = \frac{\mu}{h^2} , \tag{1.30}$$

which has the solution

$$u = \frac{\mu}{h^2}[1 + e\cos(\theta - \omega)] \tag{1.31}$$

with e and ω arbitrary constants. This is of the same form as (1.24) if we set $h^2/a(1-e^2) = \mu$. In other words, Newtonian analysis of the two mass-point problem confirms Keplerian orbital motion and sets the third law in the form

$$n^2 a^3 = G(M_1 + M_2) , \tag{1.32}$$

that appears frequently in applications to binary stars.

As noted above, Kepler's third law takes a useful form as

$$a = P^{2/3}\,(M_1 + M_2)^{1/3} , \tag{1.33}$$

with a in AU, P in years and $(M_1 + M_2)$ in solar masses. Since the masses of stars can often be roughly estimated from their spectral signatures, or other indirect evidence, and the power-law dependence is rather low, (1.33) often allows a fast and reasonably fair estimate of the separation of the components of a binary star with a known orbital period from this elementary calculation.

Orbital velocity

In general, the velocity v of a mass point in a planar orbit about a fixed mass centre can be regarded as the vector sum of two components, in the radial direction v_r and the azimuthal direction v_θ (see Fig 1.13).

In the case of circular orbits $v_r = 0$, and it is easily shown from balancing the centrifugal force with the gravitational attraction that $v^2 = v_\theta^2 = GM/a$. Since $v_\theta = 2\pi a/P$, in this circular case, $a^3 = P^2 \times \text{const} (= GM/4\pi^2)$, thus satisfying Kepler's third law directly. If r changes from a with orbital motion, a natural generalization for the foregoing formula for v^2 is to write $v^2 = GM(2/r - 1/a)$, which follows from the conservation of energy principle, and tends to the same formula for the circular case when $r \to a$; v increasing when $r < a$ and *vice versa* when $r > a$ in accordance with Kepler's areal law.

Using the foregoing formula (1.24) for the ellipse, we would then have

$$v^2 = \frac{GM}{a(1 - e^2)} \times (1 + e^2 + 2e\cos\theta) \ . \tag{1.34}$$

In other words, v^2 contains a constant term plus a term in $e\cos\theta$.

Now the velocity v at any point in the elliptical orbit, indicated in Fig 1.13 as proportional to the length AD, can be written as

$$v^2 = v_r^2 + v_\theta^2 \tag{1.35}$$

where v_r and v_θ represent the velocity components in the radial and azimuthal directions and are proportional to the lengths AB and AE, respectively.

The conservation of angular momentum allows us to put $v_\theta = h/r$, and, using formula (1.24) for $1/r$, this is seen as the sum of a constant term plus a term that depends only on the eccentricity e and $\cos\theta$. These two terms are set proportional to CD and BC in Fig 1.13. The first term is $h/a(1 - e^2)$, from the formula for r. The velocities will then be in proportion to the lengths in Fig 1.13, if CT is perpendicular to the major axis of the ellipse, for then angle ACB $= \theta$, and the ratio of the lengths AC to CD are fixed in the ratio e. Since AB completes the triangle ABD, ensuring that v is the vector sum of v_r and v_θ, we can see that this elliptical motion, subject to conservation of angular momentum, can be represented as the sum of the two constant velocities w_θ and w_y. These velocities are proportional to CD and AC in Fig 1.13, the former perpendicular to the radius vector SA, and the latter perpendicular to the major axis OP.

The magnitudes of these constant velocities turn out to be GM/h and eGM/h, respectively, as can be checked using (1.24), when we find, for example at periastron, the angle θ being zero and the constant $h = r_0 v_0 = \sqrt{GMa(1 - e^2)}$. Equating the areal derivation of $h = 2\pi a^2 \sqrt{1 - e^2}/P$ to this value of $\sqrt{GMa(1 - e^2)}$, and taking squares of both sides, the terms in e drop out and we are left with $a^3 = P^2 \times GM/4\pi^2$, thus confirming Kepler's third law in its usual form. Using the cosine law for the addition of the velocities in triangle ABD, we also confirm that the result gives $v^2 = GM(1 + e^2 + 2e\cos\theta)/a(1 - e^2)$, anticipated from the conservation of energy formula. The average value of v formed by dropping the cosine term in this expression and taking the square root of the result ('RMS value') can be seen to be close to the constant circular velocity w_θ, and tends to the same value in the limit of zero eccentricity.

1.4   Bibliography

1.1 Classical views on what appears in the sky, especially the nature of celestial bodies like the Sun, Moon and stars, are summarized in Aristotle's *De Caelo* (see e.g. *The Works of Aristotle, Encyclopædia Britannica*, Chicago, 1993, pp359–408, translated by J. L. Stocks.), where early concepts about the 'incorruptible superlunary spheres' are found. E. Cassirer's *The Individual and the Cosmos in Renaissance Philosophy*, Univ. Chicago Press, 2010; reviews the contributions of leading thinkers of that later period of scientific revival, including da Vinci, Galileo & Bruno. A. Koestler's *The Sleepwalkers*, Hutchinson, 1959, readably recounts the twists and turns of Western cosmological ideas from their ancient roots up to

Newton. R. G. Aitken's *The Binary Stars*, McGraw-Hill, 1935, gives much historical background specific to the subject of this book in its first two chapters.

The life and times of John Michell are thoroughly covered in Russell MacCormach's *Weighing the World*, Springer, 2012. Z. Kopal's autobiography *Of Stars and Men*, Adam Hilger, Bristol, 1986, includes an account (Ch 6) of John Michell and his contributions to science, mentioning his move to Thornhill in 1767 and possible contact there with William Herschel (further discussed by R. Crossley, *Yorkshire Phil. Soc., Ann. Report for 2003*, p61, 2004). Michell's landmark paper referring to photometric parallaxes of the stars and the significance of their arrangement on the sky was published in the *Phil. Trans. Royal Society*, London, Vol. 57, 1767, pp234–264. His short paper referring to the possibility of a dark star giving rise to an eclipse effect, read to the Royal Society in 1783, was published in the *Phil. Trans. Royal Society*, London, Vol. 74, 1784, pp35–37.

The Introduction to K. G. Jones' *Messier's Nebulae and Star Clusters*, Faber & Faber, London, 1968; provides a background to the development of interest in the wider cosmos beyond the more familiar starry realm at the later years of the 18th and on into the 19th century. This includes, of course, the great contributions of the Herschels, enabled in no small way by their telescopes of improved size and capability. The same (1986) book of Kopal (Ch 6) refers to William Herschel's parallel interest in double stars, culminating in his 1803 publication giving an "Account of the Changes that have happened during the last Twenty-five Years in the Relative Situation of Double-stars",*Phil. Soc. Royal Society, London*, Vol. 93, pp339–382, 1803. This paper presents evidence for the gravity induced orbital motion of Castor A and B, as well as the components of γ Leo, ϵ Boo and ζ Her. Bradley's early measurements of Castor are recalled on page 122 of the Edinburgh Review of 1837, Vol. 65, No. 132., communicated by P. Rigaud. The development of micrometers in this context was reviewed by R. C. Brooks, *J. History Astron.*, Vol. 22, p127, 1991.

Given the acceptance of the extension of Newton's Law of Gravitation to the realm of the stars, Herschel still appeared sceptical about the possibility of *close*, i.e. eclipsing, binaries; at least judging by his appraisal of the discovery and interpretation of the periodic variability of Algol (β Per) by the young astronomer John Goodricke of Hunsingore, Yorkshire. Goodricke's paper appears in the *Phil. Trans. Royal Society, London*, Vol. 73, 1783, pp474–482. Kopal's (1986) autobiography, again in Ch. 6, contains a review of Goodricke's highly significant, but sadly short, astronomical career.

F. W. Bessel's determination of the parallax of 61 Cyg is presented in the *Monthly Notices of the Royal Astronomical Society*, Vol. 4, p 152, 1838. The heliometer shown in Fig 1.4 was based on a design of J. Fraunhofer, but completed by G. Merz and delivered to the Königsberg Observatory in 1829. The historical context was reviewed by R. Willach in the *J. Antique Telescope Soc.*, issue 26, p5, 2004, who credited the original copperplate illustration to the Deutsches Museum, Munich. W. H. Wollaston's paper noting the presence of dark lines in the solar spectrum was published in the *Phil. Trans. Royal Society, London*, Vol. 92, p365, 1802. Information on the remarkable life and work of Joseph von Fraunhofer is collected together by T. Hockey in *The Biographical Encyclopedia of Astronomers*, Springer, p388, 2007. The paper on spectral analysis of G. Kirchhoff and R. Bunsen appeared in Vol. 22 of the *Philosophical Magazine and Journal of Science*, pp329–349, 1861.

Christian Doppler's principle was published under the auspices of the Gesellschaft der Wissenshaften in Prague in 1842. It is interesting that Doppler's original aim was to offer a possible explanation of the different colours of stars in binary systems in terms of their radial velocities. The effect is indeed present, and is included in the interpretation of modern high accuracy photometry, but its scale is very much less than that entertained by Doppler. Early discussion of the effect is also credited to H. Fizeau (Paris, 1870]. *Ann. Chimie et Phys.*, Vol. 19, p211, 1870, (Paris). G. Kirchhoff published his law of thermal radiation in the *Ann.*

Phys. Chemie., Vol. 109 (2), pp275–301, 1860. J. C. Maxwell's theory of electromagnetic waves appeared in the *Phil. Trans. Royal Society, London*, Vol. 155, pp459–512, 1865.

The development of stellar spectroscopy from such early luminaries as Airy, Secchi and Huggins to the influential work of W. W. Morgan, P. C. Keenan and E. Kellman in *An Atlas of Stellar Spectra with an Outline of Spectral Classification*, Univ. Chicago Press, 1943; is thoroughly backgrounded in J. Hearnshaw's *Analysis of Starlight*, CUP, 1986. Spectral classification was reviewed in *The Classification of Stars*, by C. Jaschek and M. Jaschek, CUP, 1987; and more recently by R. O. Gray and C. J. Corbally in *Stellar Spectral Classification*, Princeton University Press, 2009. Huggins' early radial velocity measures for 30 stars appeared in *Proc. Royal Society*, Vol. 20, p359, 1872. H. C. Vogel's paper demonstrating the binary nature of Algol was published in *Aston. Nachrichten*, Vol. 123, p289, 1890. The comparable identification of the binary nature of Mizar (ζ U Ma), by E. C. Pickering, was published in the same year 1890, in the *Amer. J. Science*, Vol. 34, p46. It has since transpired that both components of the Mizar visual binary are themselves close pairs. An interferometric study of Mizar A was carried out by J. A. Benson et al., (14 authors) Astron. J., Vol. 114, p1221, 1997.

E. C. Pickering's analysis of the light curve of Algol can be found in the *Publ. Amer. Acad. Arts & Sciences*, Vol. 16, p1, 1881. Schönfeld's data, used in this analysis, are cited from Vol. 36 of the *Jahresbericht Mannheim Verein für Naturkunde*, p70, 1870. The development of visual photometry from the late 18th through the 19th centuries is detailed in J. B. Hearnshaw's (1996) *The Measurement of Starlight*, CUP, especially Chapters 2 and 3. Chapter 5 in Hearnshaw's *Analysis of Starlight*, is devoted to the work of E. C. Pickering and his energetic team of female assistants at Harvard College Observatory over the last two decades of the 19th and first two of the 20th centuries.

Stebbins' historic observations of Algol were published in the *Astrophys. J.*, 1910, Vol. 32, p185, and recalled by G. E. Kron in Chapter 2 of *Photoelectric Photometry of Variable Stars.*, Willmann-Bell, Richmond, Va; eds. D. S. Hall & R. M. Genet, 1988. Also in 1910, the departure from a regular sinusoid in the radial velocity curve of an eclipsing binary system during eclipse phases was discussed by F. Schlesinger at the end of his paper on δ Lib, in the *Publ. Allegheny Observatory*, Vol. 1, p123, 1910.

History of the analysis of the light curves of eclipsing binary systems is well covered in the bibliographical notes on Chapter 6 of Kopal's (1959) *Close Binary Systems*, pp443–447. The papers of H. N. Russell and H. Shapley that set the stage for a great deal of subsequent data analysis were published in 1912 in the *Astrophys. J.*, Vol., 35, p315 and Vol. 36, p54 (Russell) and Vol. 36, p239 and p385 (Russell & Shapley). Shapley's thesis was published as the *Princeton Contr.* No 3, 1915.

The evolution of ideas that led to the appearance of the 'Herzsprung-Russell Diagram', as it has become known, received input from a number of sources. The subject is detailed in D. H. DeVorkin's *The Origins of the Hertzsprung-Russell Diagram*, published in *The HR Diagram, In Memory of Henry Norris Russell*, IAU Symp. 80, 1977, p61, eds. A. G. Davis Philip & D. H. DeVorkin.

G. Kron's light curves of YZ Cas were given in the *Lick Obs. Bull.*, No 499, 1939 and *Astrophys. J.*, Vol. 96, p173, 1942. H. L. Johnson's specifications for the well-known *UBV* magnitude system appeared in the *Astrophys. J.*, Vol. 114, p522, 1951. Further bibliographical notes on the development of stellar photometry can be found at the end of Chapters 3 & 4 in the *Introduction to Astronomical Photometry* of E. Budding & O. Demircan, CUP, 2007.

Charge-coupled devices (CCDs) have grown into very widespread use in astronomy since their introduction at by W. Smith and G. E. Boyle at Bell Laboratories in the late 1960s, with most astronomical instrumentation nowadays employing solid-state imaging devices one way or another. The growth of this technology was reviewed by M. Lesser in the *Publ. Astropn. Soc. Pacific*, Vol. 157(927), 2015.

IAU Symposium 88 on *Close Binary Stars*, edited by M. Plavec, D. M. Popper & R. K. Ulrich, was published by Reidel in 1980 and contained over 100 papers arranged over 16 related fields. Fred Hoyle's provokingly worded early presentation of interactive binary evolution was in his popular 1955 book *Frontiers of Astrononmy*, Heinemann, London, pp195–202.

The idea of overcoming the problems of observing through the Earth's atmosphere through the use of space-based observatories is old, but it was not until the formation of government-supported space facilities from the late 1950s that this idea could be realized. OAO2 was an early example (1968) of a space telescope concentrating on cosmic UV sources. The development of such resources was reviewed in J. A. Angelo's (2014) *Spacecraft for Astronomy*, Infobase Publishing, p20. A recent review of space-based photometry of close binary systems, and related topics, including pulsation effects, multi-eclipsers, heartbeat stars and exoplanets, was published by J. Southworth, arXiv211003543S, 2021. J, A. Eaton's 1975 paper on Algol in the UV range was published in *Publ. Astron. Soc. Pacific* Vol. 87, p745. The International Ultraviolet Explorer (IUE) satellite was also highly productive facility for the UV range from the last two decades of the twentieth century, reviewed by A. Boggess et al. (10 authors) in *Nature*, Vol. 275, p372, 1978.

K-Y Chen & E. G. Reuning's IR light curve appeared in the *Aston. J.* Vol. 71, p283, 1966. C. M. Wade & R. M. Hjellming's findings of the microwave emission from Algol were published in *Nature* Vol. 235, p270, 1972. The discovery of the binary neutron star PSR B1913+16 was reported by R. A. Hulse and J. H. Taylor in the *Astrophys. J.*, Vol. 195, L51, 1975. This topic is revisited in Section 3.4.3.

The X-ray spectrum of Algol was reported on by N. E. White, R. H. Becker, S. S. Holt, R. F. Mushotzky, E. A. Boldt and P. J. Serlemitsos in the *Bull. Amer. Astron. Soc.*, Vol. 11, p782, 1979; and in *Astrophys. J.*, Vol. 239, L69, 1980; from where Fig 1.8 derives. The early development of X-ray astronomy is reviewed by K. Pounds in Frontiers of X-ray Astronomy, by A. C. Fabian, K. Pounds & R. Blandford, CUP, p1, 2004. J. Tomkin & D. L. Lambert's historic identification of the secondary spectrum of Algol was published in the *Astrophys. J.*, Vol. 222, p119, 1978.

After a lengthy process of negotiation, ESA accepted the proposal for what became the HIPPARCOS satellite in 1980. Background on the motivation for this facility can be found in the article of J. Dommanget in the proceedings of the Conference on Astrometric Binaries, published in *Astrophys. Space Sci.*, Vol. 110, p47 , 1985.

The historic discovery of 51 Peg-b by M. Mayor and D. Queloz appeared in *Nature*, Vol. 378, p355, 1995. Three years before that, also in *Nature*, Vol. 355, p145, 1992, was the report on the planetary system around the millisecond pulsar PSR1257 + 12 of A. Wolszczan, & D. A. Frail. The Kepler-1 transit, seen in Fig 1.9, was picked up by F. T. O'Donovan, D. Charbonneau et al. (20 authors) as part of the Transatlantic Exoplanet Survey (TrES), and published in *Astrophys. J. Letters*, Vol. 651, L61, 2006.

The Hubble Space telescope (HST), launched in 1990, remains a major resource for astronomical science, involving the collaboration of both NASA and ESA in a long-term commitment to high-quality space-based observations. It is easy to find up-to-date information on the facility, and its long record of impressive results, from a number of internet locations, for example, www.spacetelescope.org.

The scientific aims of the Kepler Mission were detailed by W. J. Borucki et al., (14 authors) in *Scientific Frontiers in Research on Extrasolar Planets*, Eds. D. Deming & S. Seager, *ASP Conf. Ser.* 294, p427, 2003; while W. J. Borucki et al. (69 authors), *Astrophys. J.*, Vol. 736, p19, 2011, provided an early summary of interim results.

Insights into physical processes involved in interacting close binary systems like Algol arising from Doppler tomography were demonstrated by M. T. Richards, M. I. Agafonov, & O. I. Sharova in their paper in *Astrophys. J.*, Vol. 760, p8, 2012. VLBI observations of

Algol in the microwave range, leading to a picture of energized coronal loops around the cool subgiant component were shown in the *Astrophys. J.*, Vol. 328, p232, 1988, paper of J-F. Lestrade, R. L. Mutel, R. A. Preston, and R. B. Phillips. The interesting, very high resolution, map of the Algol system, including also the wide component Algol C, produced using the CHARA interferometer by F. Baron and colleagues (15 authors), was published in the *Astrophys. J.*, Vol. 752, p20, 2012.

1.2 Most of the basic terms and their usages raised in this section are detailed in more general books on astronomy, e.g. *The Cambridge Encyclopedia of Astronomy* of S. Mitton, Crown Publishers, 1981; J. M. Passachoff & A. V. Fillipenko's *The Cosmos*, Thomson-Brooks/Cole, 2007; or M. Zeilik's *Astronomy: The Evolving Universe*, CUP, 2002; and others. A good introduction to terms particular to the context of binary and multiple stars are explained in the opening chapter of A.H. Batten's *Binary and Multiple Systems of Stars*, Pergamon Press, 1973; as well as Chapter 1 of R. W. Hilditch's *An Introduction to Close Binary Stars*, CUP, 2001.

1.3 A background to much of the material in this section, especially spherical trigonometry and Keplerian motion, can be found in W. M. Smart's *Text-Book on Spherical Astronomy*, (rev. R. M. Green), CUP, 1977. The subject is also covered in detail in Chapter 2 of Hilditch's *Introduction* (op. cit.). See also A. E. Roy's *Orbital Motion*, CRC Press, 1978. A summary of the inverse data modelling problem appeared in E. Budding's *Introduction to Astronomical Photometry*, CUP, 1993, Chapter 7, together with a brief discussion of the fitting of light-curves. The subject gives the theme of the comprehensive reviews: Inverse *Problems in Astronomy*, by I. J. D. Craig & J. C. Brown, Adam Hilger, 1986; and *Definitive Parametrization of Inverse Problems in Astrophysics* of A. V. Goncharski, S. Yu. Romanov & A. M. Cherepashchuk, Moscow Univ. Publ., 1991.

2

Statics

2.1 Equipotential Surfaces

Study of the interactions of material bodies led to the notion of a 'potential energy', whose spatial gradient is associated with a force. Such ideas arose from the context of motion under gravity; for example, with planetary bodies orbiting a massive central object. When further out from the central object, a body could be considered to have more potential energy. If falling to a position closer in, while following an eccentric orbit, say, a force must have acted, whose result is the conversion of some of the potential into 'kinetic energy': the work done by the attractive force corresponding to the increase of kinetic energy. Conversely, the extra kinetic energy supplied to the body that would move it back to its original more distant position accounts for the work done against the attractive force on the return journey.

The concept of an equipotential surface arises from such considerations. This relates to the existence of macroscopically static situations, in which a large body remains more or less in the same condition for long periods of time. In such a state, there is local balance of the net forces in their combined line of action. The plane perpendicular to this direction coincides tangentially with the local equipotential surface.

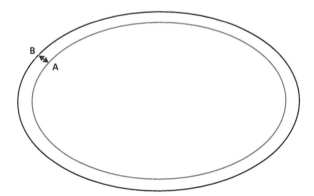

Figure 2.1   Adjacent equipotential surfaces.

DOI: 10.1201/b22228-2

In Fig 2.1, we sketch two such closed and connected surfaces A and B lying close together within a fluid body of uniform composition at rest with regard to the net action of forces. Consider a small cylinder of matter with unit cross-sectional area and of length δs joining two such surfaces along the local normal. Let us set the density in this narrow space as ρ, so the mass becomes $\rho \delta s$. The pressure at A is δP greater than at B, i.e. a negative pressure differential in the outward direction, implying an outward force. This force is balancing a 'body force' experienced by the matter in the cylinder, associated with the difference in the potential energy $\delta \Phi$ between the two surfaces. A function Ψ can be introduced here as the potential energy per unit mass, such that $\rho \delta s \, \delta \Psi = \delta \Phi$. The gravitational force on the small material cylinder $-\rho \delta s . d\Psi/ds$ (negative for an attractive force), is balanced by the pressure difference, of essentially kinetic nature at the microscopic level, so that, from A to B, we write

$$\delta P = -\rho \delta \Psi \ . \tag{2.1}$$

The difference $\delta \Psi$ is constant for all locations over the two equipotentials, so the local body force on unit mass will be inversely proportional to their local separation δs, while the force balance over these surfaces implies the pressure difference δP, over that same (varying) spatial increment δs must also be constant: the pressure *gradient* being steeper in locations where the potential gradient is also steeper by the same proportion. This, in turn, requires constancy of ρ between the equipotentials, or that the variables P, ρ and (through the equation of state, for a gas of given mean molecular weight) the temperature T can all be regarded as functions of the function Ψ in this idealized* static configuration.

In classical Newtonian terms, this function Ψ, referred to generally as the 'potential', operates wherever there is matter to exert a corresponding field of force. Consider a small unit of mass located at a distance r from an attracting massive particle of mass m. The classical field potential for the mass unit is given by $\Psi = -Gm/r + A$, where G is Newton's gravitation constant, and A is an arbitrary integration constant.

Let us introduce here an essentially similar function V applying to unit mass at a point M (Fig 2.2), external to a spherical accumulation of matter centred at distance r from M, where a typical point on a shell of radius r' is labelled M$'$, through the expression

$$V = G \int \frac{dm'}{R} \ , \tag{2.2}$$

where R is the distance between M and M'. The mass element dm' is given, in a naturally applicable spherical polar co-ordinate system, by

$$dm' = \rho r'^2 dr' \sin\theta' d\theta' d\phi' \tag{2.3}$$

with

$$R^2 = r^2 + r'^2 - 2rr' \cos\gamma \ , \tag{2.4}$$

r, θ, ϕ, being the spherical polar co-ordinates of the point M, and then

$$\cos\gamma = \cos\theta \cos\theta' + \sin\theta \sin\theta' \cos(\phi - \phi') \ . \tag{2.5}$$

The radial integration extends from the centre ($r' = 0$) to M ($r' = r$) and the angular ones (θ', ϕ') extend to cover a complete spherical surface.

---

*Although close observation shows that stellar surfaces are not in strict dynamical equilibrium, this idealization has proven a useful first approximation.

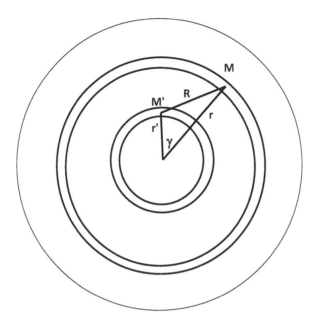

Figure 2.2   Distribution of matter in spherical layers. The geometry is relevant to expansions like (2.6).

The function V is understood to mean that which when differentiated gives the gravitational force on unit mass, although this meaning differs from the previously used Ψ, which is negative for bound matter. But the derivative of Ψ requires a minus sign for an attractive force, so the end result is the same. We will retain this positive V in (2.2) and follow a customary usage of the word potential for this context, while keeping in mind that a higher potential energy would go with smaller numerical value of V.*

In a spherically symmetrical situation, there is no loss of generality in setting the direction of M to be at $\theta = 0$, so that the denominator in (2.2) becomes $r\sqrt{1 + t^2 - 2t \cos \theta'}$, where $t = r'/r$. A direct integration of (2.2) in the angular variables then gives $V = 4\pi G \int \rho r'^2 dr'/r$; in other words, the potential at M exterior to a uniform spherical shell is the same as though all its matter were concentrated at the central point and depends only on r. In a similar way, if the unit mass at M was located internal to M' the corresponding integral over the whole sphere would vanish. These well-known results establish a conceptual 'undisturbed' situation, in which the equipotentials for a uniform mass of gas subject only to gravitational self-attraction are concentric spheres about the centre of gravity. The visible surface of such an idealized accumulation would be one such spherical equipotential.

A development of this picture occurs with the introduction of a relatively low centrifugal force associated with a uniform rotation of the fluid mass. The configuration then becomes distended, the equipotentials pushing out in equatorial regions to form oblate spheroids. As the separation of adjacent equipotentials becomes greater towards the equator, the net inward force decreases, reflecting the centrifugal action's opposition to gravity, equivalent to the reduced net potential gradient. The combination of effects is dealt with by making use of the additive property of the potential functions: the ellipsoidal equipotentials of the rotating body deriving from adding a centrifugal potential function to the gravitational one, that becomes spherically symmetric in the absence of rotation.

---

*The term 'binding energy' is also used in this connection.

Consider now the series expansion

$$\frac{1}{R} = \frac{1}{r}\frac{1}{\sqrt{1 - 2t\cos\gamma + t^2}} = \frac{1}{r}\sum_{0}^{\infty}P_n(\cos\gamma)t^n \ , \tag{2.6}$$

where P_n are Legendre polynomials in $\cos\gamma$ and $t < 1$. The integral (2.2) can thus be expressed as the sum of a convergent series in n of terms:

$$V = \sum_{0}^{\infty}r^{-(n+1)}V_n \ , \tag{2.7}$$

where each of the V_n is an integral of the form

$$V_n = G\int_0^r r'^n P_n(\cos\gamma)dm' \ . \tag{2.8}$$

A closely comparable form exists for a mass element situated internal to a surrounding mass accumulation, i.e. by reversing the arrangement of M and M' in Fig 2.2. The expansion for t (< 1) proceeds with the ratio r/r', so that we have corresponding to (2.7)

$$U = \sum_{0}^{\infty}r^n U_n \ , \tag{2.9}$$

with

$$U_n = G\int_r^{r_1} r'^{-(n+1)} P_n(\cos\gamma)dm' \ , \tag{2.10}$$

where r_1 locates the outer boundary of the matter accumulation under consideration.

When the relative scale of pertinent forces have a clear ordering, as in the case where the centrifugal contribution can be regarded as additive perturbations upon an original spherically symmetric form due to gravitation, distinct inroads into the analysis of the shapes of the fluid masses involved are possible. The integrals formed by combining Eqns (2.3) and (2.8) or (2.10) become tractable with the equipotentials in the form of spherical harmonic functions $Y_j(a, \theta', \phi')$, including the Legendre polynomials $P(\cos\theta')$. This is due to the integrability of the relevant products, i.e. the orthogonality conditions applying to products of spherical harmonics in an integral. The radial part of these Y-functions incorporates a mean radius a, whose local shift to r' is a small fraction of a. For the perturbed radius r' we then utilize the convergent series

$$r' = a\{1 + \sum_{j=2}^{\infty} Y_j^i(a, \theta', \phi')\} \ , \tag{2.11}$$

where a represents the mean radius of an equipotential whose perturbation from sphericity is given through the tesseral harmonics Y_j^i. This leads to (2.8), and similarly (2.10), becoming components of series of integrals in powers of a, where the mixed products of different order harmonics vanish. The usual practice of forming the summation from the second harmonic ($j = 2$) in (2.11) ensures that the zero order radius is the mean value a and the the configuration is centred on the average position of its particles.

The development thus far has referred only to a body's own distribution of matter and its gravitational self-attraction. For a body subject to other effects than self-attraction, with no net motion of any constituent particle in a suitably chosen frame of reference, this can be regarded as in balance with some other source of 'disturbing potential' V_d, expressible as

$$V_d = \sum_{i,j}^{\infty} c_{i,j} r^j P_j^i(\theta, \phi) \ , \tag{2.12}$$

the factors $c_{i,j} r^j$ having the dimension of energy per unit mass. The coefficients $c_{i,j}$ are supplied according to the circumstances of the problem under study. A key point in this problem is to relate the perturbation of the radius expressed in (2.11) through the tesseral harmonics Y_j^i, of initially unspecified scale but dependent on the structure of the body, to a given disturbing potential associated with forces acting in opposition to the underlying self-attraction.

In the case of a centrifugal action, for example, if the mass accumulation has a uniform angular rotation ω and we set the z-axis to be the axis of symmetry, neglecting any sectorial dependence, we would expect the relevant potential (whose derivative gives the centrifugal force) to be proportional to $\frac{1}{2}\omega^2 r^2 \sin^2\theta$. This, together with the potential associated with the disturbed shape of the body, should add up to some constant value over a particular equipotential surface. This would result in $c_2 = -\omega^2/3$ for the relevant coefficient in the disturbing function, keeping in mind that $P_2(\theta) = (3\cos^2\theta - 1)/2$. The increase of the disturbing potential towards the equator (along an equipotential) here accounts for a decrease in the corresponding gravitational contribution on the equipotential.

By balancing the coefficients in the full expansion for the combined potentials (2.8) and (2.10), with the corresponding perturbation from (2.12), since each equipotential surface is characterized by only one value of the total potential with a corresponding value of a and independently of θ or ϕ (i.e. regardless of whereabouts on the surface we may locate a test particle), we arrive at Clairaut's formulation of the first order conditions satisfied by the (internal) equipotential having a mean radius a_0. After some manipulation*, we find

$$Y_j^i \int_0^{a_0} \rho a^2 da - \frac{1}{(2j+1)a_0^j} \int_0^{a_0} \rho \frac{\partial \left(a^{j+3} Y_j^i\right)}{\partial a} da$$
$$- \frac{a_0^{j+1}}{2j+1} \int_{a_0}^{a_1} \rho \frac{\partial \left(a^{2-j} Y_j^i\right)}{\partial a} da \ = c_{i,j} a_0^{j+1} P_j^i(\theta, \phi)/4\pi G \ . \tag{2.13}$$

With further rearrangement, the number of integrals in Eqn (2.13) can be reduced to two and (2.13) cast in the form

$$\left(j\frac{Y_j^i}{a_0^{j+1}} + \frac{1}{a_0^j}\frac{\partial Y_j^i}{\partial a}\right)\int_0^{a_0}\rho a^2 da - \int_{a_0}^{a_1}\rho\frac{\partial}{\partial a}\left(\frac{Y_j^i}{a^{j-2}}\right)da = \frac{c_{i,j}}{4\pi G}(2j+1)P_j^i(\theta, \phi) \ . \tag{2.14}$$

The second integral comes from the internal potential U and it vanishes at the outer surface of the body, where the mean radius of an internal equipotential a_0 attains its surface value a_1.

Noting that

$$\int_0^{a_1} \rho a^2 da = \frac{m_1}{4\pi} = \frac{\bar{\rho} a_1^3}{3} \ , \tag{2.15}$$

for the full mass of the accumulation m_1, the following formula can be produced:

$$c_{i,j} a_1^j P_j^i = \frac{Gm_1}{a_1} \frac{Y_j^i}{\Delta_j} \ . \tag{2.16}$$

---

*More details on the intermediate steps leading up to Eqn (2.13) are given in chapter 2.1 of Kopal's (1959) book.

This deceptively simple version of Clairaut's equation (2.16) manifests the relationship between the disturbing and disturbed effects: the c s to the Y s. We expect the key structural coefficient Δ_j, introduced in (2.16) to be a purely numerical quantity of order unity, but > 1. The harmonic functions Y_j^i follow the same angular variations as the disturbing function, but the material self-attraction scales the response of m_1 to the disturbing potential. Making use of (2.14 and 2.16), Clairaut's result for the external surface can be cast in the form

$$\Delta_j = \frac{(2j+1)}{j + \eta_j(a_1)} \quad , \tag{2.17}$$

where $\eta_j(a_1)$ is the surface value of the logarithmic derivative of the disturbed potential of the star

$$\eta_j(a) = \frac{a}{Y_j^i} \frac{\partial Y_j^i}{\partial a} \quad . \tag{2.18}$$

The reduction to only the index j for Δ and η anticipates that the relevant disturbing potentials will be expressed, in practice, with an appropriate co-ordinate choice, in terms of only zonal Legendre polynomials.

From differentiating Eqn (2.14), we find

$$a^2 \frac{\partial^2 Y_j^i}{\partial a^2} + 6\frac{\rho}{\bar{\rho}} \left(a\frac{\partial Y_j^i}{\partial a} + Y_j^i \right) = j(j+1)Y_j^i \quad , \tag{2.19}$$

and, using (2.18), this can be written as

$$a\frac{d\eta_j}{da} + 6\frac{\rho}{\bar{\rho}}(\eta_j + 1) + \eta_j(\eta_j - 1) - j(j+1) = 0 \quad . \tag{2.20}$$

This first order differential equation for the structurally dependent variable η_j has been called *Radau's equation*. The dependence of η_j on a as a result of the variation of density ρ through the star is studied for different models of stellar structure. The surface of the star, whose form is derivable from analysis of appropriate observations, should then provide information to relate to such modelling.

Clairaut's equation enables us to specify the response to a set disturbing potential of a matter accumulation m_1 through the rearrangement of its equipotential surfaces. The outcome is that the disturbed potential, let us denote it V_1, external to the mass accumulation m_1, and located at at r, θ, ϕ, becomes

$$V_1 = \frac{Gm_1}{r} + \sum_{j=2}^{\infty} \frac{a_1^{2j+1}}{r^{j+1}}(\Delta_j - 1)c_j P_j(\theta, \phi) \quad , \tag{2.21}$$

where the $(\Delta_j - 1)$ factors amplify the role of the mass dispersion in the accumulation on the potential field around it.

2.2   Stellar Structure and Radau's Equation

The mathematical behaviour of the function η_j has been studied in some detail. If the stellar envelope density falls quickly away to zero, i.e. $\rho \to 0$, Eqn (2.20) could clearly be solved by $\eta_j = j+1$, in accordance with the radial component of Y_j^i having the form $c'a^{j+1}$. The coefficient Δ_j would then revert to unity, which accords with an intuitive expectation that, in the absence of matter, the disturbing and responsive potentials directly match up, i.e. $c'_j \equiv c_j/(Gm_1/a_1)$.

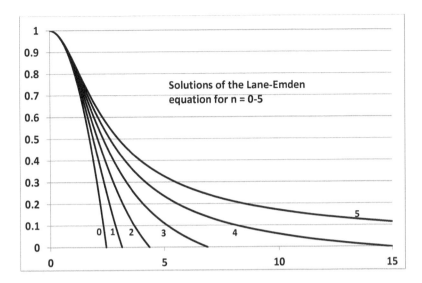

Figure 2.3 The diagram shows the degree of central condensation of an idealized gas-sphere in response to different values of the index $n = 1/(\gamma - 1)$ in the Lane-Emden structural equation. The ordinates τ are normalized to the same unit central value, while the abscissae r are proportional to the model's radius.

A finite density $\rho > 0$ has the effect of reducing η_j in (2.20) in order to balance the terms in η_j with the constant $j(j+1)$. As mentioned above, this entails a diminution of the denominator in (2.17) and an amplification of the surface distortion through corresponding increase of the coefficient Δ_j. For a body of uniform density, (2.20) can be seen to be satisfied by $\eta_j = j - 2$, so that $\Delta_j = (2j+1)/(2j-2)$. But this would be the maximum increase of Δ_j feasible for a regular astrophysical body in equilibrium.* For bodies with some degree of central condensation, like stars, η_j tends rather quickly towards $j + 1$, so that $\Delta_j \to 1$ in a similar way. $\Delta_j = 1$ thus holds for the centrally condensed 'Roche' approximation that we come to in a later chapter.

Let us now return to Equations (2.1-3) in the undisturbed situation at internal radius a so that

$$\frac{a^2}{\rho}\frac{dP}{da} = -Gm_a \ , \tag{2.23}$$

where m_a is the spherically symmetric mass internal to a. Let us suppose that P is of the form

$$P = c\rho^\gamma \ , \tag{2.24}$$

c being a constant. We will consider physical implications of this later. Eqn (2.23) can now be rearranged as

$$\left(\frac{\gamma c}{4\pi G}\right)\frac{1}{a^2}\frac{d}{da}\left(a^2\rho^{\gamma-2}\frac{d\rho}{da}\right) = -\rho \ . \tag{2.25}$$

---

*From examining limiting cases of Radau's equation, Kopal showed that

$$1 \leq \Delta_j \leq \frac{2j+1}{2(j-1)} \tag{2.22}$$

for density distributions that decrease outward. For the cases $j = 2, 3, 4$ that become of special interest in practical modelling, the Δ_j coefficients have to be between 1 and 5/2, 7/4 and 3/2, respectively.

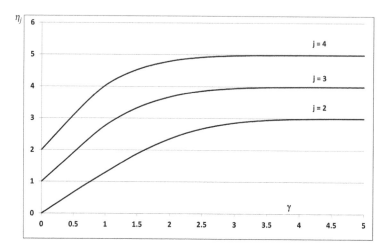

Figure 2.4 Solutions of Radau's equation for the surface values of η_j ($j = 2, 3, 4$) for different values of the index n in the Lane-Emden structural equation.

An idealized gas law of the form $P = R\rho T$ will connect the pressure P, density ρ and temperature T, R being the appropriate gas constant. Eqn (2.25) can thus be recast as a somewhat simpler differential equation for the temperature,

$$\frac{1}{r^2}\frac{d}{dr}\left(r^2\frac{dT}{dr}\right) = -T^n \;, \tag{2.26}$$

where the radial variable r represents a scaled by the constant $\sqrt{\gamma c^{3-\gamma}R^{\gamma-2}/(4\pi G)}$ and $n = 1/(\gamma - 1)$. This can be further normalized by introducing the variable $\tau = T/T_0$, so that ρ is given in terms of its central value as $\rho = \rho_0\tau^n$ and the scaling constant becomes $\sqrt{(n + 1)c\rho_0^{\gamma-2}/(4\pi G)}$. Eqn (2.26) is in the form known as the Lane-Emden equation (or sometimes just Emden's equation) with index n. It has played an important part in the development of understanding of stellar structure.

The form of Eqn (2.24) is strongly reminiscent of the connection between pressure and density in a gas undergoing adiabatic changes. In fact, if the outward flow of heat within a star is being transferred by a convective process such a condition is expected to hold to a good approximation within the convective region. An appropriate value of γ can be shown to be close to 5/3, making $n = 3/2$. On the other hand, a predominantly radiative heat transfer process can be regarded as depending on the flow of a 'photon gas', for which $\gamma = 4/3$, or $n = 3$, becomes more relevant. Corresponding degrees of central condensation can be assessed from the corresponding curves in Fig 2.3. As a generalization from such contexts, γ in (2.24) is termed the polytropic index.

Table 2.1   Structural coefficients for polytropic models.

n	k_2	β
0	0.75	0.4
1	0.2599	0.2614
2	0.0739	0.1548
3	0.0144	0.0755
4	0.0012	0.0226
4.5	0.0001	0.0094

R. A. Brooker and T. W. Olle tabulated values of the solutions $\eta_j(a_1)$ for polytropic models of stellar structure, with $j = 2, 3, \dots 7$; and 14 values of the corresponding

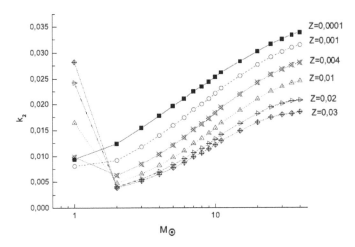

Figure 2.5 Results of numerical quadrature of Radau's equation for detailed stellar model calculations with varying metal-content in their composition. For Main Sequence stars more massive than the Sun the effective central condensation tends to increase with increasing metallicity, passing through a maximum mear the solar mass. For lower mass stars increased metallicity increases k_2 values, i.e. the stars' effective central condensation is reduced.

(Lane-Emden) index n in the range $0 \leq n \leq 5$. Their data (Fig 2.4) show a rapid increases of η_j towards $j + 1$ with increasing index n, i.e. central condensation. For most realistic models of stars, therefore, $\eta_j \sim j + 1 - \delta$, where δ is relatively small. This means that $\Delta_j \approx 1 + \epsilon$, where $\epsilon = \delta/(2j + 1)$ can be considered a small quantity, as can the alternative structural coefficients $k_j = (\Delta_j - 1)/2$. These latter coefficients are frequently used in discussions of the *apsidal motion* observed in eccentric close binary systems. Brooker and Olle's (1955) results, cited in Kopal's (1959) book, have been used in modelling the distortion of stars subject to rotational and tidal forces and form a useful basis of comparison for more recent studies.

The quadrature of Eqn (2.20) for numerical stellar models is an easy programming exercise when regarded as the pair of difference equations

$$\begin{aligned} \Delta\eta_{j,i} &= (\Delta a/a)_i f(\rho, \overline{\rho}, a, \eta_j)_{i-1} \\ \eta_{j,i} &= \eta_{j,i-1} + \Delta\eta_{j,i} \ . \end{aligned} \tag{2.27}$$

The first of these equations determines the increment for η_j at the i'th layer of the stellar model in terms of the underlying layer, and the second equation then updates η_j in this currently calculated layer. The function f comes from re-arranging Eqn (2.20) so that all terms except the first derivative are moved over to the right of the equation with their numerical values taken from the $(i - 1)$th layer. The quadrature can proceed through all the layer increments for which a given model is tabulated; from $i = 1$ (from the centre, where $\rho = \overline{\rho}$ locally, and $\eta_{j,0} = j - 2$) through to the surface, from where we emerge with the sought output value of $\eta_j(a_1)$. The stability and accuracy of the quadrature can be tested, for example, by reproducing the classical results of Brooker & Olle. Results of such numerical integrations for contemporary stellar models are shown in Fig 2.5. This diagram shows that the central condensation, which decreases with increasing k_2, becomes lower in more massive stars, and more so at lower metallicity. However, after passing through a minimum in the vicinity of one solar mass, the opposite happens. This reflects the increasing role of convection in the envelopes of lower mass stars. which starts at higher mass for stars of higher metal content.

2.3   The Potential Function and Potential Energy

The above mention of tidal forces takes us back to the disturbing potential for unit mass within the accumulation m_1 as specified by (2.12). The relevant form is reasonably anticipated from that of (2.7-8). The $j = 0$ term, arising at an external point distant r' from the centre of a body of mass m_2 and producing a tidal effect per unit mass within m_1, is readily given as

$$V_{t0} = \frac{Gm_2}{r'} \ . \tag{2.28}$$

The term in $j = 1$ vanishes as a result of choosing r' to be measured from the centre of gravity of the disturbing mass m_2. The term in $j = 2$ can be shown to be of the form

$$V_{t2} = \frac{G}{r'} \left(\frac{A + B + C - 3I}{2r'^2} \right) \ , \tag{2.29}$$

where A, B, C, are the principal moments of inertia of m_2 and I is its moment of inertia about the line joining the centres of m_1 and m_2. The quantities $(A - I)$, $(B - I)$, $(C - I)$ would vanish in the disturbance-free spherically symmetric case considered at the outset, and from the form of the leading $(j = 0)$ term in the action of m_1 on m_2, using (2.12) and (2.16), we can deduce that the term in question is of order $(Gm_2/r')(a_2/r')^6$. Retaining this order of contribution in the disturbing potential entails taking into account the action of the tides from one star on those of the other and the problem becomes convoluted. Since a stepwise approach to the analysis was implicit from the derivation of the Clairaut equation (2.13), it is logical that we deal primarily with the 'first order' problem, retaining terms in the series expansion for the disturbing potential up to and including order $(a_1/R)^5$, i.e. the radii of the binary components expressed as a fraction of the mean separation of their mass centres. For the source of the disturbing potential operating from m_2 on m_1 in the first-order problem we then have only

$$V_t = V_{t0} \ = \ \frac{Gm_2}{r'} = \frac{Gm_2}{\sqrt{R^2 - 2aR\lambda + a^2}}$$

$$= \ \frac{Gm_2}{R} \sum_0^\infty \left(\frac{a}{R} \right)^j P_j(\lambda) \ . \tag{2.30}$$

From comparing (2.30) with (2.12) we can write

$$c_{tj} = \frac{Gm_2}{R^{j+1}} \ , \tag{2.31}$$

with j effectively running only from 2 to 4 in the present application. Using (2.16), we then have the harmonic functions for the tidal distortion at the surface as,

$$Y_j(a_1) = q\Delta_j \left(\frac{a_1}{R} \right)^{j+1} P_j(\lambda) \ , \tag{2.32}$$

where q is the mass ratio m_2/m_1. From a practical point of view and to have some 'feel' for the relevance of this to the scale of observable effects of proximity in close binary systems, for a typical component having $a_1/R \sim 0.2$, the error of this first order description of the radial distortion of surface caused by neglecting terms higher than $(a_1/R)^5$ would be $\lesssim 0.02\%$.

The discussion after Eqn (2.12) referred also to the centrifugal action experienced by matter in a rotating co-ordinate system, where we anticipated a contribution to the net disturbing potential of the form

$$V_r = -\frac{\omega^2 a_1^2}{3}(P_2(\nu) + \text{const.}) \ , \tag{2.33}$$

where $\nu = \cos\theta$, θ being the zonal angle relative to the rotation axis, and the constant term determining only the numerical value of the potential to be assigned to a given equipotential surface. Again, comparison with (2.12) enables us to write

$$c_{r2} = -\omega^2/3 \tag{2.34}$$

If the rotation rate is synchronized to the mean orbital angular revolution rate ω_o, we can use Kepler's law

$$\omega_o^2 R^3 = Gm_1(1+q) \ , \tag{2.35}$$

where $q = m_2/m_1$. Let us suppose that in general $\omega^2 = \kappa\omega_o^2$, where κ is a quantity of order unity, so as not to impede the general scheme of approximation followed. We then have

$$c_{r2} = -\kappa(1+q)\frac{Gm_1}{3R^3} \ . \tag{2.36}$$

Notice that we are still considering the distortion of the mass accumulation m_1, that is subject to the ($j = 0$) tidal disturbance from m_2 and *its own* centrifugal action. The mass m_2 would have a similar distortion, but that does not enter into the effects presently considered..

A generalization of the rotation is to allow that its axis is inclined to that of the orbit. No special requirements for the axis directions of reference were imposed with (2.3), except that θ is taken to be a zonal angle and ϕ a sectorial one. In view of the discussion after (2.12), a natural choice would be to make the orbit axis and line of centres two of the principal axes of reference, so that the direction cosines λ, μ, ν of a radius vector are naturally given as $\lambda = \sin\theta\cos\phi$, $\mu = \sin\theta\sin\phi$, and $\nu = \cos\theta$. The argument of the relevant Legendre polynomial $P_2(\nu)$, when rotation and orbit axes are aligned is then simply $\cos\theta$. In the 'first order' problem, the disturbances due to tides and rotation are taken to be independent and simply additive. V_t then remains in the form of (2.30) wherever the rotation axis is located. For rotation, Eqn (2.33) keeps the same form: the issue is to relate the relevant zonal direction cosine of the inclined rotation axis, ν' say, to the adopted frame of reference.

Let the rotation axis be displaced to z', say. This is effected in a general way by a rotation about the original, undisturbed z-axis by the angle β, followed by a rotation about the new y-axis through angle α. A point in the original axis-arrangement with co-ordinates λ, μ, ν takes up the coordinates λ', μ', ν', where one system is referred to the other by

$$\boldsymbol{\lambda'} = \mathbf{R}_\eta(\alpha)\mathbf{R}_\zeta(\beta)\boldsymbol{\lambda} \ . \tag{2.37}$$

Here, in order to inter-relate the coordinate systems, we have used the appropriate 3×3 rotational transformation matrix $\mathbf{R}_\upsilon(\theta)$, where θ is the angle of rotation about the υ-axis. The rotation matrix $\mathbf{R}_\xi(\theta)$ takes the form

$$\mathbf{R}_\xi(\theta) \quad = \quad \begin{vmatrix} 1 & 0 & 0 \\ 0 & \cos\theta & \sin\theta \\ 0 & -\sin\theta & \cos\theta \end{vmatrix}$$

with corresponding forms for $\mathbf{R}_\eta(\theta).\boldsymbol{\xi}$ and $\mathbf{R}_\zeta(\theta).\boldsymbol{\xi}$. Multiplying out (2.37), we then have:

$$\begin{aligned} \nu' &= \lambda\sin\alpha\cos\beta + \mu\sin\alpha\sin\beta + \nu\cos\alpha \\ &= \cos\theta\cos\alpha + \sin\theta\sin\alpha\cos(\phi - \beta) \ , \end{aligned} \tag{2.38}$$

so the direction cosine ν' now includes the sinusoidally varying factor $\cos(\phi - \beta)$. If the hydrodynamic restoration time for the fluid masses under consideration are short compared to the rotation period, as is usually assumed for close binary stars, we may regard the material displacements due to tides and rotation as superposed contributions from their separate equilibrium configurations.

The functions V_t and V_r approximate fields giving rise to the attraction of m_2 on unit mass of the accumulation m_1, and the role of the centrifugal force in lifting unit mass away from the rotation axis. They furnish the c coefficients to be used in determining the matter-dependent, first-order, responsive disturbance from either component, as in (2.21). The corresponding potential energy can be obtained from integrating these field functions over the relevant masses. Derivatives of this potential energy then provide the forces affecting the motion of the stellar pair.

At this point we may to return to the discussion at the beginning of the chapter where we referred the work done on an increment δm, or the loss of its potential energy in being gravitationally drawn into a mass accumulation such as m_1 from some distance. The potential energy of gravitationally bound matter is here regarded as negative, so that zero potential energy corresponds to an initial state in which all the matter is dispersed, but there are different conventions about this so care is required in handling the signs of terms.

Consider a spherically symmetric situation with $m(a)$, say, filling the accumulation out to radius a. We can write for the work required to add a spherical increment δm to the condensation

$$\delta\Phi = -Gm(a)\delta m \int_a^\infty \frac{dr}{r^2} = \frac{-Gm(a)\delta m}{a} \ . \tag{2.39}$$

The total potential energy loss of $m(a)$, with the accumulation of such increments settled into a static spherical configuration, becomes

$$\Phi(a) = -G \int_0^a \frac{m(r)dm(r)}{r} = -\frac{G}{2} \int_0^a \frac{1}{r} dm^2(r) \ . \tag{2.40}$$

Now (2.2), applied to this spherical accumulation, yields

$$V(r) = G\frac{m(r)}{r} \ , \tag{2.41}$$

with the meaning that V applies to the region external to the matter already accumulated as $m(r)$. Eqn (2.40) can then be written as

$$\Phi(a) = -\frac{Gm^2(a)}{2a} + \frac{1}{2} \int_0^a m(r)\frac{dV(r)}{dr} \, dr \ . \tag{2.42}$$

The latter integral is integrated by parts to give

$$\Phi(a) = -\frac{Gm^2(a)}{2a} + \left[\frac{V(r)m(r)}{2}\right]_0^a - \frac{1}{2} \int_0^a V(r)dm(r) \ . \tag{2.43}$$

The first two terms on the right cancel each other out, so the net potential energy involved in the condensation is given by the third term formed by integrating the field V of the accumulated matter over the mass range 0 to $m(a)$. The positive notation $W = -\Phi$ is often used for this condensation energy, W keeping the positive sign, as with V. The drop $\Phi(a)$ in potential energy from the original $\Phi(\infty)$, presumably appears as the thermal energy of the heated accumulation.

The potential energy associated with self-attraction W_s is essentially unaffected by binarity within the adopted scheme of approximation. This can be deduced from (2.40), where the two vectors dV/dr and dr would be aligned in the case of disturbance-free, spherical accumulation. In the disturbed state, this product involves the cosine of the angle between the local directions $\hat{n}$ and $\hat{r}$, where $\hat{n}$ refers to the direction perpendicular to the local equipotential surface. From (2.11) we deduce that these two directions are inclined by a small quantity of order the highest term in the surface distortion $\arcsin(Y_2)$, or less, whose cosine differs from unity only by terms of higher order than the first. The implication is that stars condensing into bound binary systems are essentially similar to those of single stars of the same mass and composition.

Integrating the disturbed external field (2.21) over m_2, we obtain the potential energy associated with the tidally and rotationally perturbed mass m_1. In the approximation adopted, this is given at the distance R from m_1 by the $j < 5$ components of V_1 multiplied by m_2. The relevant potential energy is

$$
\begin{aligned}
W_{b2} &= \frac{Gm_1m_2}{R}\{1 + \sum_{j=2}^{4} q(\Delta_j - 1)\left(\frac{a_1}{R}\right)^{2j+1} \\
&- \frac{1}{3}\kappa_1(1+q)(\Delta_2 - 1)\left(\frac{a_1}{R}\right)^{5} P_2(\cos\Theta_1)\} \ ,
\end{aligned}
\tag{2.44}
$$

Θ_1 being the angle between $\hat{R}$ and the axis of bodily rotation for m_1.

It is seen that the first member on the right has the expected form, whose derivative provides the main point-mass-like component of force from m_1 on m_2, producing its acceleration about the system centre. The higher order terms in the potential function V_1, operating on m_2, should account for departures from Keplerian motion as a result of the distortions of figure from rotation and tides, within the framework of this classical perturbations treatment. We consider this in the next chapter.

The total interactive potential energy of the binary system includes $W_{b2}(R,\Theta_1)$ and a corresponding form $W_{b1}(R,\Theta_2)$, arising from the potential field from the aspherical mass m_2 operating on m_1; m_2 having its own tidal and rotational distortion. The relevant coefficients are then altered with the mass ratio q being reversed, the Δs and κs changed as applicable to m_2, and the mean fractional radius a_1/R replaced with a_2/R. The full potential energy should also include, beside an arbitrary constant of integration, terms corresponding to the self-attraction of the two mass-centres, of the form $W_s = G \int_0^{a_1}(m(a)/a)\,dm(a)$ for either star, although those are regarded as effectively constant in the adopted approximation. But for a given total loss of potential energy in the condensation process, the interactive potential $W_{b1} + W_{b2}$ detracts from that of the self-attraction, which must affect the star formation process to some extent.

2.4 Surface Gravity

The variation of surface gravity over stars in close binary systems turns out to have significant observational consequences. The extent of such variation can be readily evaluated from the formalism developed so far. Recalling the discussion at the beginning of this chapter, we can write

$$
g = \frac{dV}{dr} \ .
\tag{2.45}
$$

The question of the variation of surface gravity can be approached most simply by regarding the surface of a component in a close stellar system as an equipotential. Regarding the derivative on the right of (2.45) as the limiting ratio of two increments, the numerator for two such close equipotentials, as in Fig 2.1, would then be a constant over the surface. Paying attention to just the jth harmonic function of the series making up the total potential, we see the denominator, as the limiting form of $\delta r = \delta \{a(1 + Y_j)\}$, would vary in response to the changing value of the harmonic over the surface. Regarding δa_1 as a fixed value, to which the surface equipotentials' separation would revert in the absence of distortion, and so in a constant proportionality to the undisturbed gravity g_0, we may write

$$\delta r = \delta a_1 \left((1 + Y_j) + a_1 \frac{\partial Y}{\partial a} \right) \ , \tag{2.46}$$

or

$$g = g_0 \left((1 - Y_j) - a_1 \frac{\partial Y_j}{\partial a} \right) \ . \tag{2.47}$$

The gravity decrement $g - g_0$ over the surface is given as

$$g - g_0 = -g_0 Y_j (1 + \eta_j(a_1)) \ . \tag{2.48}$$

In regions where Y_j increases (e.g. along a sectorial arc towards the stellar equator) the local surface gravity would correspondingly fall.

2.5   Moments of Inertia

Let us consider the moments of inertia introduced in Eqn (2.29), directing our attention now to the primary star and allowing that its companion acts as mass point in producing the disturbance from sphericity, while regarding the rotational and tidal effects as separately additive. The moment of inertia about the principal axis coinciding with the line of centres was given as I in (2.29). But this is the x-axis in the body co-ordinates, i.e. $I = A$ in the symmetric model considered. We then have

$$A = \int_0^{r_1} \int_0^{\pi} \int_0^{2\pi} \rho r^4 (\sin^2 \phi \sin^2 \theta + \cos^2 \phi \sin^2 \theta) \sin \theta d\phi \, d\theta \, dr \ , \tag{2.49}$$

where we have set the x-axis along the line $\theta = 0$ in polar coordinates and the azimuthal $\phi = 0$ is measured from the z-axis.

In order to make use of Clairaut's analysis, we transform the kernel integrand in r to one in the mean radius a and allow higher order terms in the radial expansion to become negligible. Using Eqn (2.11) and limiting the harmonic expansion to its leading term, we can write

$$\rho r^4 \frac{\partial r}{\partial a} = \frac{\rho}{5} \frac{\partial r^5}{\partial a} = \rho a^4 + \rho \frac{\partial (a^5 Y_2)}{\partial a} \ . \tag{2.50}$$

Transforming the integrand in r in Eqn (2.49) to one in the mean radius a recalls the second integral in Eqn (2.13) with $j = 2$, where it was found that

$$\int_0^{a_1} \rho \frac{\partial (a_1^{j+3} Y_j)}{\partial a} = a^j Y_j (j + 1 - \eta_j) \int_0^{a_1} \rho a^2 da \ . \tag{2.51}$$

Utilizing the approximations considered in the paragraph before Eqn (2.30), we then deduce that

$$A = \int_0^{a_1} \int_{-1}^{1} \int_0^{2\pi} \rho a^4 (1 - x^2)(\sin^2 \phi + \cos^2 \phi) \, d\phi \, dx \, da$$

$$+ \frac{5(\Delta_2 - 1)c_2 a_1^5}{4\pi G} \int_{-1}^{1} \int_0^{2\pi} P_2(x)(1 - x^2)(\sin^2 \phi + \cos^2 \phi) \, d\phi \, dx \, da \ , \tag{2.52}$$

where we have abbreviated $\cos\theta$ to x. Working out the trigonometric integrals we find

$$A = \frac{8\pi}{3} \int_0^{a_1} \rho a^4 \, da - \frac{2(\Delta_2 - 1)m_2 a_1^5}{3R^3} \,, \tag{2.53}$$

where we have substituted for c_2 using (2.31).

It can easily be seen that if ρ was a constant in (2.53) the first term would revert to $\frac{2}{5}m_1 a_1^2$, i.e. the familiar formula for the moment of inertia of a homogeneous sphere about its diameter. In general, we can write $\beta m_1 a_1^2$ for this term, where β is evaluated for a given stellar model in a similar way to the calculation of the structural coefficients as discussed above. Table 2.1 lists β values for a few polytropes in the range $n = 0 - 5$.

Eqn (2.53) gives the decrement in the moment of inertia of the distorted spheroid about its long axis as a result of the tidal action of the companion. The corresponding quantities B and C are determined from replacing the trigonometric part of the integrand in (2.52) with $x^2 + (1 - x^2)\cos^2\phi$ and $x^2 + (1 - x^2)\sin^2\phi$. These two integrals have the same value so that

$$B = C = \beta m_1 a_1^2 + \frac{(\Delta_2 - 1)m_2 a_1^5}{3R^3} \,. \tag{2.54}$$

The moments of inertia due to the rotational distortion, which produces an oblate spheroid about the z-axis are found in a similar way. In that case, the two symmetric trigonometric integrands are $(1 - x^2) + x^2 \sin^2\phi$ and $(1 - x^2) + x^2 \cos^2\phi$ and the coefficient c_2 is given as $-\omega^2/3$. The greatest moment of inertia C', say, about the z-axis becomes

$$C' = \beta m_1 a_1^2 + \frac{2(\Delta_2 - 1)\omega^2 a_1^5}{9G} \,, \tag{2.55}$$

with

$$A' = B' = \beta m_1 a_1^2 - \frac{(\Delta_2 - 1)\omega^2 a_1^5}{9G} \,. \tag{2.56}$$

Combining (2.55) and (2.56) we have

$$C' - A' = \frac{(\Delta_2 - 1)\omega^2 a_1^5}{3G} \,, \tag{2.57}$$

and so the mechanical ellipticity ϵ_m is given by

$$\epsilon_m = \frac{C' - A'}{C'} = \frac{2k_2(1 + q)\omega^2 a_1^3}{3\beta \, G(m_1 + m_2)} \,, \tag{2.58}$$

where k_2 is the structural coefficient $(\Delta_2 - 1)/2$ and q is the mass ratio m_2/m_1.

If ω^2 is written as $\kappa\omega_{\mathrm{syn}}^2$, where ω_{syn} is the angular velocity synchronized to the orbital revolution (a circumstance that can be expected to occur frequently in close binaries) we have

$$\epsilon_m = \frac{2\kappa k_2(1 + q)r_1^3}{3\beta} \,, \tag{2.59}$$

where r_1 is the fractional radius of the primary $(= a_1/R)$. For the tidal deformation we similarly have

$$\epsilon_m = \frac{2k_2 q r_1^3}{\beta} \,. \tag{2.60}$$

In typical close binary situations κ and q are of order unity, $k_2/\beta \sim 0.1$, and $r_1 \sim 0.2$. Mechanical ellipticities are therefore likely to be small: of order 1% or less.

2.6   Bibliography

2.1 A comprehensive bibliography to this chapter is found in Z. Kopal's *Close Binary Systems* Chapman & Hall, 1959, pp122–123 (II.1 and 2). The key equation (2.13), that comes from balancing the coefficients of the potential function, using the trial solution on the right and the differential conditions it must satisfy on the left, is in A. C. Clairaut's original *Théorie de la Figure de la Terre, tirée des Principes de l'Hydrostatique* of 1743. The neater form (2.19) was developed by Clairaut himself, and with slight generalizations by A. M. Legendre & P. S. Laplace. The topic has been discussed and developed in many subsequent essays, with various applications in mind. The logarithmic form (2.20) was of particular interest to R. Radau (*Comptes Rendus Acad. Paris*, Vol. 100, p972, 1885). It was presented as *Radau's equation* in Kopal's book (op. cit.). The behaviour of η in the first-order differential equation (2.20) is clearly dependent on the density distribution, allowing the body's distortion of figure to map this density distribution, or the other way round in an inverse problem where the surface distortion is observed.

2.2 There are so many classical discussions of this subject referring to the polytropic structure equation (2.25), or (2.26) (see Kopal, op. cit. p122) that it is appropriate to give this approach some regard, particularly as available numerical data in the literature allow checks on the relatively simple procedure of numerical integration of a given structural model (S. Chandrasekhar, *Mon. Not. Royal Astron. Soc.*, Vol. 93, p449, 1933; R. A. Brooker & T. W. Olle, *Mon. Not. Royal Astron. Soc.*, Vol. 115, p101, 1955). Many elegant mathematical treatments were devised to speed up this process in pre-computer times, but the 'brute-force' power of modern electronics to deal directly with a simple formulation such as (2.27) obviates a direct need for such expedients. The results shown in Fig 2.5 were obtained by G. İnlek & E Budding in *Astrophys. Space Sci.*, Vol. 342, p365, 2012; with further developments for a wider range of models from G. İnlek, E. Budding & O. Demircan in *Astrophys. Space Sci.*, Vol. 362, p1671, 2017;

2.3 This Section leads up to the result (2.44) for the function that will appear in the Lagrangian treatment of binary dynamics in the next chapter. A distinction should be made between the related usages of the *potential energy* in a mechanical system, and the *potential function* as in the lead up to Clairaut's equation. The relationship between the potential function and energy in the context of stellar condensation, as discussed in Eqns (2.39-43), is detailed in S. Chandrasekhar's classic *An Introduction to the Study of Stellar Structure*, Univ. Chicago Press, 1939, (Chapter 3).

2.4-5 will also be used in subsequent chapters. Their development from the hydrostatics of Sections (2.1) and (2.2) is relatively straightforward and self-contained.

3

Dynamics

3.1 Kinetic Energy

In the previous chapter, classical theory provided a useful approach to characterizing the equilibrium forms of stars considered as fluid masses of varying density. In order to study the dynamical interactions of such bodies in the presence of gravitational forces we continue the same classical approach, utilizing the Lagrangian formulation, which requires specification of the system's kinetic and potential energies.

In a general way, we can write for the kinetic energy T of a system of particles, of mass δm_i, where v_i is the velocity of each particle in an inertial frame of reference,

$$T = \frac{1}{2} \sum v_i^2 \delta m_i \ , \tag{3.1}$$

but, as it stands, this is not specific enough to furnish the sought results directly.

Our route to such results is through giving attention to a few classical examples from binary star situations. It can be shown that the general motion of a body whose particles stay at fixed separations from each other consists of a translation of the center of mass with velocity v_0 and a rotation about the center of mass, with all mass elements rotating with the same instantaneous angular velocity ω. The total kinetic energy of the binary stars then follows as

$$T = T_o + T_r + T_v \ , \tag{3.2}$$

where T_o refers to the translational motion of the components in their orbits, T_r the rotation of each component, and T_v arises in consequence of changes in the separation of individual mass elements as a result of non-rigidity of the stars: referred to as 'vibrational' motions.

T_o can be usefully set out by arranging a 3-dimensional x, y, z inertial frame of reference centred on the binary's centre of gravity. We call one star the 'primary', usually the more massive, and identify it by suffix 1, the other becoming the 'secondary' with suffix 2. It follows from the definition of kinetic energy that:

$$T_o = \frac{1}{2} m_1 \left(\dot{x_1}^2 + \dot{y_1}^2 + \dot{z_1}^2 \right) + \frac{1}{2} m_2 \left(\dot{x_2}^2 + \dot{y_2}^2 + \dot{z_2}^2 \right) \ , \tag{3.3}$$

where the coordinates x_1, y_1, z_1 refer to the centre of mass of the primary and similarly x_2, y_2, z_2 for the secondary. This can be rearranged to allow just one set of spatial coordinates – the so-called 'reduced mass' form – by shifting the origin to the centre of mass of the primary star. The secondary then has coordinates x, y, z, where $x = x_2 - x_1$, and

DOI: 10.1201/b22228-3

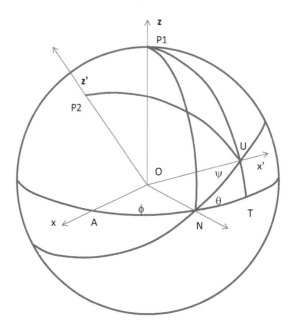

Figure 3.1   Coordinate system shifts: the Eulerian angles ϕ, θ and ψ (see Eqn 3.5).

similarly for y and z. But now, from the definition of the centre of gravity, $m_1 x_1$ is equal to $m_2 x_2$, with corresponding forms for y_1, y_2, z_1 and z_2. Differentiating the x_1, x_2; y_1, y_2; z_1, z_2; coordinates with respect to time, and using the foregoing equalities, we find:

$$T_o = \frac{1}{2} \frac{m_1 m_2}{m_1 + m_2} \left(\dot{x}^2 + \dot{y}^2 + \dot{z}^2 \right) \ . \tag{3.4}$$

In order to deal with the rotational kinetic energy, we refer to Fig 3.1, where we meet the 3 'Eulerian' angles ϕ, θ and ψ. Each of these angles is associated with a corresponding transformation matrix $\mathbf{R}$, that we met in connection with Eqn (2.37). The transformation of one particular coordinate system $\mathbf{x}$ to another $\mathbf{x'}$ we wrote as $\mathbf{x'} = \mathbf{R}_v(\beta)\mathbf{x}$, where β is the angle of rotation about the v-axis. Initially, this axis would be one of x, y or z, but, to generalize, we can write ξ, η, or ζ for corresponding axes at an arbitrary orientation of the system from $\mathbf{x}$. The particular co-ordinate systems in Fig 3.1 can be arranged so that $\mathbf{x}$, say, applies to the inertial frame of reference, (or a system parallel to that) while $\mathbf{x'}$ is a frame of reference fixed in the rotating body. We then have, after the three-angle (ϕ, θ, ψ) sequence:

$$\mathbf{x'} = \mathbf{R}_\zeta(\psi)\mathbf{R}_\xi(\theta)\mathbf{R}_\zeta(\phi)\mathbf{x} \ . \tag{3.5}$$

By multiplying out the relevant matrix products it is found that a point with a body co-ordinate x', for example, is related to the inertial frame of reference by

$$
\begin{aligned}
x' \ = \ & (\cos\phi\cos\psi - \sin\phi\cos\theta\sin\psi)x \\
+ \ & (\sin\phi\cos\psi + \cos\phi\cos\theta\sin\psi)y \\
+ \ & \sin\theta\sin\psi\, z \ ,
\end{aligned}
\tag{3.6}
$$

with similar expressions for y' and z'.

The choice of the body co-ordinate system $\mathbf{x'}$ can be arranged, without loss of generality, to coincide with the principal axes of inertia of the body. This allows the rotational kinetic energy to reduce to the diagonalised form

$$T_r = \frac{1}{2}\left(A'\omega_{x'}^2 + B'\omega_{y'}^2 + C'\omega_{z'}^2\right) \ , \tag{3.7}$$

involving only the moments of inertia A', B' and C' about the respective axes x', y' and z' and to avoid the appearance of off-axis products of inertia that would complicate the equations.

As can be seen from Section 1.4, the Eulerian angles have connections with conventional parameters. Thus, ϕ would be identified with the 'nodal' angle between a given reference direction on the sky and the direction where the plane of the relative orbit crosses the local sky-plane. Similarly, θ indicates the inclination of the orbital plane to that of the local sky. Setting ψ as the angle between the node and the x-axis identifies the corresponding principal moment of inertia as A'.

While the Eulerian angular velocities $\dot{\phi}$, $\dot{\theta}$ and $\dot{\psi}$ are also relevant, they are not the angular velocities directly required in the specification of T_r. Such connections are made by resolving the components $\dot{\phi}$, $\dot{\theta}$, $\dot{\psi}$ in the $\hat{x}'$, $\hat{y}'$ and $\hat{z}'$ directions. Since $\dot{\phi}$ is in the direction $\hat{z}$, we need to use the third row of the transformation (3.5). $\dot{\theta}$ is already in the $x'y'$ plane, so we need only the transformation $\mathbf{R}_\zeta(\psi)$, while $\dot{\psi}$ is given directly in the body co-ordinates, so it can simply be added to whatever other component is in the $\hat{z}'$ direction.

In this way, we find

$$
\begin{aligned}
\omega_{x'} &= \dot{\phi}\sin\theta\sin\psi + \dot{\theta}\cos\psi \\
\omega_{y'} &= \dot{\phi}\sin\theta\cos\psi - \dot{\theta}\sin\psi \\
\omega_{z'} &= \dot{\phi}\cos\theta + \dot{\psi} \ ,
\end{aligned}
\tag{3.8}
$$

so that, for the main rotational component of energy, using (3.7), together with (2.56) to set $A' = B'$,

$$T_r = \frac{1}{2}\left(A'(\dot{\phi}^2\sin^2\theta + \dot{\theta}^2) + C'(\dot{\phi}\cos\theta + \dot{\psi})^2\right) \ , \tag{3.9}$$

Within the framework of the 'first order' theory,[*] Kopal (1959) considered also the contribution of the tidal bulge to the total kinetic energy T_t. The perturbations to the undistorted moments of inertia due to the bulge are represented by the small additions on the right of Eqns (2.53) and (2.54). If we call these A_t, B_t and C_t, we can write (since $A_t = -2B_t = -2C_t$,

$$T_t = \frac{1}{2}C_t\left(-2\omega_{x_t}^2 + \omega_{y_t}^2 + \omega_{z_t}^2\right) \ . \tag{3.10}$$

The Eulerian transformation (3.5) would apply to the line of centres if we insert the orbital parameters, i.e. $\phi = \Omega$, $\theta = i$, $\psi = u$, where $u = \nu + \omega$ (true anomaly + longitude of periastron). The relevant angular velocities ω_{x_t}, ω_{y_t}, ω_{z_t}, follow as in (3.8) if the Eulerian angles and their rates of change are correspondingly identified. This will be the case for many close binaries whose rotations and tides are synchronized to the (circular) orbit, when the principal axes will remain clearly identified. However, interest also attaches to non-synchronized, unaligned or eccentric orbital configuration and the corresponding dynamical effects, though this brings into question applicability of the local hydrostatic equilibrium formulation of the tidal and rotational displacements. These questions have been usually progressed not through a full frontal attack on the equations of motion in their most general form, but by considering specific small developments from the two mass-point problem presented in Chapter 1. This is our approach in what follows.

---

[*]The effect of tides on tides is neglected; see the discussion after Eqn (2.29).

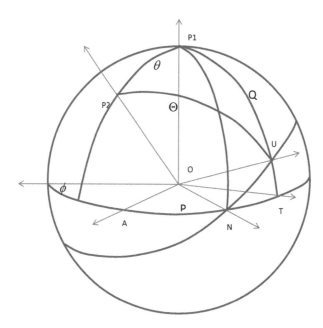

Figure 3.2   The angles Q and P relate the secondary, at U and travelling in the orbital plane from N, to the inertial reference frame centred on O, with the sectorial angle P measured from OA to OT, and Q the zonal arc P_1OU. The angle ϕ in (3.13) remains the Eulerian angle between OA and ON.

Polar coordinates have been found useful in analysis of the orbital dynamics. Let us refer to Fig 3.2, where the plane ANT represents the local plane of the sky. The relative orbit intersects this plane at the nodal line ON. The initial reference direction is indicated by OA, so the angle AN is the nodal angle indicated by Ω. The arc NU corresponds to the angle $\nu + \omega$ that we have designated as u. The polar co-ordinate system r, Q, P is selected such that P corresponds to the sectorial angle between OA and OT, and Q to the zonal angle OP_1U. The cosine rule in the triangle P_1NU yields $\cos Q = \sin i \sin u$, while the 4-parts rule in the same triangle gives $\tan(P - \Omega) = \cos i \tan u$. More directly, we can write

$$
\begin{aligned}
x &= r \cos P \sin Q \\
y &= r \sin P \sin Q \\
z &= r \cos Q \ .
\end{aligned}
\tag{3.11}
$$

In these co-ordinates the orbital kinetic energy T_o is given, using (3.4), by

$$
T_o = \frac{1}{2} \frac{m_1 m_2}{m_1 + m_2} \left(\dot{r}^2 + r^2 (\dot{Q}^2 + \dot{P}^2 \sin^2 Q) \right) \ ,
\tag{3.12}
$$

which can be seen by resolving the velocity v at U into three components in the directions of $\hat{r}$, $\hat{P}$ and $\hat{Q}$, or by differentiating the expression for x, y and z in (3.11) and using them in (3.4).

The angle between OP_2 and OU, Θ, say, which relates the position of the moving star to the basic reference plane, proves to be of general relevance. It is given by

$$
\cos \Theta = \cos \theta \cos Q + \sin \theta \sin Q \sin(\phi - P) \ .
\tag{3.13}
$$

Noting that the arc P_1P_2 is subtended about the nodal axis perpendicular to the plane P_1P_2O, the angles θ and ϕ in Fig 3.2 are seen to be the polar counterparts of the Eulerian pair θ and ϕ of Fig 3.1. The form of Eqn (3.13) may be compared with that of (2.38), with zonal angle Q playing the role of α and sectorial angle $P + \pi/2$ as β.

Regarding the vibrational component T_v, we may note from general properties of stable oscillating systems that the mean value of the kinetic energy of vibrational elements δm_i within m_1 is given by $\langle T_v \rangle = \langle W_v \rangle$, where the associated potential energy $\langle W_v \rangle$ depends on the displacements δa_i of the δm_i from their equilibrium positions of rest to the extent

$$\langle W_v \rangle \sim \frac{GM}{r} \sum_i \delta m_i \left(\frac{\delta a_i}{r} \right)^2 . \tag{3.14}$$

It is implicit that elements $\delta a_i/r$ are comparable to, or less than, the tidal or rotational distortions of a expressed by (2.11) in order for the first-order discussion of Chapter 2 to be meaningful. The vibrational kinetic energy therefore turns out to be a quantity of second order, allowing us to neglect the role of T_v in (3.2) in the first approximation.

Having thus formulated relevant and usable approximations for the potential and kinetic energies of a close pair of stars, analysis of the possible motions of the components within the framework of Newtonian mechanics becomes approachable.

3.2 Equations of Motion

The Lagrangian equations of motion for a system with specified kinetic T and potential W energies are written as

$$\frac{d}{dt} \left(\frac{\partial T}{\partial \dot{q}} \right) - \frac{\partial T}{\partial q} = \frac{\partial W}{\partial q} , \tag{3.15}$$

where q is any generalized coordinate involved in the formulation of T and W. For T_o, from (3.12) the coordinates are only r and Q: the time derivatives include also $\dot{P}$ as well as $\dot{r}$ and $\dot{Q}$. The rotational kinetic energy T_r in (3.9) involves only the coordinate θ, though includes the three time-derivatives $\dot{\phi}$, $\dot{\theta}$ and $\dot{\psi}$. The kinetic energy T_t of the tidal bulge, indicated in (3.10), if consistent with an equilibrium approach, can be treated along parallel lines to the rotational effect, though it will be found to have a more complicated appearance. In what follows, we consider the applicability of the results of classical motion studies to a few selected and possibly observable effects.

From (2.44), the combined potential energy W_b is written as

$$
\begin{aligned}
W_b(R, \Theta) = & \frac{Gm_1 m_2}{R} \{ 1 + \sum_{i=1}^{2} \sum_{j=2}^{4} q_i k_{i,j} r_i^{2j+1} \\
& - \frac{1}{3} \sum_{i=1}^{2} \kappa_i (1 + q) k_{i,2} r_i^5 P_2(\cos \Theta_i) \} ,
\end{aligned} \tag{3.16}
$$

where the symbols have the same meanings as in the previous chapter.

Note that we have replaced the radial variable r of (3.11), and subsequently in Section (3.1), by R the radius of the relative orbit. The variable r_i is henceforth used to represent the surface radius of the ith component given in units of the separation of the component centres (a). In (3.16), it is the tidal terms that first add into the parenthesized terms to the right, followed by the rotational terms for both stars. These are factored by the small structural coefficients $k_i = (\Delta_i - 1)/2$. Applied to the orbital energy, using (3.12),

Eqn (3.15) then gives:

$$\frac{d}{dt}\left(\frac{\partial T_o}{\partial \dot{R}}\right) - \frac{\partial T_o}{\partial R} = \frac{\partial W_b}{\partial R} \ ,$$

$$\frac{d}{dt}\left(\frac{\partial T_o}{\partial \dot{P}}\right) = \frac{\partial W_b}{\partial P} \ ,$$

$$\frac{d}{dt}\left(\frac{\partial T_o}{\partial \dot{Q}}\right) = \frac{\partial W_b}{\partial Q} \ , \tag{3.17}$$

leading to ,

$$\frac{d^2 R}{dt^2} - R\left(\dot{P}^2 \sin^2 Q + \dot{Q}^2\right) = \frac{\partial W_b}{\partial R} \ ,$$

$$\frac{d}{dt}\left(R^2 \dot{P} \sin^2 Q\right) = \frac{\partial W_b}{\partial P} \ ,$$

$$\frac{d(R^2 \dot{Q})}{dt} - \left(R^2 \dot{P}^2 \sin Q \cos Q\right) = \frac{\partial W_b}{\partial Q} \ , \tag{3.18}$$

W_b is seen to be made up of the predominating term $Gm_2 m_2/R$ together with the higher order perturbing terms that can be ignored in a zero-order formulation. In that case, (3.18) reverts to the two mass-point orbit problem, that can be scaled to an equivalent one mass-point, central force version. If we were to set the reference axes such that $Q = \pi/2$ coincided with the plane in which the motion is now constrained to take place in this zeroth order simplification, the first of (3.18) gives the radial acceleration, the second the conservation of angular momentum and the third becomes identically zero on both sides. Non-zero perturbations in the right hand members then lead to disturbance from the basic Keplerian motion that are dealt with by the classical procedure of successive approximations.

Aside from considerations of the orbit, the full kinetic energy includes the rotational component (3.9). Lagrangian equations that are formally similar to (3.17) but apply to the rotational energy, can be written down, thus,

$$\frac{d}{dt}\left(\frac{\partial T_r}{\partial \dot{\theta}}\right) - \frac{\partial T_r}{\partial \theta} = \frac{\partial W_b}{\partial \theta} \ ,$$

$$\frac{d}{dt}\left(\frac{\partial T_r}{\partial \dot{\phi}}\right) = \frac{\partial W_b}{\partial \phi} \ ,$$

$$\frac{d}{dt}\left(\frac{\partial T_r}{\partial \dot{\psi}}\right) = \frac{\partial W_b}{\partial \psi} = 0 \ . \tag{3.19}$$

Such a formulation does not avoid the interdependence of the variables, implicit through (3.13), apart from further interdependence associated with the energetics of the tidal bulge. Note that the last of (3.19), by asserting no dependence of the potential on ψ, allows a constant of the motion (to within the adopted scheme of approximation) to be given, namely that the angular momentum $\partial T_r/\partial \dot{\psi} = C'(\dot{\phi}\cos\theta + \dot{\psi}) = C'\omega_z' = $ const. This allows the rotational kinetic energy of (3.9) to be somewhat simplified to

$$T_r = \frac{1}{2}\left(A'(\dot{\phi}^2 \sin^2 \theta + \dot{\theta}^2) + C'\omega_{z'}^2\right) \ . \tag{3.20}$$

Even so, the full problem for fluid stellar masses being generally regarded hitherto as out of reach of formal solution, progress has been achieved by confining attention to cases where physical restrictions are imposed and linearizations around the vicinity of such cases possible. As an example, we consider the net motion of a rapidly rotating primary star with its companion in a circular orbit. The rotation distorts the star into an oblate spheroid whose axis is inclined to that of the orbit, so the spheroid experiences a restoring torque. The star may also have a tidal deformation affecting the instantaneous distribution of the stellar matter as a whole, but we neglect this in the first instance, assuming that a separable formulation can apply to the time-averaged shape of the predominant rotational distortion.* This would be valid for the equilibrium tide: for dynamical tides, however, situation calls for further analysis that would become complicated in detail.

3.3   Precession of a Spheroid

Let us consider the case where the accumulation m_1 has a substantial rotation speed, and is subject to the influence of a companion star that behaves, in the approximation we adopt, as a mass point in a circular orbit. The strong, non-synchronized, bodily rotation goes with its axis being generally inclined at θ to that of the orbit.

The angular momentum associated with the secondary's relative orbit is $m_2\omega_z R^2$, while that of the primary rotation we set as $\beta m_1 \omega_{z'}(Rr_1)^2$, where β is a numerical scaling constant, so that the ratio of the former to the latter $\sim q/(\sqrt{\kappa_1}\beta r_1^2)$. We can reasonably expect β to be small, in view of the central condensation of matter in stars. The ratio would be thus relatively high for a close pair of comparable mass: say, a young Main Sequence binary, in view of the small denominator. We therefore assume, for the present discussion, that the orbital plane remains relatively undisturbed, so that we can set it as the reference plane, for which $Q = \pi/2$.

In this case, the angular component of W_b in (3.16) reduces to

$$W_{b1} = -\frac{2Gm_1m_2}{3R}\kappa_1(1+q)k_{1,2}r_1^5 P_2(\cos\Theta) \ . \tag{3.21}$$

The angle Θ in (3.21) is modulated by $\sin u = \sin(P-\phi)$ (see (3.13)) and varies in magnitude between $\pm\arccos(\sin\theta)$. Eqn(2.58) can be used to rewrite the potential energy W_b as

$$W_{b1} = -\frac{Gm_1m_2}{R}\beta\epsilon_m r_1^2 P_2(\cos\Theta) \ , \tag{3.22}$$

ϵ_m being the mechanical ellipticity. We may notice that this disturbing potential is but a small fraction $\sim qk_{1,2}r_1^3/(1+q)\beta$ of the spin energy of the star, which suggests the precession rate will be a similar low fraction of the star's rotation rate.

The first of (3.19) can now be written, using (3.20), as

$$\frac{d}{dt}\left(A'\dot{\theta}\right) + C'\omega_{z'}\dot{\phi}\sin\theta = -\frac{3Gm_1m_2}{2R}\beta\epsilon_m r_1^2\frac{\partial\cos^2\Theta}{\partial\theta} \ . \tag{3.23}$$

Let us assume that a steady average motion, i.e. a uniform *precession* is possible, for which we neglect oscillatory effects associated with $\ddot{\theta} \neq 0$. From (3.13) we now have $\cos\Theta = \sin\theta\sin(\phi - P)$, so that (3.23) reduces to

$$C'\omega_{z'}\dot{\phi}\sin\theta = -\frac{3Gm_1m_2}{R}\beta\epsilon_m r_1^2\cos\theta\sin\theta\sin^2(\phi - P) \ . \tag{3.24}$$

With the precession rate $\dot{\phi}$ being very slow compared to the mean rate of orbital motion, the final member of the right-hand expression, i.e. $\sin^2(\phi - P)$ is fairly represented by its

---

*This situation pertains well to exoplanet host stars.

mean value of $\frac{1}{2}$. We then find,

$$\dot{\phi} = -\frac{3Gm_1m_2\beta\epsilon_m r_1^2 \cos\theta}{2C'\omega_{z'}R} \quad . \tag{3.25}$$

Now using Kepler's Law for the mean orbital angular revolution rate n, i.e. $n^2 = G(m_1 + m_2)/R^3$, we obtain

$$\frac{Gm_1m_2\beta r_1^2}{RC'} = \frac{n^2 q}{(1+q)} \quad , \tag{3.26}$$

so that

$$\frac{\dot{\phi}}{n} = -\frac{3}{2}\frac{nq}{(1+q)}\frac{\epsilon_m \cos\theta}{\omega_{z'}} \quad . \tag{3.27}$$

If we use astronomical units n becomes the number of complete revolutions per year. The small factor of the mechanical ellipticity ϵ_m in the numerator of (3.27) justifies the underlying adoption of $\dot{\phi}$ as a small quantity. We may think of the third fraction on the right in (3.27) as a property of the body, the motion of whose axis of rotation is under consideration. The second fraction pertains to the disturbing agent. The general result is that the precession rate is of the order of the mechanical ellipticity times the rotation rate. In practical terms, this would indicate precession periods of at least $\sim 10^3 \times P_{\text{orbit}}$.

3.3.1   Nutation

Suppose now that a third body is also in orbit around the rapidly rotating primary star, for which there is a corresponding contribution to the potential associated with the distorted figure of the primary due to its rotation (again neglecting tidal effects). For simplicity regarding this new extension to the discussion, let the third body's orbit be circular and in a plane at inclination i to the primary-secondary orbital plane adopted as the reference. We write, in analogy with (3.22),

$$W_{b1,3} = -\frac{Gm_1m_3}{R_3}\beta\epsilon_m r_1^2 P_2(\cos\Theta_3) \quad . \tag{3.28}$$

so that the main difference between the new and old contributions to the potential comes from the mass m_3 and distance R_3 of the third body, and also the new projection angle Θ_3, whose cosine is determined from the scalar product $\hat{R}_3.\hat{\omega}_3$. Thus,

$$\cos\Theta_3 = \sin\theta\sin P + \sin i \cos\theta \sin(P - \phi') \quad , \tag{3.29}$$

where ϕ' arises from the different longitudes of the nodes of the rotation plane ϕ and third body orbit ϕ_3 with the reference plane, i.e. $\phi' = \phi - \phi_3$. The consequent non-constant term in W_{b2} becomes

$$\begin{aligned}
\frac{3}{2}\cos^2\Theta_3 &= \frac{3}{2}[\sin^2\theta\sin^2 P + \\
&+ \quad 2\sin i \sin\theta\cos\theta(\sin^2(P)\cos\phi' - \sin P\cos P\sin\phi') + \\
&+ \quad \sin^2 i \cos^2\theta\sin^2(P-\phi')]
\end{aligned} \tag{3.30}$$

The terms in $\sin^2 P$ in this expression average to $1/2$ as before, while the mixed term in $\sin P\cos P$ has mean value zero. Matters are again simplified if we have the planes of the two orbits at low inclination to each other, as often observed. We can then neglect the third term on the right in (3.30), and we end up with

$$W_{b1,3} = -\frac{3Gm_1m_3}{4R_3}\beta\epsilon_m r_1^2(\sin^2\theta + i\sin 2\theta\cos\phi') \ . \tag{3.31}$$

The first term in the parentheses on the right is of the same form as (3.22). The scalar nature of the gravitational potential then implies that there will be a corresponding precessional effect

$$\dot{\phi}_3 = -\frac{3}{2}\frac{n_3^2 q_3}{(1+q_3)}\frac{\epsilon_m\cos\theta}{\omega_{z'}} \ . \tag{3.32}$$

The net precession from these two sources then appears as

$$\frac{\dot{\phi}}{n} = -\frac{3}{2}\frac{\epsilon_m\cos\theta}{\omega_{z'}}\frac{nq}{(1+q)}\left(1 + \frac{n_3^2}{n^2}\frac{q_3(1+q)}{q(1+q_3)}\right) \ . \tag{3.33}$$

In a typical hierarchical arrangement of stars, where we could expect $1 > q > q_3$ and $n > n_3$, the second term in the large parentheses would be small compared to unity.*

The smooth precessional components on the right and left sides of (3.23) balance, according to the foregoing result. We are left with the previously neglected term on the left of (3.23) to match the new term on the right of (3.31). It is instructive to write the augmented version of (3.23) out in full as

$$A'\ddot{\theta} - \frac{1}{2}A'\sin 2\theta\,\dot{\phi}^2 + C'\omega_{z'}\sin\theta\,\dot{\phi} = -2\epsilon_m a_3\sin 2\theta - 2\epsilon_m b_3\cos 2\theta\cos\phi', \tag{3.34}$$

where

$$a_3 = \frac{3}{8}\frac{n^2 q}{(1+q)}\left(1 + \frac{n_3^2}{n^2}\frac{q_3(1+q)}{q(1+q_3)}\right)C' \ , \tag{3.35}$$

and

$$b_3 = \frac{3}{4}\frac{n_3^2 q_3}{(1+q_3)}iC' \ . \tag{3.36}$$

In a similar way, the second of (3.19) now has a new term on the right with an acceleration term on the left:

$$\frac{d}{dt}\left(A'\dot{\phi}\sin^2\theta + C'\omega_{z'}\cos\theta\right) = -\epsilon_m b_3\sin 2\theta\sin\phi' \ . \tag{3.37}$$

A recognition that certain terms are very small compared with others permits separation and linearization procedures for approximate solutions. This was already implicit regarding the neglected tidal terms or the assumed constant location of the primary orbital plane. The precessional motion presented in (3.27) or (3.32) will thus have a counterpart for the third orbit and reference plane, so that the angle ϕ' in the coupled equations (3.34) and (3.37) will not be constant but it will steadily decrease in accordance with the trend expected from (3.27) and (3.33). In fact, if we regard the close reference orbit as a smoothed ring of matter whose mechanical ellipticity becomes simply $1/2$, subject to the perturbing effect of the wide-orbit companion, the same approach as for (3.27) yields an orbital precession rate $\dot{\Omega} = -2\pi/P_{pw}$, where $P_{pw} = 4P_w^2(1+q_w)/(3q_w\cos i)$. P_c, P_w and P_c are the orbital periods of the wide (w) and close (c) components, respectively, and $q_w = m_w/(m_w + m_c)$.

---

*This is not the case for the lunisolar precession, where the second term is $\sim$1.86 times greater than unity.

The sinusoidal right hand members on the right sides of Eqns (3.34) and (3.37) have the role of forcing terms, that create an additional slow oscillatory motion affecting the Eulerian angles θ and ϕ known as *nutation*. The scale of this nutation can be estimated by linearizing to the order of the small mechanical ellipticity ϵ_m, for which we write:

$$\theta = \theta_0 + \epsilon_m \theta_1, \quad \dot{\theta} = \epsilon_m \dot{\theta}_1, \quad \ddot{\theta} = \epsilon_m \ddot{\theta}_1;$$
$$\phi = \epsilon_m \phi_1, \quad \dot{\phi} = \epsilon_m \dot{\phi}_1, \quad \ddot{\phi} = \epsilon_m \ddot{\phi}_1. \tag{3.38}$$

If we substitute these forms into (3.34) and (3.37), writing also $-\Omega t$ for ϕ', we obtain, to the first order in ϵ_m,

$$A'\ddot{\theta}_1 + C'\omega_{z'} \sin\theta_0\,\dot{\phi}_1 = -2\epsilon_m(a_3 \sin 2\theta_0 + b_3 \cos 2\theta_0 \cos \Omega t) \tag{3.39}$$

and

$$A' \sin^2\theta_0 \ddot{\phi}_1 - C'\omega_{z'} \sin\theta_0\,\dot{\theta}_1 = \epsilon_m b_3 \sin 2\theta_0 \sin \Omega t \ . \tag{3.40}$$

The relative magnitude of the coefficient of the terms in $\dot{\theta}_1$ and $\dot{\phi}_1$, taking account of the smallness the frequency of the forcing term compared with the primary's rotation frequency $\omega_{z'}$, suggests the approximate solutions

$$\dot{\phi}_1 \approx -\frac{4\epsilon_m a_3 \sin 2\theta_0}{C'\omega_{z'}} - \frac{2\epsilon_m b_3 \cos 2\theta_0 \cos \Omega t}{C'\omega_{z'} \sin\theta_0} \tag{3.41}$$

and

$$\dot{\theta}_1 \approx -\frac{2\epsilon_m b_3 \cos 2\theta_0 \sin \Omega t}{C'\omega_{z'}} \ . \tag{3.42}$$

Upon integration of these equations and using (3.38), we find (setting $\phi(0) = 0$)

$$\begin{aligned}\phi(t) &= -\langle\dot{\phi}\rangle t - \delta\phi \sin \Omega t \\ \theta(t) &= \theta_0 + \delta\theta \cos \Omega t \ ,\end{aligned} \tag{3.43}$$

with $\langle\dot{\phi}\rangle$ specified by (3.33), and the nutational amplitudes given by

$$\delta\phi = \frac{3}{2} \frac{iq_3 n_3^2}{\Omega(1 + q_3)} \frac{\epsilon_m \cos 2\theta_0}{\omega_{z'} \sin\theta_0} \tag{3.44}$$

and

$$\delta\theta = \frac{3}{2} \frac{iq_3 n_3^2}{\Omega(1 + q_3)} \frac{\epsilon_m \cos\theta_0}{\omega_{z'}} \ . \tag{3.45}$$

Table 3.1   Dynamical parameters for the system λ Tau.

Masses	Periods (d)	Other parameters		Primary P & N
7.2	$P_c = 3.953$	$\theta_0 \sim 10°$	$\epsilon_m \sim 0.0033$	$\langle\dot{\phi}\rangle \sim -23°$ y^{-1}
1.9	$P_w = 33.03$	$i \sim 5°$	$\epsilon_c = 1/2$	$\delta\phi \sim 0.12°$
0.7	$P_{pw} = 5208$	$\kappa_1 = 1.5$	$q_w = 0.077$	$\delta\theta \sim 0.02°$

It seems reasonable to expect that the inclination of the stars' equatorial rotation plane to that of the orbit would normally be small, rendering (3.44) to have a low denominator. In turn, this points to the oscillation of the nodal position due to nutation to be relatively large compared with wobbles of the polar axis on the period P_{pw}.

An hierarchical system such as the well-known triple star λ Tau may allow us some feel for the scale of these effects in practice. Thus, with the known results of analysis of the system in the first two columns of Table 3.1, together with plausible surmises for other parameters in the third and fourth columns, using the above formulae, we can estimate a relatively fast precessional motion of the primary if its known non-synchronous rotation is unaligned to the orbital axis. Although the forced nutation effects are small, increased use of high-precision techniques, particularly involving space-based equipment, may permit detection of such mechanisms in modern observations.*

3.4   Apsidal Motion

The set of equations (3.18) correspond to Keplerian elliptic motion when the bodies involved reduce to mass points and where the equations governing bodily motions coming after (3.18) have no relevance. In many real situations this becomes almost true when the finite sizes of a pair of stars are negligible in comparison to their separation, or when their rotational or other motion energies are small in comparison to those of their orbits. In such cases, Keplerian motion indeed obtains with a high degree of fidelity.

3.4.1   Simplified Treatment

Small departures from Keplerian motion are in line with the relative scale of bodily departures from sphericity, or rotational and tidal effects. The precessional motion expressed by Eqn (3.27) is thus factored by the body's mechanical ellipticity ϵ_m. H. N. Russell produced an expression similar to (3.27), but for the apsidal motion $\dot{\varpi}$ of a mass particle moving as a satellite in an eccentric orbit of semi-major-axis A with mean rate of angular motion n around the central plane of a rapidly rotating, and thereby distorted, massive body of mean radius R_b. We can write Russell's basic formula as

$$\frac{\dot{\varpi}}{n} = 2r_1^2 k \epsilon'_m \ , \tag{3.46}$$

where k is a structural constant for the body equivalent to the k_2s given in Table 2.1. The quantity $\epsilon'_m = \epsilon_m/\Delta$, where Δ was introduced in (2.16) and ϵ_m in (2.58). For real stars $\Delta < 1.3$, and often $\Delta < 1.03$, so ϵ'_m simulates the previously defined mechanical ellipticity. It is instructive to relate the calculation to the perturbation W of the conventional potential energy. Russell's formula suggests that the relevant proportion of the binding energy at the equator of the body is given as $W = 2\epsilon' = n^2 R_b^3/Gm_1$, which, using Kepler's Law, is $(1+q)r_1^3$. The extra factor $r_1^2 = (R_b/A)^2$ (not in Eqn 3.27) enters (3.46) because the forward torque here applies to the angular momentum of the orbit rather than the body, as in (3.27). Russell extended (3.46) so as to apply to a pair of close stars, writing, for the combined apsidal advance due to the rotational distortion of both stars,

$$\frac{\dot{\varpi}}{n} = (1+q) \left(k_1 r_1^5 + \frac{k_2 r_2^5}{q} \right) \ . \tag{3.47}$$

Eqn (3.47) ignores the tidal distortion. From equations (2.59) and (2.60) it can be seen that the radial extension due to the tides along the line of centres is $3q/(1+q)$ that of the rotation alone. Russell found the net perturbative effect of the tides to be $6q/(1+q)$ times that of rotation and, adding this in with the rotational effect, he arrived at the formula

$$\frac{\dot{\varpi}}{n} = k_1 r_1^5 (1 + 7q) + k_2 r_2^5 (1 + \frac{7}{q}) \ . \tag{3.48}$$

---

*P and N in the fourth column of Table 3.1 stand for precession $\langle \dot{\phi} \rangle$ and nutation $\delta\phi$, $\delta\theta$.

More detailed analysis, that we consider next, supports Russell's simplified treatment on the basis of his adopted premises.

3.4.2   Lagrange Formulation

We can regard small perturbations to the relatively simple formulae that govern the motion of mass points in a Newtonian gravity field as producing corresponding small deviations from Keplerian orbital motion. We will apply this concept to the full set of parameters characterizing that disturbance. Let us start with the orbital acceleration formulae in rectangular coordinates (see Eqn 1.24).

$$\ddot{x} + \frac{\mu x}{R^3} = \frac{\partial V'}{\partial x}; \;\; \ddot{y} + \frac{\mu y}{R^3} = \frac{\partial V'}{\partial y}; \;\; \ddot{z} + \frac{\mu z}{R^3} = \frac{\partial V'}{\partial z}. \tag{3.49}$$

The right hand members in (3.49) are now the derivatives of the disturbing function V'. If there were no disturbing function V' would revert to zero, resulting in simple Keplerian motion. We would then have the solutions:

$$\begin{aligned} x &= f_1(c_i) \;\; ; \;\; \dot{x} = g_1(c_i) \\ y &= f_2(c_i) \;\; ; \;\; \dot{y} = g_2(c_i) \\ z &= f_3(c_i) \;\; ; \;\; \dot{z} = g_3(c_i) \;, \end{aligned} \tag{3.50}$$

where c_i ($i = 1, 2, ...6$) stands for the 6 classical constants of integration, or the parameters of the relative orbit, for this 6th order system of equations.

Notice that $g_k = \partial f_k / \partial t$, where $k = 1, 2, 3$. The idea that as the disturbing function departs from zero the constants c_i stop being constant and become functions of the time is at the basis of the *method of variation of the constants* associated with J. Lagrange. In this method, the perturbed motion takes the form

$$\frac{\mathrm{d}x}{\mathrm{d}t} = \frac{\partial f_1}{\partial t} + \sum \frac{\partial f_1}{\partial c_i}\frac{\mathrm{d}c_i}{\mathrm{d}t} \;, \tag{3.51}$$

with similar equations for y and z. But here we have just three equations helping to define the six quantities c_i; we are free to select three more and the choice that commends itself is to set

$$\begin{aligned} \sum \frac{\partial f_1}{\partial c_i}\frac{\mathrm{d}c_i}{\mathrm{d}t} &= 0 \\ \sum \frac{\partial f_2}{\partial c_i}\frac{\mathrm{d}c_i}{\mathrm{d}t} &= 0 \\ \sum \frac{\partial f_3}{\partial c_i}\frac{\mathrm{d}c_i}{\mathrm{d}t} &= 0 \;, \end{aligned} \tag{3.52}$$

so that the equations (3.52) reduce to

$$\frac{\mathrm{d}x}{\mathrm{d}t} = \frac{\partial f_1}{\partial t} = g_1 \;, \tag{3.53}$$

with similar equations for $\mathrm{d}y/\mathrm{d}t$ and $\mathrm{d}z/\mathrm{d}t$. This means that at any given time, both the co-ordinates and the velocities will correspond, instantaneously, to the Keplerian motion with the given set of orbital parameters c_i for that moment — the *osculating elements*.

If we now differentiate (3.52) with respect to time we obtain

$$\ddot{x} = \frac{\partial^2 f_1}{\partial t^2} + \sum \frac{\partial g_1}{\partial c_i} \frac{dc_i}{dt} \quad , \tag{3.54}$$

and again with essentially similar forms for y and z. Substitution of these forms into (3.50) leads to

$$
\begin{aligned}
\frac{\partial^2 f_1}{\partial t^2} + \frac{\mu f_1}{R^3} + \sum \frac{\partial g_1}{\partial c_i} \frac{dc_i}{dt} &= \frac{\partial V'}{\partial x}, \\
\frac{\partial^2 f_2}{\partial t^2} + \frac{\mu f_2}{R^3} + \sum \frac{\partial g_2}{\partial c_i} \frac{dc_i}{dt} &= \frac{\partial V'}{\partial y}, \\
\frac{\partial^2 f_3}{\partial t^2} + \frac{\mu f_3}{R^3} + \sum \frac{\partial g_3}{\partial c_i} \frac{dc_i}{dt} &= \frac{\partial V'}{\partial z},
\end{aligned} \tag{3.55}
$$

The first two members on the left of Eqns (3.55) equate to zero for unperturbed motion, implying that the third terms on the left are balanced by the derivatives of the disturbing potential on the right. Combining (3.55) and (3.52) then leads to a set of 6 equations that we may write in a short form as

$$
\begin{aligned}
\{\partial \boldsymbol{x}\}.|\frac{1}{\partial \boldsymbol{c}} \frac{d\boldsymbol{c}}{dt}| &= 0 \\
\{\partial \dot{\boldsymbol{x}}\}.|\frac{1}{\partial \boldsymbol{c}} \frac{d\boldsymbol{c}}{dt}| &= |\frac{\partial V'}{\partial x}| \quad ,
\end{aligned} \tag{3.56}
$$

where $\{\boldsymbol{a}\}$ denotes a vector set in row form and $|\boldsymbol{a}|$ is a column vector. $\partial \boldsymbol{x}$ then stands for the 3 numerator parts in the partial derivatives of x, y and z, while $(1/\partial \boldsymbol{c})(d\boldsymbol{c}/dt)$ has 6 components corresponding to each of the constants c_i.

If the first 3 rows of this array are multiplied by $-\{\partial \dot{\boldsymbol{x}}\}.|1/\partial \boldsymbol{c}|$ and the second three rows multiplied by $\{\partial \boldsymbol{x}\}.|(1/\partial \boldsymbol{c})|$ and the results all added together, we obtain an array of equations in the form

$$\boldsymbol{L}|\frac{d\boldsymbol{c}}{dt}| = |\frac{\partial V'}{\partial \boldsymbol{c}}| \quad . \tag{3.57}$$

$\boldsymbol{L}$ is an antisymmetric square matrix, whose elements are the so-called Lagrange brackets $[c_i, c_j]$, defined as

$$[c_i, c_j] = \frac{\partial x}{\partial c_i} \frac{\partial \dot{x}}{\partial c_j} - \frac{\partial x}{\partial c_j} \frac{\partial \dot{x}}{\partial c_i} + \frac{\partial y}{\partial c_i} \frac{\partial \dot{y}}{\partial c_j} - \frac{\partial y}{\partial c_j} \frac{\partial \dot{y}}{\partial c_i} + \frac{\partial z}{\partial c_i} \frac{\partial \dot{z}}{\partial c_j} - \frac{\partial z}{\partial c_j} \frac{\partial \dot{z}}{\partial c_i}. \tag{3.58}$$

The Lagrange brackets are seen to satisfy $[c_i, c_i] = 0$ and $[c_i, c_j] = -[c_j, c_i]$, and it can also be shown that they are independent of time and depend only on the formulae of Keplerian motion. In a formal sense, Eqn (3.57) solves the problem of disturbed motion, but the evaluation of the Lagrange brackets is not obvious at this point.

It can be seen that the Lagrange brackets have the character of an angular momentum per unit mass relative to pairs of angular parameters of the relative orbit, or a linear momentum if the size parameter (a) is involved. The classical procedure involves following through the variation of the form of the brackets in transferring from the fundamental reference plane to the plane of the orbit while referring to a parameter set established from the solar system context. It can be shown that the Lagrange bracket for any particular pair of parameters $[p, q]$ reduces to

$$[p, q] = \frac{\partial(\varepsilon - \varpi, L)}{\partial(p, q)} + \frac{\partial(\varpi - \Omega, G)}{\partial(p, q)} + \frac{\partial(\Omega, H)}{\partial(p, q)} \quad , \tag{3.59}$$

where ε measures the mean longitude at the reference epoch, ϖ is the longitude of the periastron, including the longitude in the reference plane of the ascending node Ω as well as the longitude measured from the node in the orbital plane ω, and $L = (\mu a)^{1/2}$; $G = L(1 - e^2)^{1/2}$; $H = G \cos i$.

Use is frequently made of Kepler's law so that $\sqrt{\mu a} = na^2$. We then find, for the three parameters appearing in L, G and H, the six remaining partial derivatives:

$$
\begin{array}{cccc}
 & L & G & H \\
\partial/\partial a & na/2 & nae'/2 & nae' \cos i/2 \\
\partial/\partial e & 0 & -na^2 e/e' & -na^2 e \cos i/e' \\
\partial/\partial i & 0 & 0 & -na^2 e' \sin i \ ,
\end{array}
\tag{3.60}
$$

where we have written $e' = \sqrt{1 - e^2}$. The six corresponding antisymmetric pairs of Lagrange brackets can now be filled in, the set of equations (3.57) set out, and, due to the triangular form of the matrix $\boldsymbol{L}$, this is easily rearranged to yield

$$
|\frac{\mathrm{d}\boldsymbol{c}}{\mathrm{d}t}| = \boldsymbol{L}^{-1}|\frac{\partial V'}{\partial x}| \ .
\tag{3.61}
$$

Now, with regard to apsidal motion, the relevant line, in this set of 6 equations, is the one dealing with the time derivative of the longitude of periastron ϖ, that reads,

$$
\frac{\mathrm{d}\varpi}{\mathrm{d}t} = \frac{e'}{na^2 e}\frac{\partial V'}{\partial e} + \frac{\tan \frac{1}{2}i}{na^2 e'}\frac{\partial V'}{\partial i} \ .
\tag{3.62}
$$

In the relaxed close binary configuration considered, the perturbation is symmetrical about the orbital plane wherein the secular motion is confined, and the second term on the right in (3.62) reverts to zero. We require the form of the potential V' and its dependence on the eccentricity e in order to deal with the apsidal motion formulated as (3.62).

We have, in Eqn (2.44), an expression for the potential energy at m_2 arising from the tidally and rotationally perturbed field from m_1 that we can rewrite as

$$
\begin{aligned}
W_{b2} = \ & \frac{Gm_1 m_2}{R}\{1 + \sum_{j=2}^{4} q(\Delta_j - 1)r_1^{2j+1} \\
& - \frac{1}{3}\kappa_1(1 + q)(\Delta_2 - 1)r_1^5 P_2(\cos \Theta_1)\} \ ,
\end{aligned}
\tag{3.63}
$$

For the disturbed accelerative potential in the approximation considered, with Θ_1 the angle between $\hat{R}$ and the axis of bodily rotation for m_1, we may then write:

$$
V_{b2} = \frac{a_1^2}{R^3}\left(\sum_{j=2}^{4} Gm_2(\Delta_j - 1)r_1^{j+1} + \frac{1}{6}\kappa_1 n^2 a_1^3(\Delta_2 - 1)\right) ,
\tag{3.64}
$$

where a_1 is the mean radius of the primary.

Eqn (3.64) reflects Russell's point about taking into account the significance of the tidal term, i.e. the first member in the large parentheses, particularly when the rotation is close to period synchronism with the orbit ($\kappa_1 \approx 1$), as often found. The lowest order tidal effect then produces a disturbing term some $6m_2/(m_1 + m_2)$ times that of the rotational one.

Russell's discussion of the apsidal motion would regard the terms in the large parentheses in (3.64) as effectively constant on average, the variation in potential within the orbit affecting only the factor in R^{-3}. The corresponding acceleration, for the lowest order tidal term, appears as

$$\frac{\partial V_{b1}}{\partial R} = \frac{-6a_1^2 k_2}{R^4} \left(Gm_2 r_1^3 + \frac{1}{6}\kappa_1 n^2 a_1^3 \right), \tag{3.65}$$

However, a closer inspection shows that the tidal term in (3.64) contains R^6 in its denominator, since $r_1 = a_1/R$. The layout of (3.65) suggests the appropriate form for the effective disturbing potential in the orbital plane to be

$$V_{b1}' = a_1^5 k_2 \left(\frac{Gm_2}{R^6} + \frac{\kappa_1 n^2}{3R^3} \right), \tag{3.66}$$

The step of relating this disturbing function to the secular component in apsidal motion through (3.62) brings in consideration of the relatively simple equation of the ellipse $A = R(1 - e\cos E)^{-1}$, where E is the eccentric anomaly, and we have written A for the semi-major axis to avoid confusion with lower case a that we have been applying hitherto in this section to the stellar radii. The two particular powers of the variable R appearing in (3.66) are -3 and -6, and these terms are operated on, from (3.62), by $\sqrt{1-e^2}/e\,(\partial/\partial e)$. $(A/R)^j$ can be directly developed as a series in powers of $e\cos E$,

$$\begin{aligned} (A/R)^j &= (1 - e\cos E)^{-j} \\ &= \sum_0^\infty \frac{(j+m-1)!}{(j-1)!\,m!} e^m \cos^m E \ . \end{aligned} \tag{3.67}$$

The procedure to find the secular advance of periastron longitude implies we are now looking only for non-periodic terms in this expansion. In fact, the odd powers of $e^m \cos^m E$ always include the periodic function $\cos E$ in the development, but non-periodic terms in e^2 or higher powers remain in the constant $e^m m!/(2^m [(m/2)!]^2)$ with m even. Factoring this into (3.67) we have, for the required series:

$$\begin{aligned} \left\langle \left(\frac{A}{R}\right)^3 \right\rangle &= \sum_{i=0}^\infty \frac{(2i+1)!}{1!\,2^{2i}\,(i!)^2} e^{2r} \\ \left\langle \left(\frac{A}{R}\right)^6 \right\rangle &= \sum_{i=0}^\infty \frac{(2i+4)!}{4!\,2^{2i}\,(i!)^2} e^{2r}. \end{aligned} \tag{3.68}$$

It turns out that the relatively large effect of the factorial in the numerator, that increases markedly with the power of (A/R), amplifies the role of tides in comparison to the rotation above the value derived by Russell, who neglected the full effects of eccentricity.

The series in powers of e^2 from (3.68), after differentiation and division by e, must be multiplied by the expansion for $\sqrt{1-e^2} = \sum_{j=0}^\infty -1^j(-e^2)^j(-1/2)_j/j!$ to retrieve the individual terms in e^2. So, for the first series in (3.68), for the rotational effect, we have

$$
\begin{aligned}
3G(e) &= \left(3 + \frac{15}{2}e^2 + \frac{105}{8}e^4 + \frac{315}{16}e^6...\right) \times \left(1 - \frac{1}{2}e^2 - \frac{1}{8}e^4 - \frac{1}{16}e^6...\right) \\
&= 3(1 + 2e^2 + 3e^4 + 4e^6...) ,
\end{aligned}
\tag{3.69}
$$

while the expansion for the tidal effect goes similarly as,

$$
\begin{aligned}
15H(e) &= 15\left(1 + 7e^2 + \frac{105}{4}e^4 + \frac{1155}{16}e^6...\right) \times \left(1 - \frac{1}{2}e^2 - \frac{1}{8}e^4 - \frac{1}{16}e^6...\right) \\
&= 15\left(1 + \frac{13}{2}e^2 + \frac{181}{8}e^4 + \frac{465}{8}e^6...\right) .
\end{aligned}
\tag{3.70}
$$

We can now return to (3.62), where we find, after a little manipulation,

$$
\frac{\dot{\varpi}}{n} = \frac{k_2 r_1^5}{1+q} \left(\kappa_1(1+q)\, G(e) + 15q\, H(e)\right) ,
\tag{3.71}
$$

where G and $H \to 1$ as $e \to 0$.

We have given attention so far to the departure from sphericity of the primary, or star 1, associated with its rotation and the tidal action on it from star 2. This distortion, that produces a consequent disturbance to the potential field external to star 1, gives rise to the deviation from Keplerian motion of star 2, relative to the system centre. This is scaled to the relative orbit by multiplying by the factor $(m_1 + m_2)/m_1$, which removes the denominator on the right of (3.71). There is a corresponding perturbation to the potential field external to star 2 disturbing the motion of star 1 in the same sense. The net advance of periastron longitude for the relative orbit from these actions is then:

$$
\begin{aligned}
\frac{\dot{\varpi}}{n} &= k_{2,1} r_1^5 \left(\kappa_1(1+q)\, G(e) + 15q\, H(e)\right) \\
&+ k_{2,2} r_2^5 \left(\kappa_2(1+q)\, G(e) + 15\, H(e)\right)/q .
\end{aligned}
\tag{3.72}
$$

Higher terms in the tidal effect on the motion of the apse could be taken in into account, within the framework of the 'first order' theory presented in Chapter 2, reaching to terms in r^{2j+1} in the relative radius, where j is the order of the harmonic in the development of (3.64) ($2 \leq j \leq 4$. Proceeding along the lines of the foregoing we would then have additional contributions

$$
\begin{aligned}
\frac{\dot{\varpi}_j}{n} &= k_{j,1} r_1^{(2j+1)}(j+1)(2j+1)q H_j(e) \\
&+ k_{j,2} r_2^{(2j+1)}(j+1)(2j+1) H_j(e)/q .
\end{aligned}
\tag{3.73}
$$

with $j = 3$ and 4.

It can be seen that the perturbing term in the tidal potential is $O(k_j r^{2j+1})$. The relative radii r_j in eccentric close binary systems that occur in studies of apsidal motion would be typically ~ 0.2, while the structural constants k_l are of the order of 0.01, so that successive contributions from higher power terms quickly become very small. The accuracy with which r_i^5 (the leading term) can be specified has been typically only within $\sim 10\%$ of its value, rendering the inclusion of terms higher than $j = 2$ in the specification of the tidal contribution to $\tilde{\omega}$ superfluous, at least with current data. Generally speaking, it has been, and continues to be, standard practice to refer only to Eqn (3.72) in comparing expected values of the apsidal motion with observations. With a typical component fractional radius of 0.2 in a young close binary with components of comparable mass, (3.72) points to the scale of the apsidal period being on the order of $\sim 10^5$ that of the orbit, so probably somewhat longer than the previously considered precession period, although we may note the high sensitivity to r_1, and also e, in (3.72).

Since the observed apsidal motion adds together the contributions of both stars, the structural constants $k_{2,j}$ cannot be separately specified, and the formula

$$\bar{k}_2 = \frac{c_1 k_{2,1} + c_2 k_{2,2}}{c_1 + c_2} \tag{3.74}$$

is used to combine the effects, where the constants c_1 and c_2 are the factors multiplying $k_{2,1}$ and $k_{2,2}$ in (3.72). Stars that can be investigated well from observations of eccentric close binary systems are often relatively unevolved massive early type pairs of fairly comparable mass and internal conditions, so the averaging process indicated by (3.74) is, in itself, not a great impediment to such investigations, compared with other sources of difficulty or inaccuracy.

If we compare (3.72) with Russell's formula (3.48), we can see that it is of basically the same form, though the coefficient of the tidal term is more than double that of Russell. This means that if this formula structure is used in inverted form, to determine an 'observational' value of $\bar{k}_2$ from an observed result for $\tilde{\omega}$, the required value of $\bar{k}_2$ to fit the data would be lower with the more detailed analysis that gives higher multiplying factors on the right. This finding at first improved support for theoretical models of stellar structure, though later there appeared a tendency for theoretical values of $\bar{k}_2$ to be too high. The effects of stellar evolution, in this context, generally make for a central condensation within the star. When such trends are taken into account the comparison between theory and observations improves.

Stellar evolution is not the only factor bearing on the interpretation of observations of apsidal motion in eccentric close binary stars. Although the tidal effect, considered in the foregoing, is thought to be the main agent, there are other causes. We may now consider that arising from the general theory of relativity (GTR).

3.4.3 GTR and Apsidal Motion

Following Eddington's (1918) presentation of the GTR theory of gravitatation, we find that the general conditions for a source-free gravitational field are equivalent to the vanishing of the fourth rank Riemann-Christoffel tensor (large to write out in full), but this is reduced, in Einstein's formulation, to the second rank symmetric tensor equation

$$G_{\mu\nu} = 0 \ , \tag{3.75}$$

the indices μ and ν running from 1 to 4. The Einstein tensor, $G_{\mu\nu}$ depends only on the metric elements $g_{\mu\nu}$. This is the four-dimensional counterpart to Laplace's equation for a classical field in general curvilinear coordinates. The 10 separate conditions thus expressed provide a basis for the GTR formulation of gravitation in vacuo. Any set of the 10 metric coefficients $g_{\mu\nu}$ that satisfy (3.75) correspond to a possible set of co-ordinates applying to a source-free field.

These reduced Riemann-Christoffel conditions allow, after some manipulation, the possibility of a metric of the form

$$ds^2 = -\gamma^{-1}dr^2 - r^2 d\theta^2 - r^2 \sin^2\theta d\phi^2 + \gamma dt^2 \ , \tag{3.76}$$

where ds is the separation of events in a four-dimensional space-time manifold with $\gamma = 1 - b/r$, b being a constant of integration. The spatial co-ordinates are chosen so as to reduce to normal spherical polars as $r \to \infty$. This metric's form is suitable to characterize conditions in a source-free volume surrounding a mass-point at the origin. The resemblance of γ to a gravitational potential becomes clearer when we set $b = 2GM/c^2$ and use conventional units. The metric at distance r from the mass centre M then becomes

$$ds^2 = c^2(1 - \frac{2GM}{rc^2})dt^2 - dr^2/(1 - \frac{2GM}{rc^2}) - r^2 d\theta^2 - r^2 \sin^2 \theta \, d\phi^2 \ . \tag{3.77}$$

Eqn(3.77) expresses the well-known Schwarzschild metric, that figures in many practical tests of GTR. Since there are no changes of angular momentum in this central force model, there are no effects on ds^2 associated with differential changes in the increments $d\theta$ and $d\phi$. In turn, this implies that one of these angular variables, θ say, can be eliminated: the orbital motion of a mass element remaining in the 'equatorial' plane $\theta = \pi/2$ with no rotation about a horizontal axis. The constant angular momentum vector about the axis $\theta = 0$ maintains constancy to

$$r^2 \frac{d\phi}{ds} = \text{const.} = \eta, \ \text{say}, \tag{3.78}$$

whereupon, after dividing by ds^2 and rearranging, (3.77) reduces to the form

$$\gamma^{-1} \left(\frac{dr}{ds}\right)^2 + r^2 \left(\frac{d\phi}{ds}\right)^2 = c^2 \gamma \left(\frac{dt}{ds}\right)^2 - 1 \ , \tag{3.79}$$

This can be further tailored to something like the normal energy of a Keplerian orbit by moving what, in typical contexts, would be a small part of the term in the rate of angular variation over to the right side, and writing $ds = c \, dt_0$, i.e. regarding it as proportional to 'proper time' for the moving mass element. We then find, with the angular momentum $\eta = r^2 \, d\phi/dt_0$ in conventional units,

$$\gamma^{-1} \left(\frac{dr}{c dt_0}\right)^2 + \gamma^{-1} r^2 \left(\frac{d\phi}{c dt_0}\right)^2 = \left[\gamma \left(\frac{dt}{dt_0}\right)^2 - 1\right] + (\gamma^{-1} - 1)\frac{\eta^2}{c^2 r^2} \ ,$$

which, when multiplied through by γc^2 gives

$$\left(\frac{dr}{dt_0}\right)^2 + r^2 \left(\frac{d\phi}{dt_0}\right)^2 = \gamma c^2 \left[\gamma \left(\frac{dt}{dt_0}\right)^2 - 1\right] + \frac{2GM\eta^2}{c^2 r^3} \ . \tag{3.80}$$

The left side of (3.80) now appears as the regular kinetic term v^2. On the right, the observed orbital time variation (dt/dt_0) is affected by the Lorentz factor $\beta = \sqrt{1 - v_0^2/c^2}$, as well as the gravitational field, and so

$$\left(\frac{dt}{dt_0}\right)^2 = \frac{\beta^2}{\gamma^2}$$

Writing now $v_0^2 = GM/a$ for the mean orbital velocity and inserting the full expression for γ, (3.80) becomes

$$v^2 = GM \left[\left(\frac{2}{r} - \frac{1}{a}\right) + \frac{2\eta^2}{r^3 c^2}\right] \ , \tag{3.81}$$

which looks like the familiar Keplerian form for the orbital velocity, except for the small additional term involving the angular momentum constant η on the right.

Now, v^2 in (3.81) can be rewritten in terms of the angular variable ϕ, so that

$$\left(\frac{\eta}{r^2}\frac{dr}{d\phi}\right)^2 + \frac{\eta^2}{r^2} = GM\left[\left(\frac{2}{r} - \frac{1}{a}\right) + \frac{2\eta^2}{r^3 c^2}\right] , \qquad (3.82)$$

and then choosing $u = 1/r$ as the variable, and dividing by η^2:

$$\left(\frac{du}{d\phi}\right)^2 + u^2 = \frac{GM}{\eta^2}\left(2u - \frac{1}{a}\right) + 2\frac{GMu^3}{c^2} . \qquad (3.83)$$

Differentiating this last equation and taking out the common factor $du/d\phi$ gives

$$\frac{d^2u}{d\phi^2} + u = \frac{GM}{\eta^2} + 3\frac{GMu^2}{c^2} . \qquad (3.84)$$

Apart from the last member on the right, (3.84) looks like the classical differential equation for Keplerian motion. Comparison of the form of those two right hand members shows that the second one, given that close binary stars are typically orbiting with speeds of several hundred km per second, is of order 10^{-6} that of the main term, so the classical formula $u = GM(1 + e\cos(\phi - \varpi))/\eta^2$ can be substituted into (3.84) to develop a solution by successive approximations. Equation (3.84), in this 'post-Newtonian' approach, then becomes;

$$\frac{d^2u}{d\phi^2} + u = \frac{3(GM)^3}{c^2\eta^4}\left[1 + 2e\cos(\phi - \varpi) + \frac{e^2}{2}(\cos 2(\phi - \varpi) + 1)\right] . \qquad (3.85)$$

The term in $\cos(\phi - \varpi)$, although small, produces an ongoing effect in the orbit through its resonance with the main orbital argument. Restricting attention to this term, the others being very small, we observe that

$$\frac{d^2u}{d\phi^2} + u = \frac{3(GM)^3}{c^2\eta^4}2e\cos(\phi - \varpi) \qquad (3.86)$$

has a particular integral

$$u_1 = \frac{3(GM)^3}{c^2\eta^4}e\phi\sin(\phi - \varpi) .$$

Adding this to the classical orbit formula we find, using the small angle approximation, to the first order in $\delta\varpi$:

$$u = \frac{GM}{\eta^2}(1 + e\cos(\phi - \varpi - \delta\varpi)) , \qquad (3.87)$$

where $\delta\varpi = 3GM\phi/[c^2a(1 - e^2)]$, and so with each orbital cycle the increment $2\pi\delta\varpi$ is added to the longitude of periastron measured from a fixed reference point.

The foregoing model of a fixed mass centre with an orbiting mass element corresponds with the relative orbit in a close binary system if we replace the central mass M with the sum of the component masses $m_1 + m_2$. In summary, the rate of apsidal advance in a close binary system arising from the general theory of relativity is given, to the first order in small quantities, as

$$\frac{\dot{\varpi}}{n} = \frac{3G(m_1 + m_2)}{c^2a(1 - e^2)} . \qquad (3.88)$$

Eqn(3.88) implies that the relativistic apsidal motion rate relates to the orbital period P_o by a fraction of about the same order as a stellar Schwarzschild radius relates to the size of the orbit: so typically on the order of $\sim 10^{-6} \times P_o$. This is about a tenth that of typical apsidal advance rates associated with the torques arising from non-sphericity of the component stars. For young Main Sequence close binary systems these would be of the order of $1°$ per year.

However, binary systems sometimes contain collapsed stars, of comparable masses to those in typical close binaries, but with periods and separations that may be less by an order of magnitude or so in either case. A famous example is the Hulse-Taylor pulsar (PSR 1913+16) that is made up of a radio-pulsating neutron star and a radio-quiet neutron star companion. These objects move in eccentric orbits around a common centre of mass in a period of about 8 h. Both stars are of mass around 1.4 times that of the Sun. The semi-major axis a is close to 2.8 $R_\odot$ and eccentricity $e \sim 0.62$. The orientation of periastron advances by about 4.2 degrees per year and, in this case, it must be dominated by the GTR component, since the relative radii $r_{1,2}$, used in the formula for the classical effect (3.73) are so small.

Accurate timing of pulses from the active component of the binary* furnishes two other quantities that constrain the parametrization of the system: a sinusoidal variation in the frequency of pulse arrival times, that enables a value of the mass-function $f(m) = m_2^3 \sin^3 i/(m_1 + m_2)^2$ to be derived; and a steady decline of the orbital period $\dot{P}_o$ associated with the loss of orbital energy due to the emission of gravitational waves. This is accounted for in more detailed GTR analysis of close binary orbits (see bibliography). The relevant result is given (in the post-Newtonian approximation) as

$$\frac{\mathrm{d}P_o}{\mathrm{d}t} = -\frac{192\pi}{5} \left(\frac{2\pi GM}{c^3 P_o}\right)^{5/3} \frac{\mu}{M}\phi(e) \ , \tag{3.89}$$

where $\phi(e)$ is a function of the eccentricity* that increases strongly with e (≈ 0.6171 for PSR 1913+16). μ is here the reduced mass (1.23) and M the total mass of the system.

The three quantities $\dot{\varpi}/n$, $f(m)$, and $\dot{P}_o$, two of them coming from the formulae of GTR, allow the masses of the components and the inclination of the orbital axis to the line of sight to be determined. The evolving decline of P_o with time according to (3.89) can be compared with observations to check for consistency with the GTR prediction. The special circumstances whereby PSR 1913+16, and similar objects, can provide checks on GTR, or alternative theories of gravity, is of great scientific interest and no doubt was influential in the award of the Nobel Prize for physics to R. A. Hulse and J. H. Taylor in 1993.

3.4.4   Third Body Effects

The complexity of the general problem of the dynamics of a three-body configuration in a gravitational field is well-known. As it happens, observational data very often relate to limiting or restricted conditions for which analytical methods have been explored in some detail in classical papers. The motion of the Moon or other planetary satellites form relevant examples. In these cases, two of the bodies form a close pair, while the third body is at a relatively large distance.

In the previous account of nutation we introduced an hierarchical arrangement of this kind and applied relevant formulae to the relatively close triple of λ Tau. We deduced that the disturbance to the two-body precession of the equator given by (3.27), corresponding to the third body, i.e. the second term in the large braces in (3.33), is only $\sim 10^{-3}$ times the main term. The equatorial precession of a component in a close binary system might be

---

*Sometimes the designation PSR B1913+16 is used.

*$\phi(e) = (1 + 73/24e^2 + 37/96e^4)(1 - e^2)^{-7/2}$.

discernible with high precision photometry, through the effect known as gravity darkening, that would be dependent on the orientation of the rotation axis to the line of sight. But for the present context we can deduce that a third body would be unlikely to have significant effects on the precession of the body rotation axes, that would be dominated by the effects of their closer companions.

In relation to systems like λ Tau, we deduced that the node, where the plane of the close orbit intersects that of the wide one, would recede at a rate

$$\frac{\dot{\Omega}}{n_w} = -\frac{3}{4}\frac{m_3}{m_1 + m_2 + m_3}\frac{n_w}{n_c}\cos i \ , \tag{3.90}$$

n_w and n_c being the mean angular velocities of the wide (w) and close (c) components, respectively. Motions in the plane of the sky, however, that are associated with the nodal angle in spectroscopic and photometric contexts, are not detectable by those kinds of observation, which refer essentially to effects in the line of sight. Practical information on Ω must rely on the growing and improving facilities of high-precision astrometry.

Although superficially different, Eqn (3.46) for the apsidal advance, or its developed form, (3.72), are of essentially similar form to (3.27) for the recession of the nodes, in stating that

$$\frac{\dot{\alpha}}{n} \sim \left(\frac{\Delta W}{T_{\text{rot}}}\right) \ , \tag{3.91}$$

where α is the relevant angular parameter and $\Delta W \sim k r_1^5 W_0$; W_0 and T_{rot} being the relevant orbital potential and kinetic energies. This similarity is borne out by the energy transferred to the orbit, associated with the eccentric tides, being comparable to the work from the torque about the nodal axis associated with the orbit's gravity towards the invariable plane.[*] The rate of apsidal advance is thus found comparable to the rate of nodal recession in a close orbit perturbed by a wide one.

Non-Keplerian motion of a distorted component in a close binary system, such as that described by (3.27), may be regarded as deriving from a spherical mass surrounded by an equatorial girdle whose relative mass, together with the sphere's moment of inertia, are in the correct proportion to correspond to the quantity ϵ_m in (3.25). In this representation, the other component is replaced by a uniform surrounding ring of matter of the same mass. We may thus consider the relative effect of a third body, since, for that case, when $r_3(= R_2/R_3)$ raised to the third power falls below ϵ_m; i.e. $r_3 \lesssim 0.1$, or $P_2/P_3 \lesssim 0.03$, the effect of the third body on orbital precession rates becomes small in comparison to the body distortions. In Batten's (1973) sample of well-documented triple systems, about 20% satisfy the condition $r_3 \gtrsim 0.1$, though observational selection strongly favours closer companions. Only about 2% of stars in the Catalogue of Spectroscopic Binaries satisfy Batten's separate requirements to permit a reliable measurement of apsidal motion. The chances of having to include the effects of a third body in a particular example of apsidal-advance data determining an observational structure constant $\bar{k}_2$ may therefore appear minor. There are, however, a small number of historic cases of apparent disagreement between the rate of apsidal motion predicted from known properties of the component stars and the observed value. Third body effects may need to be included, in a general way, to account for such disagreements. A notable example is the previously mentioned λ Tau, about which there has

---

[*]The concept of an *invariable plane* comes from Laplace's analysis of the planetary motions in the solar system, where the planets all have low inclinations to a plane for which the system's net angular momentum vector maximises. It can become different from the plane about which orbits precess when proximity effects become significant.

been some controversy. Such examples may increase in number with the steady improvement of observational technologies.

3.5 Bibliography

3.1-2 The formulation of the kinetic energy starts along essentially similar lines to that given in Section 2.3 of Z. Kopal's *Close Binary Systems*, Chapman & Hall, 1959, (cf. also *Dynamics of Close Binary Systems*, Dordrecht Reidel, 1978, by the same author) but is not as generalized, the present analysis being directed to specific examples. Section 4.4 in H. Goldstein's Classical Mechanics, Addison-Wesley, p107, 1953, gives a thorough introduction to the Eulerian angle transformations, also mentioning various alternative forms and ambiguities that have appeared in the literature.

3.3 The first example is that of precession, where the rotational axis of a star revolving at relatively high speed shifts in response to the gravitational influence of an orbiting companion. The Earth's precession forms an obvious comparison, for which Eqns (3.23,24) can be applied to derive an approximately valid result. These equations appear as (5) and (6) in the comparable treatment of D. Brouwer, *Astron. J.*, Vol. 52, p57, 1946, see also H. Goldstein's *Classical Mechanics*, Addison-Wesley, p182, 1953. The use here of a derivation from rigid body mechanics might be questioned, but, given the long periods associated with the higher-order effects compared with the relatively short time for hydrostatic restitution of the bodies involved, it can be understood that (3.27) will apply to fluid bodies, provided they retain their mean equilibrium shapes over the precession and nutation timescales. Brouwer (op. cit.) implies the same argument can apply to the equilibrium tidal deformation, which remains in the line of centres so that the corresponding torque vanishes. The potentially important role of dynamical tides was reviewed by G. J. Savonije & M. G. Witte, *Astron. Soc. Pacific Conf. Ser.*, Vol. 229, p133, 2001.

The oscillatory motion that accompanies terrestrial precession, known as nutation, is also approached from the vantage point of the familiar luni-solar example. A step-by-step account of this is given in R. Fitzpatrick's (2011) *Gravitational Potential Theory*, http://farside.ph.utexas.edu/teaching/336k/Newtonhtml/node103.html.

3.4 H. N. Russell's treatment of apsidal motion, in the *Monthly Not. Royal Astron. Soc.*, Vol. 88, p641, 1928, though not fully sourced, seems intuitively clear and provides a useful introduction to the more detailed presentation in Section (3.4.2). Here the not obvious connections between orbital parameter variation and corresponding derivatives of the potential function, originally due to J. Lagrange, as given in D. Brouwer & G. M. Clemence's *Methods of Celestial Mechanics*, Academic Press, 1961, Chapter 11, are followed through. The application of this method to the derivation of apsidal motion in close binary stars with eccentric orbits then essentially follows the exposition of T. E. Sterne in *Mon. Not. Royal Astron. Soc.*, Vol. 99, p451, 1939. The study of the massive binary Y Cyg by D. Holmgren, G. Hill & C. D. Scarfe, *The Observatory*, Vol. 115, p188, 1995 provides a good example of how this theory relates to observational data.

The earlier apparent disagreements between observation and theoretical prediction for structural constants using observed rates of apsidal motion were discussed in A. H. Batten's *Binary and Multiple Systems of Stars*, Pergamon Press, 1973, Chapter 6. A recent review can be found in İ. Bulut, A. Bulut & O. Demircan, *Monthly Not. Royal Astron. Soc.*, Vol. 468, p3342, 2017. It would appear that the role of observational error is relatively high for low mass stars, preventing a very clear assessment of the degree of agreement between theoretical models and observed results. However, for massive stars a more significant comparison can be made and the level of agreement in more recent data is encouraging, even allowing Main Sequence ages of the stellar components to be estimated from the empirical structural coefficients $\bar{k}_2$.

The section on the general relativistic contribution to apsidal advance is based on A. S. Eddington's *Report on the Relativity Theory of Gravitation, Phys. Soc. London*, Fleetway Press, London, 1918, Chapter 5. A comprehensive discussion of this and related subjects may be found in Volume 2 of Landau and Lifshitz' *Classical Theory of Fields*, 2nd ed., Pergamon Press, Oxford, 1962. The discovery of the binary pulsar PSR 1913+16 and its implications were well reviewed in J. H. Taylor's Nobel lecture (1993), https://www.nobelprize.org/uploads/2018/06/taylor-lecture-1.pdf. The observed decline of the orbital period of PSR 1913+16, set the stage for the development of gravitational wave astronomy that started with the LIGO discovery of the source GW150914 in September, 2015 (B. P. Abbott et al. (969 authors), *Phys. Rev. D*, Vol. 94, id. 102001, 2016). More is said about this in Section 11.4. Eqn (3.89) is given in J. H. Taylor & J. M. Weisberg's review of PSR 1913+16 in *Astrophys. J.*, Vol. 253, p908, 1982. Those authors cite P. C. Peters & J. Mathews *Phys. Rev.*, Vol. 131, p435, for the underlying analysis.

The third body effects considered in (3.4.4) can be seen to develop from the considerations given in (3.3.1) with the introduction of third body term in the potential in Eqn (3.28). The derivation of the orbital precession period P_{pw} given after Eqn (3.37) essentially reappears in Eqn (3.90). These results were given in the detailed work of P. Slavenas in *Yale Trans.*, Vol. 6, p35, 1927, and are recollected by A. H. Batten (op. cit.), p128. This thorough, if somewhat dated, overview pointed out subtleties of third body effects, that may, depending on the relative inclination of the orbits, result in apsidal regression as well as advance (though the former were not confirmed). The role of third body effects in complicating the determination of apsidal motion using observed times of minima data was studied for the triple systems V 354 Lac, YY Sgr and DR Vul by M. Wolf et al. (9 authors) in *Acta Astron.*, Vol. 69, p63, 2019. The incidence of triple stars, in general, was considered by C. R. Chambliss, *Publ. Astron. Soc. Pacific*, Vol. 104, p663, 1992, as part of his collection of data on 80 known multiple stars. The ratio of P(wide) to P(close) in these bright triples varies over 6 orders of of magnitude.

4

The Roche Model

4.1   Roche Equipotentials

In preceding chapters, the 'first order' theory of surficial distortion in close binary systems was outlined. 'First order' here means that the effects of proximity in one component induced by the other are regarded as being due to a mass point, i.e. the surficial distortion of a component does not affect that of its companion. The description is self-consistent up to terms of order $(R/a)^5$, where R is a stellar radius and a is the mean separation of the two stars. For a typical close binary system this ratio may be ~ 0.2, so that the error of the description may regarded as quite less than $\sim 1\%$ of the effect. Until recently this would have been well below the measurement accuracy of proximity effects in the majority of such close pairs.

However, if the stars are regarded as acting, in principle, like mass points, then the description of their corresponding equipotential surfaces can be extended to an arbitrary level of exactness, since complications associated with the mutual interactions of near surface layers can be neglected. Given the key role of the equipotential concept in the physics of close binary systems, there is a natural interest in this approach.

We follow this by returning to the orbital co-ordinate frame presented in Fig 1.12 as ξ, with the primary star centred at the origin and the secondary at 1, 0, 0; thus normalizing the separation of mass centres to unity. We will further simplify the discussion by assuming uniform circular motion of the components, with mean angular motion n given by $n^2 = G(m_1 + m_2)$ according to Kepler's third law in the natural units of the system. The centre of gravity of the system is at $q/(1+q)$, 0, 0; or μ 0, 0; where μ is now the mass of the secondary as a fraction of that of the pair. We now refer to a co-ordinate system $\mathbf{x}$ which rotates with the stars and is centred on the primary's centre. The potential function, introduced in Section 2.1 and regarded as acting on unit mass at distance $r = \sqrt{x^2 + y^2 + z^2}$ from the origin and $r' = \sqrt{(1-x)^2 + y^2 + z^2}$ from the centre of the secondary, contains the terms: Gm_1/r and Gm_2/r' whose derivatives give rise to the gravitational attractions on the unit mass at the set location. In this rotating frame of reference there is also a centrifugal force acting on a mass element that is at rest with respect to the moving axes. This can be associated with a potential of the form $n^2 r_c^2/2$, where r_c represents the distance of the mass element from the axis of rotation. These scalar components can be simply added to give the

DOI: 10.1201/b22228-4

complete potential function, as discussed in Chapter 2, that can be written as:

$$V = \frac{Gm_1}{r} + \frac{Gm_2}{r'} + \frac{n^2}{2}\left[\left(x - \frac{q}{1+q}\right)^2 + y^2\right] \tag{4.1}$$

The derivative of (4.1) expresses the full combination of forces acting upon unit mass at x, y, z in this *Roche model* of the configuration.

Introducing now a normalized or relative potential function $\Omega = V/Gm_1$ and setting q for the mass ratio m_2/m_1, with a little rearrangement we can write

$$\Omega = \frac{1}{r} + q\left(\frac{1}{r'} - x\right) + \frac{1+q}{2}\left(x^2 + y^2\right) + \frac{q^2}{2(1+q)} \ . \tag{4.2}$$

It is seen that we need consider values of q only in the range 0-1 to quantify the equipotential contours for the binary, since secondary and primary are simply interchanged with reversal of the mass ratio. Different authors have followed different notations for other contexts, but there is some convenience in following Kopal (1959) and dropping the last constant term in q. Let us write Ω_K for this reduced form, so that

$$\Omega_K = \frac{1}{r} + q\left(\frac{1}{r'} - r\lambda\right) + \frac{1+q}{2}r^2\left(1 - \nu^2\right) \ , \tag{4.3}$$

where we have written $r\lambda$ for x and $r\nu$ for z.

The surfaces corresponding to constant values of Ω in (4.2) are generally referred to as Roche equipotentials. The form of (4.2) shows that Ω will be dominated by the first or second (gravitational) term whenever r or r', respectively, is small. The corresponding contours tend towards near spherical surfaces centred on either mass point. This accords with the expected configuration of a wide, or visual, binary system. At large r, relative to the separation of the mass centres and near the orbital plane, centrifugally dominated equipotentials would move towards cylindrical forms parallel to the z-axis, though relevant matter would, by then, be unbound to the reference frame. For lower values of Ω, however, the contours correspond to more complex shapes that may have physical relevance to very close, interacting, or common envelope binaries.

Multiplying both sides of (4.3) by r and using the fact that $1/r' = 1/\sqrt{1 - 2\lambda r + r^2} = \sum_{j=0}^{\infty} P_j(\lambda)r^j$ we can multiply through by r and rearrange (4.3) as

$$(\Omega_K - q)r = 1 + q\sum_{j=2}^{\infty} P_j(\lambda)r^{j+1} + n^2 r^3(1 - \nu^2) \ . \tag{4.4}$$

In the zeroth approximation, the second and third terms on the right of (4.4) are neglected and we have

$$r_0 = \frac{1}{\Omega_K - q} \ , \tag{4.5}$$

implying that for large Ω_K, $r \to$ constant, which results in the small near-spherical forms referred to above.

Writing now

$$r_1 = r_0 + \Delta'r = r_0\left(1 + \frac{\Delta'r}{r_0}\right) \tag{4.6}$$

and

$$(\Omega_K - q)r_1 = 1 + q \sum_{j=2}^{\infty} P_j(\lambda)r^{j+1} + n^2 r^3(1 - \nu^2) \ , \tag{4.7}$$

where we note the leading term in r on the right is of order r^3. Substituting from (4.6) into (4.7) we have

$$\frac{\Delta' r}{r_0} = q \sum_{j=2}^{\infty} P_j(\lambda)r^{j+1} + n^2 r^3(1 - \nu^2) \ . \tag{4.8}$$

We do not have the exact value of r on the right in advance, but it can be taken that $r = r_0 + \epsilon$ where ϵ is a term of order r_0^3. The relative error introduced by inserting r_0 for r on the right is then a term of order r^6. The 'first order' approximation for r_1 is therefore self-consistent up to and including terms of order r_0^5 on the right, when

$$\frac{\Delta' r}{r_0} = q \sum_{j=2}^{4} P_j(\lambda)r_0^{j+1} + n^2 r_0^3(1 - \nu^2) \ . \tag{4.9}$$

Equation (4.8) compares with (2.11) in Chapter 2, if we set the structural coefficients Δ_j to unity.

It is possible to proceed with a recursive process, as in the formulation of (4.8) we have written the relative deviation of r_1 from r_0, i.e. $\Delta' r/r_0$, in terms of r_0 on the right. We could, in the same way, produce a second order correction of r_2 from r_1, $\Delta'' r/r_0$, involving r_1 as well as r_0 on the right; and so on with higher order differences. However, evaluation and display of contours for given values of Ω by numerical iterative solution of Eqn (4.2) are not difficult programming exercises in general, though special precautions may be required in regions near to the vanishing of derivatives. We can intuitively expect that there will be some point between the two mass centres at which the net force on a mass element disappears.

To pursue this, we consider more closely the mathematical character of Eqn (4.2), so let us differentiate Ω along the x-axis with respect to x to find

$$\frac{\partial \Omega}{\partial x} = -\frac{x}{r^3} + \frac{q(1-x)}{r'^3} - q + 2n^2 x \ . \tag{4.10}$$

A differentiation with respect to y for Ω in the x-y plane yields

$$\frac{\partial \Omega}{\partial y} = -y \left(\frac{1}{r^3} + \frac{q}{r'^3} - 2n^2 \right) \ . \tag{4.11}$$

The z-derivative of Ω in the x-z plane is similarly given as

$$\frac{\partial \Omega}{\partial z} = -z \left(\frac{1}{r^3} + \frac{q}{r'^3} \right) \ . \tag{4.12}$$

Both $\partial\Omega/\partial y$ and $\partial\Omega/\partial z$ vanish along the x-axis, which means that the net force $\nabla\Omega$ disappears on this axis wherever $\partial\Omega/\partial x$ also becomes zero. It can be seen that this will happen at least once in the range $(0 < x < 1)$, since $\partial\Omega/\partial x$ becomes large and negative as $x \to 0$, while it becomes large and positive as $x \to 1$. We shall denote such a point as L_1 in anticipation of a discussion of the *Lagrangian points* of the system later in this chapter.

Along the x-axis, setting $\partial\Omega/\partial x = 0$ in (4.10) yields, after multiplying through by the denominators,

$$(1 + q)x^5 - (2 + 3q)x^4 + (1 + 3q)x^3 - x^2 + 2x - 1 = 0 \ . \tag{4.13}$$

For x in the range 0-1 the left side of (4.13) is a smooth monotonic function starting at –1, rising to $+q$ and approaching the straight line $\Omega_x = 2x - 1$ for $q = 1$. Let us write x_{L_1} as the solution of (4.13). Kopal pointed out that, for small values of q, or more accurately $\mu = q/(1+q)$, an approximate value of x_{L_1} is given by

$$x_{L_1} \approx 1 - \sqrt[3]{\frac{\mu}{3}} \; , \tag{4.14}$$

which is found by expansion of (4.13), or its equivalent in μ, about its root. This estimate can be used to get a starting value for x_{L_1} in an iterative process. Having found x_{L_1}, the corresponding potential Ω_{L_1} is directly derived from (4.2).

Eqn (4.13) can be rearranged into the form $(0 < x < 1)$

$$(1 - x^3)(1 - x)^2 = q(1 - x')^2(1 - x'^3) \; , \tag{4.15}$$

where x' measures the distance along the x-axis from the secondary point. The differences of the two cubes on both sides contain a factor of the form $(1 + x + x^2)$, which has the roots $x = 1/2 \pm i\sqrt{3}/2$, where $i = \sqrt{-1}$. Interestingly, this can be given a geometrical interpretation. The real part, for both x and x', with both sides zeroing simultaneously, refers to the midpoint between the two mass centres. If the imaginary part is regarded as an extension perpendicular to the x-axis in the $x - y$ plane, where $\partial\Omega/\partial z = 0$, we could anticipate two further zeros of the force at two points forming equilateral triangles with the line joining the two mass centres for any value of q, since from (4.11) $\partial\Omega/\partial y = 0$ at these other Lagrangian points. We return later to this.

4.2   The Contact Equipotential

The equipotential corresponding to Ω_{L_1} is of special interest. A star in a binary system starting as a near-spherical mass corresponding to a relatively high value of Ω, as indicated by (4.5), would slowly expand during the course of its evolutionary development. With a high degree of central condensation, the star would approach such a contour, that makes contact with a corresponding contour pertaining to the other mass centre. Further expansion would then entail that matter near the point L_1 would experience an onward pressure sending it beyond the region where it was gravitationally bound to the original star and move towards the other one. Although the Roche contours under consideration are only an approximation to the location of the surfaces of binary components with the same masses and separation but a more realistic density distribution, the role of such a critical surface, 'Roche limit' or 'lobe', is expected to have a topological equivalent in a more realistic model. The adoption of this topological equivalence has been a central feature in theories explaining the characteristics of evolved stars in binary systems, in terms of Roche lobe overflow (RLOF), for decades. Information on the properties of such critical surfaces is then pertinent to practical modelling of binary evolution.

Relevant data include the 'side' points $x_1, y_1, 0$ and $x_2, y_2, 0$ where the derivative dy/dx on the critical surface in the x-y plane becomes zero. These can be located by solving iteratively the simultaneous pair

$$\begin{aligned} \Omega(x_1, y_1, 0) &= \Omega(x_{L_1}, 0, 0) \; , \\ \Omega_x(x_1, y_1, 0) &= 0 \; . \end{aligned} \tag{4.16}$$

A starting value for x_1 here can reasonably be taken to be zero: for y_1, the already known value of x_{L_1} gives a guide. For the secondary, we have similar starting points $x_2 \approx 1$

and $y_2 \sim (1 - x_{L1})$. Solutions are approached using the Newton-Raphson procedure. For the second of (4.16) we then need the second derivative of $\partial\Omega/\partial x$, given by

$$\frac{\partial^2\Omega}{\partial x^2} = -\frac{1}{r^3} + \frac{3x^2}{r^5} - \frac{q}{r'^3} + \frac{3q(1-x)^2}{r'^5} + 1 + q \ . \tag{4.17}$$

The y-coordinate requires iterations that involve $\partial\Omega/\partial y$. This is formulated in (4.11).

The corresponding 'pole' coordinates x_1, 0, z_1 and x_2, 0, z_2 (improved values of x_1 and x_2 being progressively set in the iteration sequence) are given by similar implicit equations

$$\Omega(x_{1,2}, 0, z_{1,2}) = \Omega(x_{L_1}, 0, 0) \ , \tag{4.18}$$

again using the Newton Raphson procedure, where we need to use (4.12) to calculate values for $\partial\Omega/\partial z$.

In Table (4.1) we list values of the L_1, side and pole coordinates of for a variety of mass ratios. Such tables have appeared in classic literature on the Roche model. Their reproduction here, apart from application to relevant observations, serves to cross-check numerical validity, since the numbers were calculated by different methods.[*]

Table 4.1 Key co-ordinates for critical Roche surfaces: L_1; 'side' ($y_{1,2}$) and 'pole' radii ($z_{1,2}$), for different mass ratios (q).

q	x_{L_1}	Ω_{L_1}	x_1	$\pm y_1$	$\pm z_1$	x_2	$\pm y_2$	$\pm z_2$
1.0	0.50000	4.00000	−0.01134	0.37420	0.35621	1.01134	0.37420	0.35621
0.9	0.51084	3.79876	−0.01151	0.38397	0.36502	1.01115	0.36454	0.34744
0.8	0.52295	3.59475	−0.01168	0.39501	0.37491	1.01092	0.35388	0.33770
0.7	0.53663	3.38741	−0.01184	0.40767	0.38614	1.01064	0.34199	0.32677
0.6	0.55234	3.17594	−0.01198	0.42244	0.39909	1.01029	0.32853	0.31431
0.5	0.57075	2.95918	−0.01209	0.44010	0.41433	1.00984	0.31301	0.29984
0.4	0.59295	2.73524	−0.01213	0.46189	0.43278	1.00926	0.29465	0.28260
0.3	0.62087	2.50084	−0.01204	0.49015	0.45599	1.00847	0.27204	0.26123
0.25	0.63808	2.37791	−0.01189	0.50806	0.47026	1.00797	0.25837	0.24823
0.2	0.65856	2.24939	−0.01163	0.52989	0.48714	1.00735	0.24233	0.23294
0.15	0.68392	2.11287	−0.01117	0.55774	0.50781	1.00656	0.22280	0.21425
0.1	0.71750	1.96365	−0.01034	0.59609	0.53451	1.00552	0.19745	0.18991
0.05	0.76874	1.79004	−0.00859	0.65804	0.57291	1.00397	0.15979	0.15365
0.02	0.82456	1.65722	−0.00618	0.73070	0.60977	1.00245	0.11992	0.11522
0.01	0.85852	1.59915	−0.00457	0.77779	0.62867	1.00165	0.09614	0.09231
0.005	0.88635	1.56257	−0.00327	0.81807	0.64170	1.00110	0.07689	0.07378
0.002	0.91520	1.53405	−0.00201	0.86165	0.65258	1.00063	0.05708	0.05474
0.001	0.93227	1.52148	−0.00135	0.88815	0.65761	1.00041	0.04550	0.04360
0.0005	0.94598	1.51355	−0.00091	0.90992	0.66088	1.00026	0.03623	0.03472
0.0002	0.96000	1.50737	−0.00052	0.93264	0.66348	1.00015	0.02679	0.02566
0	1	1.5	0	1	0.66667	1	0	0

Table (4.1) shows that the normal appearance of the contact equipotential for intermediate values of the mass ratio consists of two elongated 'raindrop' shaped lobes that touch at the interior point x_{L_1} on the x-axis (Fig 4.1). As the mass ratio decreases to zero, the primary lobe becomes equatorially distended, so that the z-diameter approaches 2/3 that of the equatorial diameters. The secondary takes an approximately prolate form whose long diameter, along the line of centres, is 3/2 times those of the other two. These properties are of interest, for example in connection with the occurrence of close exoplanets at low mass-ratios. Given the previous estimate of the long-axis radius for the small component of $\Delta x = 1 - x_{L_1} \approx \sqrt[3]{q/3}$, we can estimate the relative volume V_2 of such a component as

$$V_2 = \left(\frac{2}{3}\right)^4 \pi q \ . \tag{4.19}$$

---

[*]Note that the potentials Ω_{L_1} listed in Table 4.1 correspond to Eqn (4.2).

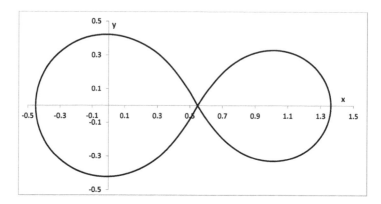

Figure 4.1 Critical Roche contour in the $x - y$ (orbital) plane for the mass ratio $m_2/m_1 = 0.6$. The maximal excursions of the contour perpendicular to the line of centres are slightly displaced from the x coordinates of the mass-points (0 and 1).

In order to check this point, we may evaluate Eqn (4.3) in a limiting form with $r = 1 - \Delta x$, whereupon we find $\Omega = 3/2 + \epsilon$, where, neglecting high order terms in Δx,

$$\epsilon = \frac{q}{\Delta x} + \frac{3}{2}\Delta x^2 - \frac{q}{2} + \frac{q\Delta x^2}{2} \ . \tag{4.20}$$

The same limiting form for ϵ is found at $r = 1 + \Delta x$, while at $x = \sqrt{1 - \Delta y^2}, y = \Delta y$, there results

$$\epsilon = \frac{q}{\Delta y} - q + \frac{q\Delta y^2}{2} \ . \tag{4.21}$$

Substituting the approximation $q = 3\Delta x^3$, combining the two expressions for ϵ, and neglecting high order terms in Δx and Δy, we are left with

$$\frac{9}{2}\Delta x^2 \Delta y = 3\Delta x^3 \ , \tag{4.22}$$

or $\Delta y/\Delta x = 2/3$. Essentially the same formulation works also in the $x - z$ plane, confirming that the limiting second order ellipsoid tends towards the prolate form with long axis $3/2$ greater than the short ones, as seen in the last two columns of Table 4.1 as $x_{L_1} \to 1$.

4.2.1   Areas and Volumes

As indicated above, a Newton-Raphson procedure was used to construct Table 4.1. The same procedure is then reasonably available to examine the solution of the functional equation $r = \Omega^{-1}\{\Omega_{L_1}\}$ for any point r, θ, ϕ on the contact surface, θ and ϕ being supplied at will, and the constant potential fixed by the condition $\Omega_{L_1} = \Omega(x_{L_1}(q))$, the mass ratio q being given and x_{L_1} found as in the discussion after Eqn (4.13). In this way, a mapping of a contact surface for a given mass ratio can be produced as in Fig 4.2, whose closeness to the continuous form of the lobe will improve as more points are included in the network used in the mapping. This applies to the numerical evaluation of the projected area of a Roche surface, when we approximate the integral $\int_0^{2\pi} \int_0^{\pi} r^2 \sin\theta \, d\theta \, d\phi$ by the summation $4\sum^m \sum^n \bar{r}^2 \sin\theta\Delta\theta\Delta\phi$, where we have utilized the quadrant symmetry in the construction and the sectoral and zonal ranges are split into m and n subdivisions. To each small segment $\Delta\theta$, $\Delta\phi$ there corresponds a mean radius $\bar{r}(\theta, \phi)$. For the volume, we similarly sum the small pyramid contributions $4/3\sum^m \sum^n \bar{r}^3 \sin\theta\Delta\theta\Delta\phi$.

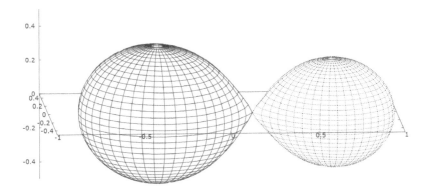

Figure 4.2 A 3d representation of the Roche lobes with the same mass ratio as in Fig 4.1, using the freely distributed GNUPLOT software (©T. Williams, C. Kelley). The origin is here placed at the system's centre of mass. The shape is formed from a network of 30 zonal and 40 sectoral divisions. The angular co-ordinates being given with the potential fixed to that at L_1, the appropriate value of the radius is determined by solving the implicit equation (4.2) with a suitable numerical procedure.

In Tables 4.2 and 4.3, we provide these areas ($A_{1,2}$) and volumes ($V_{1,2}$), together with various measures of representative radii for varying mass ratios. Table 4.4 extends this information by listing the sums of both lobe radii, according to different ways in which the representative radii might be calculated.

A range of formal properties of the Roche surfaces exist in the literature. It can be shown, for example, that in the limiting case of $q = 0$ that the radius of the primary tends to $r = 2\sin(\theta/3)/\sin\theta$. The mean radius integrated over the approximately spheroidal volume can thus be formally evaluated and the result is given in the last row of Table 4.2, along with corresponding areas and volumes.

Table 4.2   Roche surface data (primary component).

q	r_{side}	$\bar{r}$	A_1	V_1	$r_{\bar{A}}$	$r_{\bar{V}}$
1.0	0.37403	0.37285	1.80891	0.22959	0.37941	0.37986
0.9	0.38379	0.38205	1.89894	0.24691	0.38873	0.38918
0.8	0.39483	0.39238	2.00287	0.26743	0.39923	0.39967
0.7	0.40749	0.40416	2.12475	0.29217	0.41120	0.41164
0.6	0.42226	0.41779	2.27062	0.32273	0.42508	0.42552
0.5	0.43991	0.43393	2.44986	0.36165	0.44154	0.44198
0.4	0.46171	0.45361	2.67824	0.41332	0.46166	0.46210
0.3	0.48997	0.47868	2.98515	0.48632	0.48739	0.48784
0.25	0.50788	0.49428	3.18552	0.53607	0.50348	0.50394
0.20	0.52972	0.51297	3.43529	0.60034	0.52285	0.52332
0.15	0.55757	0.53627	3.76182	0.68797	0.54713	0.54764
0.10	0.59592	0.56722	4.22336	0.81857	0.57973	0.58030
0.05	0.65785	0.61409	4.98754	1.05127	0.63000	0.63078
0.02	0.73044	0.66325	5.88655	1.35018	0.68442	0.68565
0.01	0.77743	0.69116	6.45073	1.55143	0.71647	0.71815
0.005	0.81757	0.71223	6.90829	1.72258	0.74145	0.74364
0.002	0.86089	0.73173	7.36259	1.89999	0.76544	0.76834
0.001	0.88714	0.74175	7.61040	2.00024	0.77821	0.78162
0.0005	0.90857	0.74882	7.79306	2.07594	0.78750	0.79136
0.0002	0.93070	0.75500	7.95919	2.14634	0.79585	0.80021
0	1	0.76430	8.23144	2.26662	0.80934	0.81489

Table 4.3 Roche surface data (secondary component).

q	r_{side}	$\bar{r}$	A_2	V_2	r_A	r_V
1.0	0.37403	0.37285	1.80891	0.22959	0.37941	0.37986
0.9	0.36437	0.36372	1.72170	0.21321	0.37015	0.37060
0.8	0.35371	0.35360	1.62761	0.19600	0.35989	0.36035
0.7	0.34183	0.34225	1.52531	0.17784	0.34840	0.34886
0.6	0.32838	0.32935	1.41303	0.15860	0.33533	0.33579
0.5	0.31287	0.31438	1.28826	0.13809	0.32018	0.32065
0.4	0.29451	0.29656	1.14722	0.11608	0.30215	0.30261
0.3	0.27192	0.27447	0.98379	0.09221	0.27980	0.28026
0.25	0.25826	0.26104	0.89048	0.07943	0.26620	0.26666
0.20	0.24223	0.24521	0.78653	0.06595	0.25018	0.25064
0.15	0.22271	0.22586	0.66806	0.05165	0.23057	0.23102
0.10	0.19738	0.20060	0.52789	0.03630	0.20496	0.20540
0.05	0.15975	0.16282	0.34875	0.01951	0.16659	0.16699
0.02	0.11990	0.12254	0.19817	0.00836	0.12558	0.12592
0.01	0.09612	0.09840	0.12804	0.00435	0.10094	0.10124
0.05	0.07688	0.07880	0.08226	0.00224	0.08091	0.08116
0.02	0.05708	0.05857	0.04554	0.00092	0.06020	0.06040
0.001	0.04550	0.04672	0.02901	0.00047	0.04805	0.04822
0.0005	0.03623	0.03723	0.01844	0.00024	0.03830	0.03844
0.0002	0.02678	0.02754	0.01010	0.00010	0.02835	0.02846

4.2.2   Roche Lobes and Eclipses

It can be seen from Table 4.4 that the sum of average radii of the two Roche lobes is $\approx 0.75 \pm 0.01$ for a wide range of mass ratios. This fact, although reducing discriminatory information in the light curve analysis of near-contact binary systems, is a useful indicator of the likelihood of a near-contact condition when empirical evidence yields $r_1 + r_2 \approx 0.75$.

Table 4.4 Roche lobes (sum of representative radii).

q	r_{side}	$\bar{r}$	r_A	r_V
1.0	0.74806	0.74570	0.75882	0.75972
0.9	0.74816	0.74577	0.75888	0.75978
0.8	0.74854	0.74598	0.75912	0.76002
0.7	0.74932	0.74641	0.75960	0.76050
0.6	0.75064	0.74714	0.76041	0.76131
0.5	0.75278	0.74831	0.76172	0.76263
0.4	0.75622	0.75017	0.76381	0.76471
0.3	0.76189	0.75315	0.76719	0.76810
0.25	0.76614	0.75532	0.76968	0.77060
0.20	0.77195	0.75818	0.77303	0.77396
0.15	0.78028	0.76213	0.77770	0.77866
0.10	0.79330	0.76782	0.78469	0.78570
0.05	0.81760	0.77691	0.79659	0.79777

Apart from the radial data of Tables 4.2-4, the opening angle of the Roche lobes at their point of contact L_1 is of interest. The definition of the L_1 point is tantamount to the vanishing of the first order derivatives there. The second-order derivatives corresponding to (4.17) for y and z are found to be

$$\frac{\partial^2 \Omega}{\partial y^2} = -\frac{1}{r^3} + \frac{3y^2}{r^5} - \frac{q}{r'^3} + \frac{3qy^2}{r'^5} + 1 + q \;, \tag{4.23}$$

and

$$\frac{\partial^2 \Omega}{\partial z^2} = -\frac{1}{r^3} + \frac{3z^2}{r^5} - \frac{q}{r'^3} + \frac{3qz^2}{r'^5} \;. \tag{4.24}$$

The second-order mixed derivatives are similarly found as

$$\frac{\partial^2 \Omega}{\partial x \partial y} = \frac{3xy}{r^5} - \frac{3q(1-x)y}{r'^5} \;, \tag{4.25}$$

$$\frac{\partial^2 \Omega}{\partial x \partial z} = \frac{3xz}{r^5} - \frac{3q(1-x)z}{r'^5} \;, \tag{4.26}$$

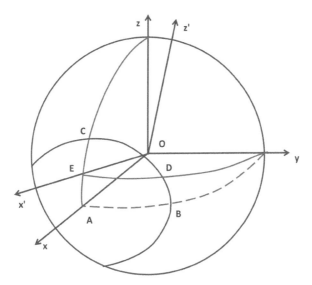

Figure 4.3 The elliptical cone presented by the Roche lobes from the L_1 point (O) projects onto the celestial sphere as the ellipse centred on the x axis at A, and containing the arc BC, where AB and AC correspond to the maximum extensions in the y and z directions as tabulated in Table 4.6.

$$\frac{\partial^2 \Omega}{\partial y \partial z} = \frac{3yz}{r^5} + \frac{3qyz}{r'^5} \ . \tag{4.27}$$

Table 4.5 Roche lobe opening angles at L_1. Observe that these stay approximately constant for a wide range of mass ratios. The opening angle in the $x - z$ plane θ_z is the complement of the angle listed as minimum inclination for an eclipse to be seen $i_{\min}$.

q	x_{L_1}	θ_y (deg)	$i_{\min}$ (deg)
1.0	0.50000	57.3122	34.4499
0.9	0.51084	57.3138	34.4494
0.8	0.52295	57.3193	34.4478
0.7	0.53663	57.3303	34.4445
0.6	0.55234	57.3491	34.4389
0.5	0.57075	57.3795	34.4298
0.4	0.59295	57.4279	34.4153
0.3	0.62087	57.5069	34.3919
0.2	0.65856	57.6440	34.3513
0.1	0.71751	57.9243	34.2690

The y and z factors in these derivatives ensure that they vanish on the x-axis. Constancy of the value of Ω around the surface near to L_1 allows us to write

$$\Delta x^2 \frac{\partial^2 \Omega}{\partial x^2} + \Delta y^2 \frac{\partial^2 \Omega}{\partial y^2} + \Delta z^2 \frac{\partial^2 \Omega}{\partial z^2} = 0 \ . \tag{4.28}$$

Table 4.6   Roche lobes: Eclipse tangency angle ϕ_1 (deg).

q	Inclination i (deg)					
	90	80	70	60	50	40
0.1	57.92	57.34	55.46	51.79	45.08	30.91
0.2	57.64	57.06	55.16	51.47	44.73	30.48
0.3	57.51	56.92	55.01	51.32	44.55	30.27
0.4	57.43	56.84	54.93	51.23	44.46	30.15
0.5	57.38	56.79	54.85	51.12	44.33	30.01
0.8	57.32	56.73	54.82	51.11	44.32	29.99
0.9	57.31	56.72	54.81	51.10	44.31	29.98
1	57.31	56.72	54.81	51.10	44.31	29.98

Substitution of the coordinates of the point L_1 into the second order derivatives in (4.17) and (4.23-24) then shows that (4.28) takes the form

$$A\Delta x^2 = B\Delta y^2 + C\Delta z^2 \ , \tag{4.29}$$

where A, B and C are determined from (4.17) and (4.23-24) to be all positive. We then find that this constraint on the second order derivatives is equivalent, in the vicinity of L_1 to the equation of an elliptical cone along the x-axis with apex at $x_{L_1}, 0, 0$; i.e.

$$\Delta x^2 = \Delta y^2/a^2 + \Delta z^2/b^2 \ , \tag{4.30}$$

where the cone's semi-major and semi-minor axes are $a = \sqrt{A/B}$ and $b = \sqrt{A/C}$, so that the opening angle θ_y in the $x-y$ plane is $\arctan\sqrt{A/B}$, and in the $x-z$ plane θ_z is given by $\arctan\sqrt{A/C}$. Note that for $q = 1$, $r_1 = 1/2$ and so $A = 34$, $B = 14$ and $C = 16$ and then $\theta_y = 57.312$ and $\theta_z = 55.550$ deg as can be seen from Table 4.5. It can also be seen from Table 4.5 that these angles vary very little over a wide range of mass ratios.* This can be understood as again a consequence of the relative insensitivity of the L_1 point to the mass ratio, which entails that $(1+q) \approx ((1/x_1{}^3 + q/(1-x_1)^3)/8$. If we substitute $x_1 \approx 1/2 + \epsilon$ in this expression, where ϵ is a small quantity we will find that $\tan\theta_z \approx \sqrt{2+1/8}$ to order ϵ^2.

We may consider next how the cone, which defines the limiting angle for possible eclipses in contact systems, bears on the eclipse phase range when viewed from remote locations with a line of sight inclined at some arbitrary angle to the rotation axis. To do this, we find the intersection of the elliptic cone given by (4.29) with the celestial sphere, which is concerned only with angular separations, and so satisfies

$$x^2 + y^2 + z^2 = 1 \ . \tag{4.31}$$

Producing the osculating variables in (4.30) to a general extension and combining with (4.31) results in the elliptical form

$$y^2/\alpha + z^2/\beta = 1 \ , \tag{4.32}$$

$(\alpha > \beta)$ where the ellipse's axis parameters α and β are determined from the foregoing values of A, B and C. The co-ordinates y and z in this scheme correspond to angular projections on the sky (see Fig 4.3), and in the normal arrangement used for phasing the orbits of eclipsing systems can be set as $y = \sin\phi$, and $z = \cos i \cos\phi$, so that

$$y^2 = \alpha \left(1 - \frac{z_0^2 \cos^2\phi}{\beta} \right) \tag{4.33}$$

---

*It can be shown that $-\frac{1}{2} \leq \cos 2\theta_y \leq -\frac{5}{12}$, and $-\frac{5}{13} \leq \cos 2\theta_z \leq -\frac{9}{25}$.

Here $z_0 = \cos i_0$ corresponds to the minimum angular separation of the viewing angle to the line of centres and will be given, in general, according to the accidental arrangement of any given system. The corresponding phase ϕ_1 at which the eclipse starts then follows as

$$\phi_1 = \arccos \sqrt{(1-\alpha)/(1 - z_0^2 \alpha/\beta)} \ . \tag{4.34}$$

Values of ϕ_1 for given mass ratios and inclinations are set out in Table 4.6. It is seen from this Table that the phase at which eclipse starts for the 'contact' geometry under consideration is quite insensitive to the mass ratio, and for higher inclination binaries an eclipse lasts for typically $\sim 30\%$ of the light cycle.

4.3   External Envelopes

Hitherto we have been concerned with closed equipotentials that could approximate the surfaces of real stars in binary systems. These will satisfy the condition $\Omega \geq \Omega(L_1)$, the equality corresponding to the critical Roche surface examined in the previous section.

But what happens if $\Omega < \Omega(L_1)$? Intuitively, we might expect a single surface surrounding the two contact forms. Higher Ω values correspond to the situation where the small denominators in the first two terms on the right of (4.2) predominate. With regard to (4.2), though, we noted before that it must become large again for large distances from the z-axis, when the centrifugal term increases indefinitely. Along the x-axis, therefore, apart from the minimum in Ω located at the L_1 point, there must be at least two further minima beyond the region between the two mass centres, i.e. where $x < 0$ and $x > 1$. In order to locate such minima we set out a more general form for $\partial\Omega/\partial x$, keeping in mind the sense of the gravitational force in relation to the variable x, thus

$$\frac{\partial\Omega}{\partial x} = \mp \frac{x}{r^3} \pm \frac{q(1-x)}{r'^3} - q + (1+q)x \ , \tag{4.35}$$

the plus sign on the first member on the right operating for $x < 0$ and the minus sign on the second member operating for $x > 1$. Otherwise, the region $0 < x < 1$ is covered by (4.10).

These generalizations alter the last three terms on the left of (4.13). In the case $x < 0$ we have

$$(1+q)x^5 - (2+3q)x^4 + (1+3q)x^3 + x^2 - 2x + 1 = 0 \ , \tag{4.36}$$

and for $x > 1$,

$$(1+q)x^5 - (2+3q)x^4 + (1+3q)x^3 - (1+2q)x^2 + 2x - 1 = 0 \ . \tag{4.37}$$

The approach used in locating x_{L_1} for small q, can be applied also to the other 'collinear Lagrangian points', L_2 and L_3, on the x-axis. Note first that if $q = 0$, (4.13) and (4.37) both reduce to the same form, which, from (4.15), we can see is satisfied by $x = 1$. Insertion of $x = 1 + \epsilon$ into (4.37), ϵ being a small quantity, produces a similar result to (4.14), except that this time we find the root beyond $x = 1$ gives $\epsilon \approx \sqrt[3]{q/3}$. This root is identified with L_2, so we see that as $q \to 0$ L_1 and L_2 coalesce, rather quickly, from either side of the second mass centre. The approximate position of L_2 is then given by $x_{L_2} \approx 1 + (q/3)^{1/3}$

Eqn (4.36), meanwhile, is satisfied by $x = -1$ at $q = 0$. If we now set $|x| = \xi$, say, in (4.35), the first member gives the largest contribution, with the next significant remaining term being $(1+q)\xi$. Putting now $\xi = 1 - \epsilon$, we can deduce that $\epsilon \to q/3$, so that for small q, x_{L_3} moves towards $-(1 - q/3)$, i.e. linearly with q. Figure 4.4 shows that, in fact, both these approximations are not bad over the whole range $0 < q < 1$.

It may be noticed that the two pentics given by (4.36) and (4.37), p_2 and p_3, say, image each other about the x-axis, in that $p_2(q, x) = -p_3(Q, x')$, where $Q = 1/q$ and $x' = 1 - x$.

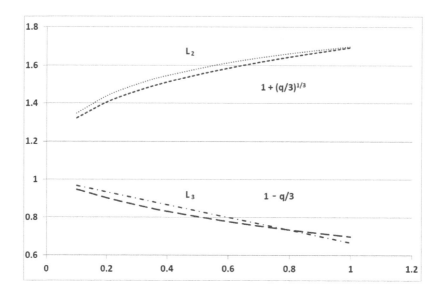

Figure 4.4   The formulae for the approximate positions of the external Lagrangian points for different mass ratios (q) as shown near the corresponding curves: in short dash for L_2 and dot-dash for L_3. These may be compared with much more precise evaluations from the Newton Raphson root-finding technique (dotted and long dash, respectively).

This can be verified by direct substitution, or by the realization that the change essentially swaps the roles of primary and secondary in the configuration, and with that interpretation the locations of L_2 and L_3 interchange. This naturally entails that $x_{L_2} - 1 = -x_{L_3}$ for $q = 1$, as can be seen in Table 4.7. Less obvious is that if $q = g_n$, $x_{L_2} = 1 + g_n (= 1/g_n)$, where g_n is the 'golden number' $(\sqrt{5} - 1)/2$, but it can be established by dividing $x^2 + x - 1$ into $p_2(x, 1 + x)$, noting that g_n is a root of $x^2 + x - 1$.

The possibility of off-axis zeros of the force field was raised in Section (4.1). This can perhaps be more directly confirmed by considering $\nabla\Omega$, where (4.2) for Ω is set in cylindrical coordinates (r, θ, z). From (4.12), it is clear that $\partial\Omega/\partial z$ vanishes only in the plane $z = 0$, so we can confine attention to that plane, writing

$$\Omega = \frac{1}{r} + \frac{q}{r'} - qr\cos\theta + \frac{(q+1)}{2}r^2 \ , \tag{4.38}$$

so that

$$\frac{\partial\Omega}{\partial r} = -\frac{1}{r^2} - \frac{q}{r'^2}\left(\frac{r - \lambda}{r'}\right) - q\lambda + (q+1)r \tag{4.39}$$

where we have written λ for $\cos\theta$. Eqn (4.39) will zero when $r = r' = 1$. For $\nabla\Omega_\theta$ we have

$$\frac{\partial\Omega}{r\partial\theta} = -\frac{q}{r'}\sin\theta + q\sin\theta \ , \tag{4.40}$$

which, if zero at $\theta = \pm\pi/3$, entails $r = r' = 1$, thus locating the other off-axis Lagrangian points at the equilateral positions referred to as L_4 and L_5.

The term 'Lagrangian points' derives from a classical analysis of the so-called restricted three-body problem, published by Joseph Lagrange in 1772. The key insight of Lagrange was

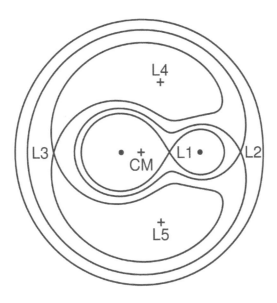

Figure 4.5 The five Lagrangian points in the $x - y$-plane are shown. (Courtesy of Philip D. Hall.)

Table 4.7 External Lagrangian point details.

q	x_{L_2}	Ω_{L_2}	x_{L_3}	Ω_{L_3}	$\Omega_{L_{4,5}}$
1.0	1.69841	3.45680	−0.69841	3.45680	2.75000
0.9	1.68108	3.30120	−0.71546	3.26570	2.61316
0.8	1.66147	3.14434	−0.73414	3.07362	2.47778
0.7	1.63904	2.98578	−0.75471	2.88050	2.34412
0.6	1.61304	2.82487	−0.77752	2.68633	2.21250
0.5	1.58238	2.66059	−0.80303	2.49109	2.08333
0.4	1.54538	2.49125	−0.83180	2.29476	1.95714
0.3	1.49917	2.31378	−0.86461	2.09739	1.83462
0.2	1.43808	2.12180	−0.90250	1.89903	1.71667
0.1	1.34699	1.89835	−0.94693	1.69982	1.60455

to reduce aspects of a purely dynamical problem, by an appropriate choice of coordinates, into the kind of statics issues considered in this chapter. And what an insight this has turned out to be! The L_1 point is now understood to have a pivotal role regarding the evolution of interactive binaries, as we shall see in later chapters.

Although matter reaching the L_1 point as a result of secular expansion of the evolving star seems likely to fall down the potential well and into the companion's Roche lobe, if that lobe is also approaching fullness, then the configuration of a 'contact', or, more aptly, 'common-envelope' binary may be attained. Alternative theoretical models for such systems exist that are supported to some extent by observational data. The subject is reviewed in Chapter 8.

Relatively little potential energy separates the L_1 and L_2 points, so in situations where some suitably directed increment of matter leaves its original confines with sufficient energy, it may attain the lower binding energies associated with the other Lagrangian points. If energized matter, from a common envelope situation, say, arrives in the vicinity of L_2, it may then be able then to escape the system entirely, since the potential energy gradient becomes continuously outward beyond this point. But here we should recall that the Roche contours apply only for the rotating coordinates, where, formally, the centrifugal force would continue to operate on matter fixed in that frame of reference.

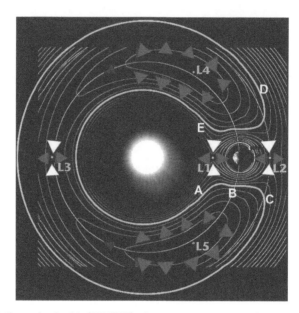

Figure 4.6 Possible 'horseshoe' orbit (ABCDE) of a mass increment in relation to the Lagrangian points, in a special case of the three-body problem. The 'external' equipotentials are shown as faint curves around the two mass centres. (Illustration courtesy of NASA.) The horseshoe configuration was discussed in connection with a natural satellite of the Earth by A. A. Christou & D. J. Asher in *Monthly Not., Royal Astron. Soc.*, Vol. 414, p2965, 2011.

In practice, a range of possibilities arises for matter energized beyond the L_2 level occur, including the curious 'horseshoe' condition. It is seen that for a low mass particle moving at a low velocity relative to the rotating frame of reference, the associated kinetic energy will be small compared with the energy separating equipotentials on the scale of those of the main binary. Thus, for example, there is a $\sim 7\%$ difference in the potentials of L_2 and L_3 at $q = 0.5$. Given conservation of the total energy, the particle's orbit would therefore remain relatively close to one of the outer equipotentials, shown as faint curves in Fig 4.6. In that diagram, such a slow-moving mass particle drifts from a moderately high potential region near L_3 along the inner side of the 'plateau' of the equilateral region around L_5, gaining speed as it approaches m_2. The slight excess of kinetic energy raises the particle back through the A B C region to the outside track, past the plateau and on towards the L_3 region, which it passes beyond, now at a slightly higher potential and pursuing its excursion around the other equilateral Lagrangian point L_4 in the opposite sense: D to E and then back along the inside track.

The equilateral Lagrangian points, L_4 and L_5, are convex stationary points for the potential, indicated by the outward-pointing grey triangles. This is unlike the others, which are all saddle points (the white triangles indicating the inward direction of the potential gradient). Matter gathering in the L_4 and L_5 regions with sufficient self-attraction may therefore tend to accumulate. This is the situation believed to hold for the 'Trojan' asteroids associated with Jupiter and the Sun, with parallels for relatively small amounts of matter at the equilateral points of other binary systems (Fig 4.7).

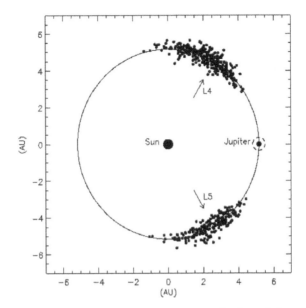

Figure 4.7   Trojan asteroids in the solar system congregating at the L_4 and L_5 points of the Sun-Jupiter system. For details of the diagram, courtesy of J. M. Matthews, see Section 4.4.

4.4   Bibliography

This chapter is based on Chapter 3 of Z. Kopal's *Close Binary Systems*, Chapman & Hall, 1959, which contains a wealth of detailed information. We have been able to take some short-cuts, given the speed and convenience of numerical procedures on modern computers.

4.1 Kopal (op. cit.) cites the original work of E. Roche in the *Mém. Acad. Sciences, Montpelier*, Vol. 1, p243 and p333; Vol. 2, p21; 1849-51, and Vol. 8, p235, 1873, as the source publications for what has become known as the Roche Model. Issues relating to the stability of the model are summarized in the book of L. Lichtenstein, *Equilibrium Figures of Rotating Fluid Masses*, Berlin, 1933, sections 24–28. A numerical study of small adiabatic radial oscillations in close binary components distorted according to the Roche model was carried out by B. Ulaş, *Astron. Nach.*, Vol. 333, p744, 2012; and comparisons made with observational data. It was shown that measured frequencies and amplitudes of surface waves can vary slightly for distorted stars in comparison to undistorted ones.

4.2 Tables, such as those given in this chapter, were set out by G. P. Kuiper in the *Astrophys. J.*, Vol. 93, p133, 1941, and in more detail by Z. Kopal in the *Jodrell Bank Ann.*, Vol. 1, p37, 1954; and in more extensive form by M. Plavec & P. Kratochvil, in the *Bull. Astron. Czech.* Vol. 15, p165, 1964. The latter paper contains some useful approximate formulae linking representative radii for Roche limiting surfaces r_L to the mass ratio q of the two components. A similar approximation having the useful short form $r_L = 0.38 + 0.2$ log q, valid for intermediate values of q was given by B. Paczyński in *Ann. Rev. Astron. Astrophys.*, Vol. 9, p183, 1971. A more accurate approximation with a greater range of validity for q, introduced by P. P. Eggleton in *Astrophys. J.*, Vol. 268, p368, 1983, as $r_L = 0.49q^{2/3}/(0.6q^{2/(3} + \ln(1 + q^{1/3}))$, is frequently cited. A JAVA interface to calculate geometrical properties of critical Roche surfaces, was published by D. A. Leahy & J. C. Leahy in *Comp. Astrophys. Cosmology*, Vol. 2(4), 2015.

The discussion of equipotentials that follows from Eqn 2.1 is based on classical treatments that include only gravitational and centrifugal terms. But there is also the pressure of radiation, which becomes significant for high temperature O and B type stars in a close binary configurations. The consequent modification to equipotentials can be approximated as a reduction of the leading gravitational term in (4.2) by a constant proportion (δ). This was pursued in the paper of H. Drechsel, S. Haas, R. Lorentz & S. Gayler, *Astron. Astrophys.*, Vol. 294, p723, 1995, with numerical calculations showing the modifications to the standard Roche lobe geometry. These effects were found consistent with analysis of the broad eclipses of the B system in the quaternary QZ Car in W. S. G. Walker, M. Blackford, R. Butland & E. Budding, *Mon. Not. Royal Astron. Soc.*, Vol. 470. p2007, 2017. Issues in matching models of massive, early type binaries to observations were discussed by J. Figueiredo, J-P. De Greve & R. W. Hilditch in *Astron. Astrophys.*, Vol. 283, p144, 1994. They must surely arise in connection with GCIRS 16SW, a massive eclipsing binary near the Galactic center examined by M. S. Peeples, A. Z. Bonanos, D. L. DePoy, K. Z. Stanek & J. Pepper, *Astrophys. J.*, Vol. 654 L61, 2007.

4.3 S. F. Dermott & C. D. Murray discuss the dynamics of horseshoe and comparable orbits in *Icarus*, Vol. 48, p1, 1981. The subject of Trojan and similar mass concentrations, relating to the properties of the Roche surfaces, was discussed in the book edited by R. Dvorak & J. Souchay, *Dynamics of Small Solar System Bodies and Exoplanets, Lecture Notes in Physics*, Vol. 790, Springer, 2010, see, in particular, P. Robutel & J. Souchay, on p197. F. Yoshida & T. Nakamura review more familiar solar system examples in the *Astron. J.*. Vol. 130, p2900, 2005. Fig 4.7 appeared in the paper of R. M. Winslow & J. M. Matthews in *Canadian Undergraduate Phys. J.*, Vol. 6, p7, 2008.

Visual and Astrometric Systems

5.1 Introduction

William Herschel had catalogued several hundred double stars by the time his third list was published in 1821. His son John went on to extend the catalogue to over 5000 examples. By the present time, the latest edition of the Washington Double Star catalogue runs to over 100000 entries. A sample from the catalogue of O. Malkov et al (2018) is shown in Fig 5.1.

Even in Herschel's time, it was strongly expected that the nearest stars must be at least of the order of a few parsecs in distance, while the great majority of stars visible through a

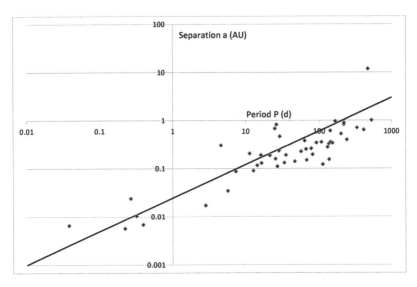

Figure 5.1 A sample of 50 visual binaries from the compilation of Malkov et al., (2012), showing the trend of separation versus period. The general validity of Kepler's third law, is shown by reference to the straight line trend $\log a = 2/3 \log P + $ const., regardless of the vertical scatter associated with the mass distribution in the sample. The intercept at $a = 1$ AU corresponds to a representative mass sum of about 3 solar masses, with the scatter suggesting a mass range of about 0.5 to 5 $M_\odot$. We can also notice that the number of binaries increases with increasing separation, at least in the separation range up to $\sim$1 AU. At greater separations the sampling becomes less complete.

DOI: 10.1201/b22228-5

telescope of the size range used by historic visual binary observers (∼20 cm diameter, say) would be within a few hundred parsec. This implies that double stars that might just be resolved with such an instrument, say at 1 arcsec, would be spatially separated by distances (a) from at least several AU for the few nearest examples to some hundreds of AU. Orbital periods (P) of such binaries should therefore range from the order of decades up to a few thousand years. Pairs that are easily seen, having separations of the order of arcminutes rather than arcseconds, could have separations running into the thousands of AU or more, entailing orbital periods of many thousands of years. It would thus be difficult to decide, from the evidence of direct measures alone, whether such pairs were real visual binaries – 'physical doubles' as they are sometimes called – or just 'optical pairs' that chanced to be in close alignment.

Our view of the occurrence of binary stars, as mentioned before, is dependent on the technique we are using to detect them. For example, it is clear that the chance of observing an eclipsing system must decline with increasing separation of the pair, so if our knowledge of binaries was limited only to photometric means we might, from superficial indications, deduce that binary incidence must require components with sizes an appreciable fraction of their separation. With such a restricted selection, our direct evidence on the properties of well separated pairs would be seriously curtailed. Advantages from combining different observational methods for a fuller understanding of multiple stars thus become apparent.

In the case of visual binaries, useful physical parametrization is dependent, not only on the separation of the two components, but also on their separation from us, i.e. their parallaxes, since that information is vital to scale an observed orbit to absolute units. Historically, parallaxes were difficult to measure directly with high precision: a point that has limited the extent to which the astrometry of visual pairs could bear on stellar astrophysics until relatively recent times.

In the introductory chapter it was mentioned that William Herschel argued that his observations of the motions of selected double stars indicated a mutual attraction between them, and thereby the universal nature of gravitation was realized. But it is quickly seen that Herschel's data fell short of an exact proof of Newton's law with the same constant, even given that the detected orbital motions were of circular or elliptical form.

Consider Castor for example; the first of the several double stars that Herschel discussed in his classic paper of 1803, and for which he estimated an orbital period of about 342 years.* Suppose that the stars involved are of closely comparable mass to the Sun: a plausible assumption in Herschel's day. The use of Kepler's law, in the same form as applies to the solar system, would entail the separation of the two main components to be about 62 AU. Herschel measured the mean separation of these stars to be close to 5 arcsec. If that separation can be taken as representative of the size of the semi-major axis of the orbit, though in general it would be less, then the distance to Castor should be at least 12 pc. In fact, modern estimates put this at close to 14 pc, corresponding to a distance modulus ($V - M_V$) of about 0.73. The average visual magnitude of the two stars in Castor is about 2.4, though the Sun's absolute magnitude is known to be about 4.8, i.e. using the Sun's luminosity as a guide, the Castor stars are much too bright. Indeed, errors of over 2 magnitudes would occur if these components of Castor were genuinely sunlike.

So what is wrong? The calculation depended on the adopted solar masses of the stars, but not so sensitively, and, as we now know, the estimated distance is not too far amiss. The brightness of a star must be more sensitively dependent on the mass to reconcile the

---

*In fact, both stars, Castor A and B, are close (unresolved) binaries with periods of a few days. But that does not really affect this discussion of the wide (AB) orbit.

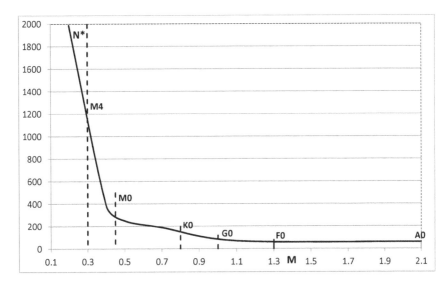

Figure 5.2 Numbers of different types of star are plotted for each 0.1 $M_\odot$ interval of stellar mass (M) using the power-law approximation $N^* \sim N_i M^{n_j}$, where N_i and n_j are constants for particular ranges of mass, as specified by A. N. Cox (1999). The distribution is arranged to compare with the ~3000 stars within the 20 pc radius considered by Eggen (1956). A selection effect in such data becomes clear, namely that the dependence on parallax evaluation for absolute parametrization must bias the information strongly towards stars of low mass. The diagram marks by vertical lines the approximate boundaries of the spectral classes of Main Sequence stars.

magnitude information however. From such considerations, the critical role of *parallax data* in the use of visual binaries to probe independently such physical issues as the 'mass-luminosity law' can be seen.

But an interesting general point arises on the basis of our knowledge of those relatively near binaries that can be assigned parallax values reliable to a few percent. Thus, O. J. Eggen (1956) counted 333 stars involved in known bound systems out to 20 pc distance from the Sun. On the other hand, C. W. Allen's (1974) catalogue of the nearest 100 stars, that should be fairly inclusive out to ~6.5 pc, implies there would be around 3000 stars in the sphere considered by Eggen. We then deduce that at least 10% of all stars must be in bound multiple systems. This is without making allowances for the difficulty of recognizing faint stars at 20 pc, which would add another 2.5 mag to the 17.4 V magnitude limit in Allen's list of nearby stars. Nor does it allow for the completeness in including faint secondaries at angular separations that may be several orders of magnitude greater than their parallax. Other surveys of this kind were made by G. Kuiper (1942), who included stars out to ~10.5 pc, and P. van de Kamp (1969), whose list was limited to ~5 pc distance.

What kinds of binary would such surveys have included? The answer is summarized diagrammatically in Fig 5.2, where we have set out a power-law mass distribution for un-evolved stars scaled to the ~ 3000 stars posited to lie within 20 pc of the Sun. Most of these stars are significantly less massive than our Sun: more than half have spectral class M. There is thus an observational selection effect that comes from the requirement for parallax information in order to specify absolute parameters of visual binaries. This selection effect renders the knowledge of stellar properties gleaned from visual binary data to be strongly biased towards stars of low mass.

Here we may note that subtle differences occur in the usage of expressions such as double star, visual binary, astrometric pair, etc. The term *visual binary* is generally applied to two stars that can be optically resolved by direct observations and their separation and position angle measured accordingly. Historical photographic data also showed some stellar images to be of atypical outline, and that such distorted but unresolved images could be seen to change their appearance over time. Knowledge of these *astrometric* pairs, sometimes having periods of the order of only a few years, has increased dramatically with the development of more sophisticated methods of observation and analysis, particularly since the introduction of space-borne observation systems. Examples are considered later in this chapter.

The historic, two-datum micrometric observations of visual binary stars included in Eggen's catalogue, given accurate parallaxes, could furnish, apart from the orbital elements, estimates of the sum of the masses of the two components, using Kepler's third law. With the use of photographic plates and precise measuring engines in the development of astrometric techniques in the early twentieth century, systems' centres of gravity could be located, and, after further reductions, the masses of the individual components assessed.

This was still only to a low precision, though, and for a rather restricted sample of stars. Yet problems of stellar cosmogony range across the whole Herzsprung-Russell diagram. The acceleration of stellar evolution with stars of higher mass then tended to divert astrophysical attention away from the kind of information that comes from historic visual binary orbit analysis, regarded in isolation. In combination with other related forms of analysis, however, the picture becomes more complete. For example, if a collection of stars sharing some common properties such as by being members of a cluster or moving group*, were to contain a visual binary it is possible to invert the use of Kepler's law, by assuming a representative value for the stellar masses, make a reasonable estimate for the parallax of the group. This point is considered further in Section (5.4.3).

An important step in the development of astrometry occurred in 1970, with the discovery, by A. Labeyrie, that very high time resolution imaging in the focal fields of large telescopes allow the problems associated with the blurry appearance of stars in classical observations to be circumvented. In effect, fast time-sampling of the focal field revealed the multitude of separate diffraction-limited images produced by the image-disturbing effects of the intervening atmospheric mass. In principle, one could then measure separations and position angles of two stars separated on the scale of the telescope's full theoretical resolving power (i.e. tens of mas) from all the separate speckles, as the small stellar images are called, and take averages as representative measurements. Of course, more automated procedures were called for to carry out this task effectively, and by the 1980s programs were available to carry out Fourier analysis of the image field and recover the sought mapping of the source.

The transfer of data to the Fourier domain recalls the general procedure of interferometry, where the received signal appears as a time-modulated effect, or *visibility*, relating to a *baseline* that separates different elements in the aperture plane. Such ideas led to the introduction of *masking* of the telescope's main entrance pupil by purpose-built diaphragm arrays. The mountings of modern large telescopes are usually of alt-azimuth type, so the image plane will slowly rotate, with respect to the axes set by the sky coordinates. The observing arrangement then resembles that of high-resolution radio-interferometry, for which the processing of visibilities to recover accurate maps of the source in the sky plane is well-established. This process can be sped up by adopting a given model for the source, e.g. a pair of point-sources. The map will then deliver the separation and position angle as well as positions in the sky co-ordinates. The accuracy of the result depends on how fully the visibilities are sampled in time as well as the coverage of the angular frequencies in both dimensions of the image plane. Due to the former factor, the technique has a greater application to stars brighter than, say, $V \sim 10$.

Conversion of the input to the Fourier domain is not a necessary requirement for source positioning analysis. Speckles, particularly those that are well-defined and comparable in

---

*A group of stars sharing a common motion through space.

size to the Airy disk images of point-like sources, can be selected and constructively shifted and added so as to produce a result comparable to the theoretical performance of the telescope objective. There are general interests supporting such technology, relating to the stabilization of imaging devices, with widespread applications. This has opened up significant opportunities for the astrometry of visual binaries in recent decades.

5.2 Recognition of Binaries

Michell's prior argument, that the fairly frequent occurrence of close pairs constitutes evidence of likely gravitational interactions, raises the question of how close the pair should be for this to become more probable than not. In turn, this brings attention to possibly close, in the sense of gravitationally interacting but transitory, encounters between stars.

Regarding the distances of double stars, let us note first that, for a given luminosity of star

$$\log \rho = 0.2V + \text{const} \ , \tag{5.1}$$

where we are simply dividing the distance modulus formula (1.2) by 5. For the relatively plentiful sunlike stars the constant $(= 1 - 0.2M_V)$ would be not far from zero. Such stars are well-catalogued up to distances of order 100 pc, i.e. magnitudes up to $V \sim 10$.

Consider now the projected angular separation of the components of a double star $\theta = a/\rho$, which is given in seconds of arc if a is in AU and ρ is in pc. Let there be some limiting separation a_0 at which it becomes likely that a wide binary would be disrupted by the accumulation of encounters with other stars. Given $\log \theta = \log a - \log \rho$ and Eqn (5.1), for star pairs of a given mean magnitude V we can expect a limiting separation

$$\log \theta_0 = \log a_0 - \text{const} - 0.2V \tag{5.2}$$

beyond which it becomes unlikely that an observed double star is physically connected. Cox (1999) gives

$$\log \theta_0 = 2.8 - 0.2V \ , \tag{5.3}$$

as a rule-of-thumb formula, suggesting a projected spatial separation a_0 of about 600 AU and period about 10000 y for a pair of mid-G type Main Sequence stars at a distance of about 100 pc. This may work out about right for the stellar types just mentioned, but earlier stars at the same apparent magnitude would be much farther away. The angular separation θ_0 would become small with increasing V for a fixed value of a_0 in (5.2), implying that, if θ_0 is determined only by the magnitude, more massive binaries can maintain their hold on each other out to significantly greater distances. For stars closer to the Earth, say $V \sim 5$, Eqn (5.3) suggests a pair would become more likely optical (i.e. not physically related) at separations greater than a few tens of arcsec, which looks to be confirmed in more detailed studies.

Along comparable lines, E. Öpik introduced a 'measure of difficulty' $C = 0.22\Delta m - \log \theta$, where Δm refers to the difference in magnitude of the two stars in a visual binary system. This means that the difficulty of binary identification will increase for stars at a given separation as the magnitude difference increases, while more difficulty attaches at fainter magnitudes for a given magnitude difference.

P. P. Eggleton and his associates argued that essentially all visual pairs brighter than magnitude 6 should by now be known and catalogued. About 5% of such stars have absolute magnitude less than –4, implying that 95% of the most well-known (visible) stars are within about 1000 pc of the Sun, a volume that should contain, according to the foregoing discussion, around 40 million stars. Given that of the order of 10^5 stars are already known and catalogued as binaries, it is reasonable to suppose that, while many of the fainter stars have unconfirmed status as bound pairs, the sampling should be fairly complete for the brightest $\lesssim 1\%$ that are in the *Bright Star Catalogue*. Even so, further checks on available data, such as that in Fig 5.3, suggest that there may be some intrinsically bright stars with companions at distances of a few AU yet to be identified.

Binaries with periods less than about a year were usually first identified from radial velocity monitoring and thereafter studied spectroscopically. These are the subject of the following chapter. But binaries with periods of several years have been increasingly investigated as visual pairs using technically developed observational equipment at higher angular resolution. These include methods such as speckle or long baseline interferometry, adaptive optics, aperture-mask aided imaging, or fast photometry of lunar occultations. The higher resolution aimed at depends, basically, on the spatial extent of the light receiving system. It is, of course, easier to initiate the set-up of such a system on the ground, but dealing with the problems posed by the complex optical properties of the intervening atmosphere involves highly specialized engineering. An *intermediate* range of binaries with periods of order a few years is open to systematic monitoring from dedicated space-based surveys, which provides a significant component of the motivation towards programmes such as those of HIPPARCOS and Gaia.

Multiple stars are also revealed indirectly from surveys of particular stellar groups, where they show up as overluminous objects. A binary composed of two similar stars would be a conspicuous $\sim 3/4$ of a magnitude above the sequence of single objects in Hertzsprung-Russell diagrams of clusters. Such sequences have thus been analysed, where close multiplicity incidence and mass ratio distributions are parametrized to fit observed vertical dispersions.

Another possible route to binary detection comes from *spectral* (sometimes *spectrum*) binaries where two quite distinct sets of absorption features are seen in the same (composite) spectral image. The occurrence of two different spectral types together may often be missed: binary identification is considerably eased when they are of comparable luminosity. An illustration is shown in Fig 5.4. Many such discoveries would, after sufficient time and coverage, turn out to be regular spectroscopic or astrometric binaries, but in cases where the orbital motion is slow the composite spectrum may alert observers to the possibility of wide binarity as with ζ Aur, 31 and 32 Cyg. Spectral binaries may have orbital axes inclined at any angle to the line of sight, so some spectral binary candidates are confirmed astrometrically if they are near face-on, while others may eventually show appropriate radial velocity changes.

Long baseline interferometry was mentioned above in the context of optical instruments such as Georgia State University's CHARA array at Mt. Wilson, or the University of Sydney's SUSI interferometer at the Paul Wild Observatory, NSW. Interferometric techniques have been a central feature in radio astronomy for decades, although the radio brightness of most stars is still low compared to currently available signal detection limits. A small number of stars are, however, 'radio bright': a notable example is the multiple star AB Dor.

AB Dor is a young, low-mass, relatively near (~ 15 pc), quadruple system whose components have deep convective envelopes that rotate at high speed: conditions associated with the generation of strong magnetic fields and active surface regions associated with starspots, flare-like phenomena and significant radio emission. The four components comprise a pair of binaries: AB Dor A-C and AB Dor Ba-Bb, the two binaries separated by close to 9 arcsec on the sky, corresponding to about 135 AU in space. AB Dor A, the primary, has a mass of close to 0.9 solar masses. AB Dor B (a+b), at a mean distance from A of 2.3 AU is a pair of 'brown dwarf' stars. These stars, typically, have masses on the order of 100 times that of Jupiter. The orbits of the A-C system were obtained by combining a variety of techniques, including data from HIPPARCOS and very long baseline radio interferometry

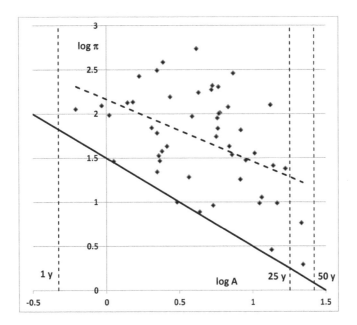

Figure 5.3 A sample of 50 visual binaries are plotted with parallax (π in mas) against component separation (a in au). The arrangement reflects a presentation by J. Dommanget (1985) (see Section 5.6). The linear trend line (dashed) is given by $\log \pi = 2.2 - 0.7 \log a$, suggesting a selection effect working in favour of detecting closer visual pairs at lower distances. The continuous line shows the limiting reach of 'classical' visual binary observations, resulting from the separation versus distance compromise. The general direction of progress in astrometric binary research is towards the bottom left corner of this display, where the conventional spectroscopic and photometric binaries are located.

(VLBI). AB Dor B was observed between 2007 and 2013 with the Australian Long Baseline Array (LBA), at 8.4GHz. Radio emission, as from point-like sources, was detected from both AB Dor Ba and Bb. This allowed parametrization of the absolute orbits, and thence determination of the individual masses as 0.28±0.05 and 0.25±0.05 solar masses (Fig 5.5).

These results, of great significance to theories of low mass stars and their pre-Main-Sequence conditions in the general context of star formation, emphasize the dividends of multiwavelength observational applications and point the way to future information gathering on stars with improved detection capabilities at radio wavelengths.

5.3 Quantification of Binary Occurrence

Let us now return to the problem considered by Michell (Chapter 1) regarding, among a certain number N of stars distributed randomly over the sky, the likelihood of of finding the nearest neighbour of a given star at an angular separation θ. For general random processes it is tempting to utilize the Poisson distribution function. For the present question, we could envisage a 'mean free area' $4\pi/N$. The problem then looks analysable on the basis that such an area is related to the linear size of the zone of exclusion about the given star (i.e. the probability of no other star within that radial extent). Such an approach was followed by Bahcall and Soneira (1981), who obtained

$$P(N, \theta) = N\theta \exp(-N\theta^2/4)/2 \ , \tag{5.4}$$

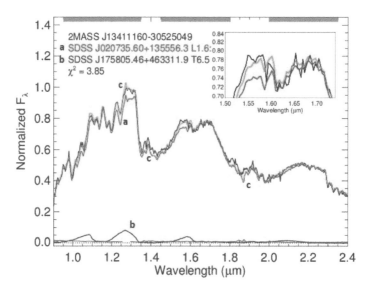

Figure 5.4 Example of a low mass spectral binary, consisting of a pair of brown dwarfs, after Bardalez Gagliuffi et al., *Astrophys. J.* Vol. 794, p143, 2014; ©AAS. In the diagram the observed spectrum (heavy black) appears to combine unresolved components in the cool dwarf object 2M1341. The light grey (a) and dark grey (b) spectra are taken from the sources labelled in the corresponding grey shades. These have been used as primary and secondary templates, scaled to their absolute magnitudes and combined into the model spectrum, that is matched with the black one. Small differences can be seen in the regions marked c. This two star template produces a significantly better fit to the original source than using other trialled single spectra (upper right inset). From such fitting experiments it is concluded that the peculiar (black) spectrum is likely to be a blend of those from two distinct stars.

for the probability density function of the nearest neighbour to a given star in a population of N stars being at angular separation θ.

It turns out that there is an alternative, more direct approach*. Place the given star at the pole of a celestial sphere and fix attention on another star, which can be placed at random over the sphere's surface. Using the same areal consideration as Michell, the probability of that other star being within an angle θ_0 of the pole is

$$\frac{\int_0^{2\pi}d\phi \int_0^{\theta_0}\sin\theta\, d\theta}{\int_0^{2\pi}d\phi \int_0^{\pi}\sin\theta\, d\theta} = (1 - \cos\theta_0)/2 \ . \tag{5.5}$$

In the same way, the incremental probability, associated with the probability density function, that the other star is in the range θ to $\theta + d\theta$ is $\frac{1}{2}\sin\theta d\theta$. The probability that it is at a *greater* separation than θ_0 is then $(1 + \cos\theta_0)/2$. Of course, there are $N - 1$ other stars, which have to be further away than our selected star if it is to be the nearest neighbour to the pole. At the same time, there are $N - 1$ possibilities to select this other star, so the combined probability density function for the nearest neighbour at separation θ is:

$$P(N, \theta) = (N - 1)\frac{\sin\theta}{2}\left(\frac{(1 + \cos\theta)}{2}\right)^{N-2} \ . \tag{5.6}$$

---

*The following presentation derives from that of Scott and Tout (1989).

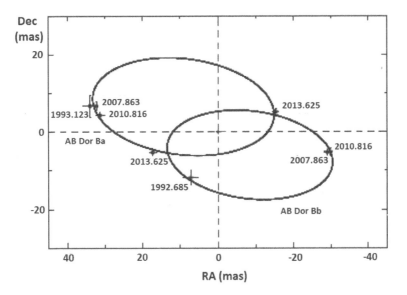

Figure 5.5 Orbits of the Ba and Bb components of the multiple star AB Dor derived from observations with the Australian Long Baseline Array. (Courtesy of R. Azulay et al., *Astron. Astrophys*, Vol. 578, A16, 2015, ©ESO).

Perhaps unexpectedly, it turns out the both Eqns 5.4 and 5.6 tend rather quickly to the same distribution, even for relatively low values of N. Integrating (5.6), the cumulative probability that the nearest neighbour is in the range 0 to θ_0 turns out to be simply

$$P_c(N, \theta_0) = \left[\left(\frac{1 + \cos\theta}{2} \right)^{N-1} \right]_{\cos\theta_0/2}^{1} , \tag{5.7}$$

i.e. 1 minus the $(N-1)$-fold probability that there is no star among the available $N-1$ closer than θ_0. Eqn (5.7) can also be written as

$$P(N, \theta < \theta_0) = 1 - \left(\cos\frac{\theta}{2} \right)^{2(N-1)} . \tag{5.8}$$

Eqn (5.8) enables us to take an alternative view of Eqn (5.3), because it indicates that, below a certain separation θ_l, the probability is very small, δ say, that such doubles are actually chance close positionings on the celestial sphere. We can then write

$$1 - \delta = \left(\cos\frac{\theta}{2} \right)^{2(N-1)} , \tag{5.9}$$

or, taking logarithms of both sides and using well-known expansion formulae

$$\delta = 2(N-1)\epsilon . \tag{5.10}$$

where ϵ is the small difference between $\cos^2(\theta_0/2)$ and unity, i.e. $\theta_0^2/8$. For large values of N this reduces to $N\theta_0^2/4 \approx \delta$.

Now Allen's (1974) table of stellar densities against magnitude allows the fair approximation

$$\log_{10} N = 0.621 + 0.500V \tag{5.11}$$

as a mean over the whole sky for brighter magnitude stars. When combined with (5.10), this gives

$$\log_{10} \theta_0 = 5.305 + \frac{1}{2} \log_{10} \delta - 0.25V \ . \tag{5.12}$$

This is of comparable form to (5.3) and would imply that 'unlikely' in Cox's rule of thumb means about 1 chance in 10^5, or that essentially no close pair satisfying $\theta < \theta_0$ of (5.3) in the Bright Star Catalogue would be an optical double.

5.4 Visual Binary Orbits

The geometrical parametrization of binary stellar orbits was already set out as the 'direct problem' in Section (1.3). It will be useful to refer back to that section regarding the meanings of parameters. These parameter-sets (or elements) grew, historically, out of ideas like those raised in Herschel's study of visual binaries, where re-application of Newton's system to explain planetary motions was expected to work also for pairs of stars. Such a pair of mass points, as they may be considered when their separation is large compared to their sizes, perform Keplerian motions in a plane containing the common centre of gravity.

In this scheme, the elliptical orbit's major axis forms a natural reference line, and the direction from the primary star to the periastron position on this line, or *apsis*, allows the orbital motion to be traced in terms of the radial separation of the stars r and the angular movement from the apsis or true anomaly ν.

Consider now an elliptical orbit lying exactly in the plane of the sky. The focus would be located at the position where it is determined to be from the observed orbit — without any shift arising from projection. If this ellipse is rotated about its minor axis, say, the apparent ellipticity would decrease; the focus of the projected ellipse, or *apparent orbit*, would move towards the centre in response to the decreasing apparent ellipticity. But the *projected* position of the focus of the *true orbit* would remain at its same relative distance along the length of the major axis. The focus of the apparent, or projected, orbital ellipse will therefore not, in general, coincide with that of the true orbit.

On the other hand, if we view three successive positions of the secondary in its true orbit about the primary, such that the area swept out at the primary between the first and second positions is the same as that between second and third, then the times taken for these two segments would be the same, according to Kepler's second law. The projections of these two segments from the plane of the true orbit to that of the apparent one would remain in the same ratio. So Kepler's areal law would still hold for the motion of the secondary in the apparent orbit. This point is made use of in observational analysis of the data for the apparent orbit.

5.4.1 Orbit Parameters: The Inverse Problem

The canonical form an ellipse is given as

$$\frac{\xi^2}{a^2} + \frac{\eta^2}{b^2} = 1 \tag{5.13}$$

A linear transformation of co-ordinate axes

$$\begin{aligned} x &= a\xi + b\eta - c \\ y &= -b\xi + a\eta - d \end{aligned} \tag{5.14}$$

when substituted into the above can be easily seen to produce the second order form

$$Rx^2 + Txy + Sy^2 + Ux + Vy + W - 1 = 0 \tag{5.15}$$

and by dividing both sides by $W - 1$ we obtain the general ellipse in standard form as

$$Ax^2 + 2Hxy + By^2 + 2Gx + 2Fy + 1 = 0 \ . \tag{5.16}$$

The essence of the inverse problem for the visual binary orbit is to relate the five coefficients A-H to the five classical orbital parameters a, e, i, ω and Ω.

The specification of A-H, can be made, in principle, by noting where the apparent orbit crosses the x and y axes that are centred, for convenience, on the primary, and correspond to cardinal directions on the sky. If the two crossing points on the x-axis are x_1 and x_2, say, then, clearly

$$
\begin{aligned}
Ax_1^2 + 2Gx_1 + 1 &= 0 \\
Ax_2^2 - 2Gx_2 + 1 &= 0
\end{aligned}
\tag{5.17}
$$

so that A and G are derived from the simultaneous solution of (5.17). The same procedure for the crossing points on the y axis determine B and F. Any point elsewhere on the ellipse, $x_0 \, y_0$ say, can be used to allow H to be calculated from

$$H = -(Ax_0^2 + By_0^2 + 2Gx_0 + 2Fy_0 + 1)/(2x_0 y_0) \ . \tag{5.18}$$

Returning to the canonical form for an ellipse and referring to the orbit's natural coordinate system ξ, η, ζ; we have, for the true orbit, with axes centred on the focus,

$$\frac{(\xi + ae)^2}{a^2} + \frac{\eta^2}{a^2(1 - e^2)} = 1 \ . \tag{5.19}$$

This would correspond directly with the apparent orbit in the 'face-on' conditions that Ω, ω and $i = 0$. The three angular parameters being set in this way, the other two, a and e, would be directly determined from the measurements of the orbit. The five coefficients of (5.16) reduce to only A and B, that fix size and ellipticity in this special case.

The linear transformation (5.14) involves rotations of the axes, without translations or changes of scale, so for a rotation θ about the ξ axis, we have (cf. 2.37) $\boldsymbol{\xi}' = \mathbf{R}_\xi(\theta).\boldsymbol{\xi}$. In general, the three additional degrees of freedom among the available coefficients permit three arbitrary rotations of the model to be determined from the information content of the data. Explicit formulae are obtained by applying the three relevant rotations about the appropriate axes in succession (cf. Section 1.3 and Eqn 3.5). In this way, with suitable definitions of the sense of angles and arrangements of coordinates,[*] we can write, following I. Ribas et al. (2002), that

$$
\begin{aligned}
x &= \frac{a(1 - e^2)}{1 + e \cos \nu} [\cos(\nu + \omega) \sin \Omega + \sin(\nu + \omega) \cos \Omega \cos i] \\
y &= \frac{a(1 - e^2)}{1 + e \cos \nu} [\cos(\nu + \omega) \cos \Omega - \sin(\nu + \omega) \sin \Omega \cos i] \\
z &= \frac{a(1 - e^2)}{1 + e \cos \nu} \sin(\nu + \omega) \sin i
\end{aligned}
\tag{5.20}
$$

---

[*]Sign ambiguities are possible concerning this co-ordinate set — see, e.g. the reference to van den Bos (1962) in Section 5.6.

The form of (5.20) can be reconciled with that of (3.5) by writing, for the position vector **x**, after 3 successive axial rotations from the natural orbit system back to the observed reference system,

$$\mathbf{x_3} = \mathbf{R}_\zeta(-\phi)\mathbf{R}_\xi(-\mathbf{i})\mathbf{R}_\zeta(-(\nu+\omega))\mathbf{x_0} \ . \tag{5.21}$$

The angle ϕ is in the sky plane, but the nodal angle Ω is measured from the y_3-axis, so that $\phi = \Omega + \pi/2$. Also, the prescription applies to a right-handed Cartesian frame viewed externally, where, the y_3-axis being directed towards increasing declination, the x_3 component of (5.21) would point in the opposite direction to that of increasing RA on the sky, so the first of (5.20) adopts the positive form as written. Alternatively, x_3, y_3 in (5.21) may be regarded as y, x in (5.20), with $\phi = \Omega$ and z increasing outwards. Data for visual binaries concern only $\rho\sin\theta$ and $\rho\cos\theta$, so there remains ambiguity, since the effect of adding π to Ω in the first and second of Eqns 5.20 can be cancelled by adding π also to ω. The z motion, if available, eliminates this ambiguity, otherwise Ω is adopted to be in the range 0 to π.

Classical methods for finding the parameters involve substituting the transformed coordinates from the inverse of (5.21) into (5.19). The coefficients of the terms in x^2, xy, y^2, x, y are then directly related to the values of A, H, B, F, and G in (5.16), which are known following the discussion around (5.17-18). Five linear equations are formed, that are inverted to yield the sought geometric parameters. There remain two time-related parameters that fix the position of the secondary with respect to the primary for any particular time, i.e. the period P and reference epoch t_0. A pair of reference points are sufficient to derive these, their true anomaly values being determined from (5.20) with the known geometric parameters. Both ν values have a corresponding mean anomaly, which, taken together, fix the values of P and t_0 (cf. Eqns 1.7-1.8).

Given the widespread availability and use of computers, however, it becomes easily practicable to deal with the parametrization of the fitting function by programmed optimization methods. Instead of solving the inverse problem by using five values of A-H determined from selected points read from a representation of the apparent orbit, the ellipse that corresponds to (5.19) is progressively matched to all the x and y data so as to produce a minimum of the residuals. The fitting functions will contain the 7 constants discussed above and involved in (5.20), as well as small translational corrections, Δx_0, Δy_0, in the position to be assigned to the origin.

Formally, we can write for the solution of the inverse problem set out in this way,

$$\mathbf{a}_{\mathrm{opt}} = [\chi^2]^{-1}\mathrm{Min}[\chi^2(\mathbf{a})] \ , \tag{5.22}$$

where $a_{j,\mathrm{opt}}$ is the vector of the best estimates of each parameter in the adjustable set $\{a_1, a_2, a_3...a_j...a_m\}$. The observed values of the variable, which could be x or y, are matched by the values of the fitting function, calculated from (5.20) in the present context. z values, that may be available in certain cases (discussed in Section 5.4.4), can be treated in the same way. The quantity χ^2 depends on the squared differences of observed and calculated quantities, and the optimal estimate for a_j is taken to occur when χ^2 is minimized. $[\chi^2]^{-1}$ expresses the idea of *inversion* of the dependence of χ^2 on **a**. An example of the results of such a method is shown in Fig 5.6.

The general usefulness of modelling procedures is circumscribed by the extent and accuracy of the orbital data. Uncertainties and ambiguities in the parametrization must surely arise with observational scatter and limited coverage of the ellipse.

5.4.2   Graphical Interpretation

It is of interest to interpret the orbital parameters in terms of the appearance of the binary system in the sky. To this end, we refer to Fig 5.7. This approach follows the graphical method of H. J. Zwiers.

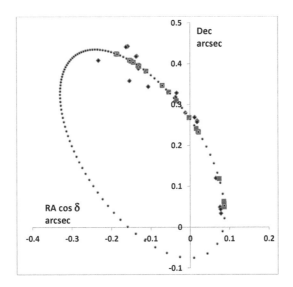

Figure 5.6 A residuals-minimizing fitting to the astrometric orbit of the close visual binary WDS J07598-4718AB, which is formed by the A and B components of the multiple star V410 Pup. The diamonds represent the observations of separation and position angle. Corresponding optimal orbital positions are indicated by the sequence of full squares, which lie on the model orbit indicated by the dots, that are spaced at uniform time intervals of about 3.6 years. This fitting has treated all observations as having the same weight, although it is clear that accuracy improved over the century of coverage.

The elliptical apparent orbit is shown as a full curve centred at C. The primary star is at P and the position angle values, with their corresponding star separations, are measured in the eastwardly sense from the direction PN. It is assumed that there are sufficient observations to allow a confident localization of the apparent orbit on the local sky plane. The line CP, containing the projections to the sky of both centre and focus of the true orbit must then be the projection of the major axis of that orbit. This line intersects the apparent orbit at T and U in Fig 5.7. The ratio CP/CT, not changed by the projection, thus corresponds to ae/a, i.e. the eccentricity e of the true orbit.

We now construct another ellipse tangential to the apparent orbit at T and U. This is done by setting an arbitrarily positioned line parallel to TU as shown at T'U' and bisecting it to locate the midpoint V. The line VC is produced to form another axis traversing the apparent orbit as shown. For any point on the apparent orbit, Q_i say, we produce another point Q_j, such that Q_iQ_j is in the direction parallel to the axis VC, the line Q_iQ_j meeting the axis PC at R. The location of Q_j on this line is fixed by ensuring the ratio Q_jR/Q_iR is the same as $1/\sqrt{1-e^2}$.

The auxiliary ellipse thus created turns out to be the projection of the eccentric circle of the true ellipse. The eccentric circle is in the same plane as the true orbit, to which it is tangential at either end of the major axis, and this circumstance is unaltered by projection. Its departure from circularity results from the inclination of the true orbit plane to that of the sky. The ratio of its (measurable) axes YC/XC = b/a directly furnishes the inclination as

$$\cos i = b/a \tag{5.23}$$

The radius of the eccentric circle gives the semi-major axis a of the true orbit, since this is equal to the semi-major axis of the auxiliary ellipse that remains unchanged in the projection. This axis CX must then also be parallel to the line of nodes since, lying already in the sky plane, that line is not affected by projection. The angle from PN round to PX'

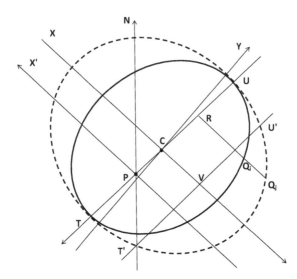

Figure 5.7   The full ellipse corresponds to the apparent orbit, centred at C and with the primary star at P. The north direction then aligns with PN. The dashed ellipse is the *auxiliary ellipse*, which corresponds to the projection of the eccentric circle of the true orbit.

thus defines the nodal angle Ω. In this way, 4 of the 5 geometrical parameters of the problem are measured directly from this geometrical construction.

The remaining angle ω is evaluated from considering the meaning of the angle X'PU. This is the projection of the angle between the node and periastron, i.e. ω, onto the sky plane, λ say. Spherical trigonometry applied to this projection shows that

$$\tan\omega = \tan\lambda \sec i \tag{5.24}$$

Now, for any point on the apparent ellipse there is a measured value of the position angle θ which, in correspondence with the previous equation, satisfies

$$\tan(\nu + \omega) = \tan(\theta - \Omega)\sec i \tag{5.25}$$

so that the true anomaly will be defined for any position angle θ. For any true anomaly ν there is a corresponding mean anomaly M, that can be specified, in principle, from iterative use of the *equation of the centre* $\nu = \nu(e, M)$. The role of M is as a suitable phasing of the elapsed time t in terms of the period P and offset from the time of periastron passage t_0, i.e. $M = 2\pi(t - t_0)/P$. In this way, from a pair of timed position angle values the two required time parameters t_0 and P complete the basic modelling set of seven.

Parameters estimated from the foregoing construction may allow a preliminary set for computerized optimal modelling, though attention is required in matching the construction's angle directions to the formulae in (5.20). With sufficient orbit coverage, a robust optimizer will usually 'home in' on a set of model points that are in reasonable accord with the observations. The finding of an optimized set satisfying (5.22) does not, however, complete the orbit specification regarding uncertainties and uniqueness issues. While such matters are pursued in further data-modelling examples later, it is worth recalling here A. H. Batten's caveat that "... a consistent set of elements may be obtained, of which the only disadvantage is that it is completely wrong."

5.4.3 Dynamical Parallaxes

The application of Kepler's third law to visual binaries was apparent in the explanation of Fig 5.1 above. An alternative application occurs in the dynamical parallax Π_d of a visual binary considered by F. H. Seares (1922) as

$$\Pi_d = \frac{\alpha}{(M_1 + M_2)^{1/3} P^{2/3}} \ , \tag{5.26}$$

where P is the period in years, α is the apparent semi-major axis of the secondary about the primary in arcsec, determined from the analysis of a sufficient ρ, θ data-set. M_1, M_2 are the masses of the two stars. From Kepler's third law the denominator is seen to be the spatial semi-major axis a in AU. If the distance to the system is d (parsec), the separation a is equal to αd, consistent with (5.26).

Although the derivation of stellar masses in a key aim in visual binary data analysis, in working with (5.26), $M_1 + M_2$ may be unknown in advance, and a mean value of 2 M$_\odot$ could be adopted, for example, if several binaries in a known cluster were involved. Since the dependence on mass is as a low power, the error in such estimates does not have a relatively big effect on Π_d. A parallax of ~0.025 arcsec was found in this way for the Hyades cluster. Better estimates for the masses may be made based on spectral properties and inferences of the likely evolutionary status of the binary components. Dynamical parallaxes have been used to check expected (V) magnitudes against the apparent ones. Initially adopted masses may then be adjusted so as to allow better agreement between dynamical and photometric parallaxes.

5.4.4 Astrometry of the R CMa System

The aim of extending the range of double star astrometry to closer systems of relatively shorter period was mentioned above. For systems like R CMa that may have a wide orbit companion to a close eclipsing binary, historic astrometric data covering the ~100 y period associated with the wide orbit have been used to help place recent developments into context. The detailed study of I. Ribas et al. (2002) collected together relevant high quality astrometric observations of R CMa over the last century. The study provides an interesting example combining different modes of data analysis.

The idea of a wide orbit companion to the close binary R CMa is supported by the differences between observed and calculated (O – C) times of minimum light for the eclipsing pair (Fig 5.8), interpreted as a light travel time (LTT) effect. The z coordinate in Equations (5.20) would then relate directly to the LTT values — in light-seconds, say. LTT data are often analysed separately, but they could be expected to fit in naturally alongside x and y in an astrometric setting. In order to compare numbers, z are converted from light seconds to an equivalent number of mas on the sky by multiplying by $c\pi/A_0$, where c is the velocity of light, π is the parallax in mas, and A_0 is the astronomical unit. The amplitude of the z variation works out at about 145 mas if regarded in this way. The LTT data yield complementary information to that of historic astrometry, though the variation in z cannot separate the semi-major axis a and inclination i parameters, and makes no reference to the nodal angle Ω.

Fig 5.8 shows the O – C differences fitted to Eqn (5.21) for z, utilizing the timings gathered by Ribas et al., and a Levenberg-Marquardt type fitting optimizer. The form of the O – C curve indicates a short time of recession followed by a much longer approach, pointing to a significant eccentricity to the presumed elliptical orbit. A value $e = 0.49$ was adopted by Ribas et al., who also inferred that the orbit's major axis should lie close to the plane of the sky, as indicated by the periastron longitude ($\omega = 10.5$ deg). The asymmetric form of Fig 5.8 places the periastron passage within the relatively short recession range.

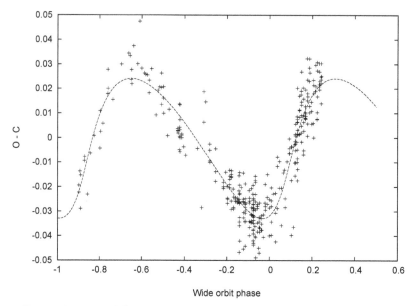

Figure 5.8   Times of minima differences, between observed and calculated (O – C) values (in days), plotted against orbital phase. The phase values are calculated using a model of the z motion based on the third of Eqn (5.20). Starting values for the model fitting were taken from the work of I. Ribas et al., (2002).

The astrometric wide orbit in the sky plane is derived, in principle, by first taking out the mean proper motions from accumulated positional data, then checking for the residuals in x and y to come into approximate alignment with the period P obtained from the LTT variations. For R CMa a period of around 93 y was initially inferred from the O – C data. Systematic shifts appear discernible in the historic astrometric data, but with an appreciable scatter ($\Delta\theta$). In fact, the median datum error in RA (x) is ∼154 mas, while the amplitude of the fitted orbital model is only ∼136 mas. For the declination direction (y) the situation is worse, with a median error of ∼ 173 mas and a derived variation amplitude of 116 mas.

With 5 of the 7 basic orbital parameters determined from the variations of z (though a and $\sin i$ are combined), the x and y data, even if relatively inaccurate, can be used to bear on just the two remaining ones: Ω and i. The astrometric data listed by Ribas et al. (2002) were used to do this, including also certain HIPPARCOS positions for the epoch 1991.25 (ESA, 1997), relating all the measures to the mean equinox of 2000.0. Two additional parameters (Δx_0, Δy_0) arise in the fitting of the x and y data. These are the shifts that optimize the position of the centre of the orbit model and serve as a reference point for the position of the system at 2000.0.

Although the orbital trends in x and y are only of the same order as individual datum accuracies, a model orbit resulting from analysis of the combined O – C and astrometric data is shown in Fig 5.9, with adopted parameters listed in Table 5.1. The agreement on optimal Ω and i is not very consistent for the RA and declination data taken separately, but the latter has a more definite diminution in χ^2 around the optimal fitting. The derived parameters are also not very consistent with those of Ribas et al., who did not find much movement in declination, and did not report a shift of the historic reference point. The RA data show rather large inherent inconsistencies in the early points, that render the curve-fitting less sensitive to the parameter values, though the swing of ∼200 mas found by Ribas e al. is supported.

In carrying out these procedures the included HIPPARCOS satellite positions were given high weights in accordance with their measurement accuracy. This ensured that the various trial orbits in the fitting process would pass close to the HIPPARCOS RA and declination.

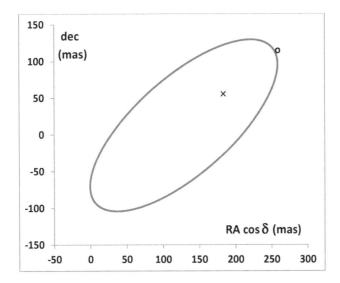

Figure 5.9 A feasible model of the sky motion of R CMa A from optimal fitting of historic astrometric measures combined with timings of mid-eclipses. The axes are marked in milliarcseconds (mas). The optimal system centre is marked with an X, which is several tens of mas removed from the historic reference position of 1894.7 at the origin. The periastron position is indicated with an O. The sense of the rotation is clockwise and periastron passage occurred at 1895.8 and 1997.3.

Table 5.1 Optimized model for the astrometric orbit of R CMa A

Parameter	Value
Δx_0 mas	184 $\pm$20
Δy_0 mas	58 $\pm$20
a mas	167 $\pm$5
i_3 deg	63. $\pm$10
Ω deg	49. $\pm$10
P y	101.5 $\pm$1.5
T_0 y	1895.8 $\pm$1.5
χ^2/ν	0.81
N	18
$\Delta\theta$ mas	150

The accurately measured proper motions from HIPPARCOS, when compared with the mean proper motions over a century of historic astrometry, also constrain the orbit modelling. The presented wide orbit model can be reconciled with this point, since the adopted value for the historic mean value of $\mu_\alpha \cos\delta = 167.8$ mas y^{-1}, is only slightly greater than the HIPPARCOS value (165.4). The close binary, in 1991.25, should then have been close to the stationary point in its orbital RA motion. In declination, there is a slight difference between the HIPPARCOS value (–136.2) and the adopted mean (–143.6), indicating a small northward relative motion a few years before the periastron passage in 1997.3. This can be seen in the model orbit shown in Fig 5.9.

The problem is, however, that although Fig 5.8 appears persuasive as a LTT effect, further research has failed to confirm the existence of the wide component in the R CMa system, at least in the form proposed by earlier studies. The system has been more recently observed with many individual images included and at higher resolution. The third body, predicted to be a late Main Sequence K-M type star, ought to have shown metallic lines and a RV signature complementary to the systemic velocity of the close pair, but these have not been detected. Effects that could be associated with the addition of an additional continuum apart from the contributions of the close pair were also not confirmed in subsequent observations.

The implication is that Fig 5.8 does not correspond to an LTT effect; though this raises the question of what other process could give rise to such an O – C variation. It seems difficult to reconcile the various strands of evidence on this puzzling system with one coherent model at the present time and further observations, astrometric and otherwise, are called for.

5.5 Space Age Developments

5.5.1 HIPPARCOS

Space-based astrometry was encountered in the previous section, where its accuracy was contrasted with that of ground-based data. It was realized already by the mid-20th century that further improvements to measurement accuracies of star positions by established procedures faced serious problems. The effects of the Earth's atmosphere were predominant, but other issues concerned instrumental stability and range. Even so, the HIPPARCOS (High Precision Parallax Collecting Satellite)* mission was able to play a complementary role with historic observations, notably in allowing parallaxes to be determined for binaries with known relative orbits. HIPPARCOS data has been able to improve the parametrization of such orbits, allowing masses to be determined for a few hundred of the nearest systems to accuracies of several per cent. For a score or so of short-period systems mass-ratios were determined from direct observations of the components' separate orbital arcs around the system centre over the satellite's functioning interval of almost four years.

The initial proposal to the French space agency CNES, submitted in 1967, envisaged stellar astrometry being progressed through the use of an advanced-technology space platform. Regarded as too expensive for a single national programme, after enlisting international support, the HIPPARCOS project was accepted within ESA's scientific programme in 1980. HIPPARCOS went on to provide the positions, parallaxes, and annual proper motions for around 100,000 stars with an accuracy of at least 0.002 arcsec. We may contrast this with data on the $\sim$3000 relatively near-by stars mentioned in Section 5.1, that characterized the situation before space-age facilities.

HIPPARCOS has proved to be one of the most productive of 20th century programmes in yielding useful knowledge of basic stellar properties from a homogeneous supply. It completed its survey of the targeted stars over the whole sky during the period 1991-4, surpassing its originally proposed aims by a significant margin. A subsidiary programme of the same mission – *Tycho* – involving an additional photoelectric device, gathered photometric information with less precise positional data but on a much larger number of stars. The final version of the Tycho Catalogue covered $\sim 2.5 \times 10^6$ stellar sources in the B and V spectral ranges of the Johnson system.

The principal detector element in the HIPPARCOS satellite was its image dissector tube. This was a photomultiplier arrangement equipped with a a grid of about 2700 finely arranged slits in the focal plane across which stellar images passed. Processing of the light modulation as a star moved over this grid allowed the positions of apparently single images to be refined to $\sim$2 mas. The situation for close double stars having separations of order 100 mas was more complex. The useful range for measurement of visual binary separations ρ was then taken to be $\sim 0.1 \lesssim \rho \lesssim 30$ arcsec, but reductions for the range $2 < \rho < 5$ arcsec

---

*The name HIPPARCOS also reflects due recognition of the ancient stellar catalogue maker and father-figure astronomer Hipparchus of Nicaea.

required a somewhat different algorithm, since both stars would illuminate the photocathode together. It was towards the short end of this separation range that new knowledge was anticipated.

Astrometry of δ Lib

δ Lib is a well-known Algol binary that has been surmised to have a relatively close astrometric companion. In this connection, T. F. Worek drew attention to irregularities in the primary's radial velocity curve over a relatively short orbital period of 2.762 y. Worek's model put constraints on the companion's mass, depending on the relative inclination of the wide orbit. This could, in principle, be checked from HIPPARCOS data that covered a $\sim$3y time interval, since the scale of the implied orbital displacements should be on the order of 5 mas i.e. greater than the $\sim$2 mas typical accuracies of the satellite's stellar positions. As with the R CMa system, the possibility of combining astrometric with other time-dependent information offers an opportunity to investigate the complete geometrical arrangement of the component stars and separate out parameters.

The measurements of HIPPARCOS are not directly in the equatorial system, but they provide angular positions along known great circles, p, say. During the course of its three and a quarter year mission HIPPARCOS took, typically, of the order of 100 measurements on each star of the survey. The variations of p arising from small differences from a given position of a star, X_1, Y_1, Z_1 and its proper motions $\dot{X}_1$, $\dot{Y}_1$, $\dot{Z}_1$, through the observation period, were tabulated as the residuals Δp, so that

$$\Delta p = \frac{\partial p}{\partial X_1}\Delta X_1 + \frac{\partial p}{\partial Y_1}\Delta Y_1 + \frac{\partial p}{\partial Z_1}\Delta Z_1 + \frac{\partial p}{\partial \dot{X}_1}\Delta \dot{X}_1 + \frac{\partial p}{\partial \dot{Y}_1}\Delta \dot{Y}_1 + \frac{\partial p}{\partial \dot{Z}_1}\Delta \dot{Z}_1. \qquad (5.27)$$

In practice, there is considerable difference in the astrometric effect of a given spatial displacement, in the plane of the sky and so the scale of components ΔX_1, ΔY_1; and in the line of sight ΔZ_1. The relative accuracy with which HIPPARCOS astrometry can fix the X and Y co-ordinates ($\sim10^{-8}$ in circular measure, so corresponding to $\sim$0.2 AU at 100 pc distance), is a number of orders greater than that with which it can specify parallax (implying a range of uncertainty of $\sim$20 pc at the same distance). Hence, although a possible correction to a star's *mean parallax* is retained in Eqn (5.27), we need not concern ourselves with the changes of Z displacements arising from orbital motion of a system such as δ Lib as far as HIPPARCOS data are concerned. For practical purposes, Equation (5.27) then reduces to:

$$\Delta p = \frac{\partial p}{\partial X_1}\Delta X_1 + \frac{\partial p}{\partial Y_1}\Delta Y_1 + \frac{\partial p}{\partial \Pi_1}\Delta \Pi_1 + \frac{\partial p}{\partial \dot{X}_1}\Delta \dot{X}_1 + \frac{\partial p}{\partial \dot{Y}_1}\Delta \dot{Y}_1, \qquad (5.28)$$

where Π_1 is the mean parallax, and $\Delta X_1, \Delta Y_1, \Delta \dot{X}_1$ and $\Delta \dot{Y}_1$ combine the effects of systematic displacements due to an orbital motion, together with constant errors in the assigned mean position X_1, Y_1, and mean proper motions $\dot{X}_1$, $\dot{Y}_1$. This is on the basis that the originally adopted HIPPARCOS mean position, parallax and proper motion values, by excluding a possible orbital motion of a source, would show greater residuals than a solution that included such effects.

With small but systematic displacements, x and y say, being supplied for a given model of the orbit in dependence on the parameters a, e, i, ω, Ω, M_0 and P through (5.20), we can rewrite (5.28) as

$$\Delta p = \frac{\partial p}{\partial X_1}(x + \delta X_1) + \frac{\partial p}{\partial Y_1}(y + \delta Y_1) + \frac{\partial p}{\partial \Pi_1}\Delta \Pi_1 + \frac{\partial p}{\partial \dot{X}_1}\Delta \dot{X}_1 + \frac{\partial p}{\partial \dot{Y}_1}\Delta \dot{Y}_1, \qquad (5.29)$$

To calculate where the star would be at any particular time, we also need the parameter t_0 that gives the epoch of periastron passage. The small constants δX_1, δY_1 make up the differences $\Delta X_1 = x + \delta X_1$, $\Delta Y_1 = y + \delta Y_1$ that apply in optimally fitting the orbit-including model to the data using (5.29).*

The natural reference frame for the wide orbit gives fairly simple expressions for the time derivatives that occur, after appropriate axis transformation, in (5.29). Using the decomposition into the constant velocities w_θ and w_y given in Section 1.3.2.5, it follows directly that:

$$\dot{\xi} = \frac{-2\pi a \sin \nu}{P\sqrt{1 - e^2}}$$
$$\dot{\eta} = \frac{2\pi a (\cos \nu + e)}{P\sqrt{1 - e^2}}$$
$$\dot{\zeta} = 0. \tag{5.30}$$

The relevant transformations of these velocities, following the discussion of (5.20-21), give:

$$\dot{x} = k(U_1 \cos \phi + U_2 \sin \phi \cos i), \tag{5.31}$$

$$\dot{y} = -k(U_1 \sin \phi - U_2 \cos \phi \cos i) \tag{5.32}$$

$$\dot{z} = kU_2 \sin i, \tag{5.33}$$

where, as before, $\phi = \pi/2 + \Omega$, $U_1 = \sin(\nu + \omega) + e \sin \omega$, $U_2 = \cos(\nu + \omega) + e \cos \omega$, and $k = 2\pi a/(P\sqrt{1 - e^2})$, P being the orbital period.

If the wide orbital period were large in comparison to the period during which the observations were gathered, then the mean proper motions derived by HIPPARCOS would be affected by the orbital motion, leading to values that could be systematically different from those coming from an average of many terrestrial observations over many years. Such deviations, for binaries of period a decade or two, could show systematic variations over the 3-4 years of the HIPPARCOS survey. In some known examples of this kind of additional orbital motion, the catalogue compilers introduced extra terms in (5.29), i.e. 7 or even 9 term solutions.

This is not the case, however, for an orbit like that of Worek's model for the third orbit in δ Lib, whose period is less than that of the survey. The instantaneous orbital velocities, as seen from Earth, are sometimes a noticeable fraction of the systematic proper motion. Over the $\sim$3 survey years, if evenly distributed in time, these movements would look like additional scatter in the proper motions data. But with correct mean proper motions determined, the orbital displacements would separate out as a discernible Keplerian motion superposed on that of the system centre of gravity. If the data were unevenly distributed, however, as appears the case for δ Lib, then the net result would be just a systematic shift of the HIPPARCOS mean proper motions from the historic trend. Small linear corrections $\Delta \dot{X}_1$, $\Delta \dot{Y}_1$, arising in this way, should have the effect of reducing the observational scatter in a revised solution that took into account the additional motion.

The five extra terms in (5.29), i.e. δX_1, δY_1, $\Delta \Pi_1$, $\Delta \dot{X}_1$ and $\Delta \dot{Y}_1$ add to the seven classical orbital parameters make up a total of 12 unknown parameters for a model optimization

---

*In principle, $\Delta \Pi_1 = z\Pi/a + \delta \Pi_1$, where z, due to orbital motion, is in mas like x and y, and $a = 2.06 \times 10^8$, but the term in z would generally be so small compared with available accuracies as to be negligible.

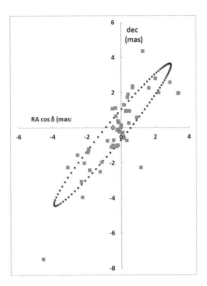

Figure 5.10 Plot of HIPPARCOS angular measurement components in the directions of RA and dec (in mas) for δ Lib, on the assumption of displacements due to T. F. Worek's (2001) 2.762 year period third body. The orbit model appears as dots at uniform intervals of phase; full squares mark the individual observations.

procedure. The five partial derivatives – $\partial p/\partial X_1$, $\partial p/\partial Y_1$, etc. in (5.29) – are supplied as numerical tabular values for each measured HIPPARCOS position with the designations $F_i(i = 1, 5)$. We can thus use (5.29) and each HIPPARCOS measurement as an equation of condition to optimize the fit of the parameters for a given model. In practice, as with many optimization problems, it is advantageous, if possible, to split the determinacy between the astrometric and other observations that involve the same parameters. This could be from radial velocity analysis, as with Worek's modelling, or (O – C) timing data, as used in the previous example of R CMa.

If the model is completely wrong, we would expect the run of residuals Δp to be worse than for the no-orbit solution. Indeed, checks were carried out by the HIPPARCOS data reduction consortia for numerous stars of suspected duplicity (including δ Lib), but often with no significant improvement over the original positional findings

Orbit model

If there were no errors of measurement in the HIPPARCOS astrometry of an example having a previously unreported orbital motion the deviations Δp would all arise from the ordered changes of position neglected in the presumed no-orbit solution in the HIPPARCOS catalogue. In reality, we construct an approximate orbital model for the situation where the changes of position due to binary motion are significantly greater than the errors of measurement. We then write, for the apparent orbital positions, $\Delta X_1 = F_1\Delta p$, $\Delta Y_1 = F_2\Delta p$, and apply a model optimization procedure to find relevant parameters.

Certain of these parameters were derived separately in Worek's spectroscopic analysis of the δ Lib system, thus: $a\sin i = 5.3$ mas, $e = 0.28$, $\omega = 198°$, $P = 2.762$ y and $T_0 = 1983.24$. The timings of HIPPARCOS observations refer to the 'orbit number' counted from the epoch $T_0 = 1989.847$, with each orbit having a period of 0.442292 days. The parameters just cited then determine that the companion's mean anomaly at HIPPARCOS' time zero was 141°. A cursory examination of the HIPPARCOS Intermediate Data Search Facility (IDSF) (http://www.rssd.esa.int/HIPPARCOS/apps/PlotHipi.html) information on δ Lib raises the possibility of a slight quasi-sinusoidal variation on a ~2.8 year time base.

After extraction of the relevant data from this source and plotting against the sky co-ordinates, we find the distribution of data points rather elongated about an axis at position angle $\sim$40° – 220°, suggesting an inclination of the wide orbit (i.e. the angle between the line of sight and the orbital axis) not far from 90°, as proposed by Worek. Fig 5.10 shows this distribution, together with a plotting of a model orbit, taken to be coplanar with the eclipsing system. The nodal angle is here $\Omega = 40°$, and the other 5 orbital parameters are taken from Worek's model.

Although the orbit model in Fig 5.10 looks feasible, since the plotted ellipse matches the general distribution of observations, it is, in fact, significantly non-optimal. The rms error is more than twice the $\sim$1.5 mas that could be expected for normal HIPPARCOS positions. Assuming a coplanar inclination and adopting all the other parameters from Worek, minimization of the sums of squares of errors by changing only Ω moves the ellipse major axis away from the apparent X, Y correlation axis, though without improving the rms error significantly.

Preliminary examination of the data suggested a period slightly longer than Worek's value. Use of such an increased period improved the modelling match, so that the scatter became more comparable to the distribution of the original residuals about the no-orbit solution.

It is possible that a still better fit could be obtained with further parameter adjustments, but, in fact, convincing support for Worek's wide orbit has not been found by matching the HIPPARCOS residuals. Worek used spectroscopic radial velocity data to construct his model for the wide orbit. While that was based on only 8 points, there seems no reason to doubt the high quality of the measures included, that certainly appear to identify a systemic RV amplitude of at least 5 km s^{-1}, with an accuracy better than 0.5 km s^{-1}. In turn, this leads to a mass function $f = P_{\mathrm{wide}}(\mathrm{y})(K/29.79)^3$ of $\sim$0.015.

Worek adopted a fairly high inclination, together with Tomkin's (1978) spectroscopically derived masses for the close pair ($M_1 + M_2 = 6.6$ M$_\odot$) to find a third mass of about 0.95 M$_\odot$. With a lower angle between the line of sight and the orbital axis (70°) and the longer period coming from the astrometric analysis, the third mass could be raised to 1.1 M$_\odot$. This scenario provides a feasible explanation of findings that include, as well as Worek's spectrometry, photometric analysis of δ Lib's eclipsing binary light curve, that introduced a third light source in the system with a luminosity slightly greater than that of the Sun.

However, this evidence is not unequivocal. The time of eclipse minima $(O - C)$ data, for example, produces no clearly decisive single set of optimal parameters on a $\sim$2.8 yr time-scale, unlike the situation for R CMa. In reality, tentative third body wide orbits produced for multiple systems like R CMa and δ Lib, from the mutual interaction of photometric, spectroscopic and astrometric data, while stimulating interesting research, await further confirmation or refutation from the more precise astrometry of Gaia.

5.5.2   Gaia

The European Gaia space telescope, launched in 2013, followed a logical development from the generally acknowledged successes of the HIPPARCOS programme. Gaia was orig-inally an acronym for Global Astrometric Interferometer for Astrophysics, since optical interferometry was an intended observational technique. The name Gaia somehow stuck even though the instrumental design has changed significantly.

The 1 m aperture-class telescope-bearing satellite is located at the Lagrangian L$_2$ point of the Sun-Earth system, around 1.5×10^6 km further out from the Sun than Earth, where it should find itself in a relatively stable environment. The orbit is carefully arranged to avoid blocking of the sunlight by the Earth that would limit the amount of solar energy the satellite could utilize, as well as disturb the spacecraft's thermal equilibrium. A 10-m diameter umbrella always faces the Sun, allowing the facility's operations to be powered from the solar panels on its surface, while keeping all telescope components under temperature control. In fact, it is the high stability of the platform together with the large (giga-pixel) CCD camera that is key to the high precision of the data.

The importance of more precise parallaxes was emphasized above in relation to the use of data on visual doubles. Gaia continues to extend and make more precise the kind of survey work referred to in this chapter. It is also significant to the general process of establishing the large distances found in astronomy: the 'cosmic ladder' that is met in estimating how far away are the remotest objects.

Gaia aims to find the positions, parallaxes, and annual proper motions of $\sim 10^9$ stars, with precisions that depend on the stellar brightness: typically, ~ 20 microarcsec (μas) at 15 mag, and 200 μas at 20 mag. This means that distances to some 2×10^7 stars will be measured with a precision of at least 1%, and measurements on an order of magnitude more than this number of (fainter) stars will be accurate to better than 10%. Distances accurate to 10% should thus be achieved as far away as the Galactic Centre. Tangential velocities to within about 500 m s^{-1} for an estimated 4×10^7 stars are anticipated.

The data are released in stages, 'DRn', that contain progressively more detailed information. The incomplete DR1 was based on 14 months of observations carried out until September 2015. DR2 appeared in April 2018 and involved data from observations of over a billion stars between mid-2014 and mid-2016. DR2's reference coordinates take into account observations of about half a million quasars. These have been checked on agreement with other fundamental positioning systems to within ~ 25 μas. The DRs are accessible to potential users via Internet from the Gaia archive or other astrophysical information sites such as the CDS.

Regarding binary and multiple stars, Gaia's observations should permit statistical studies of multiple star distribution functions in different environments of the Galaxy, exploring the distribution of binaries in open clusters and associations, as well as the general field. Gaia's high-precision astrometry of visual binary orbits is expected to yield around 10^4 new stellar mass determinations, accurate to within about 1%, significantly improving factual awareness of stellar properties. While there are better prospects for determination of the parameters of the secondaries at separations greater than a few arcsec, Gaia can, in principle, resolve binary separations down to about 20 mas. Simulations of the expected results show that binaries with periods down to ~ 4 years will be open to investigation. While Worek's model for a companion star to δ Lib, having a maximum separation of ~ 40 mas, is at the low end of the separation range it should still be possible to check if the proposed third star is really there.

It has been reported that even shorter period systems will be discovered from irregularities in precise proper motion data. While such data may not furnish a complete set of parameters they should still be useful for statistical studies of binary incidence and characteristics. Altogether of the order of 6×10^7 binaries are expected to be recovered, permitting an unprecedented amount of empirical information on the nature of stars out to distances of the order of 1000 pc. Gaia's facilities include photometric and spectrographic instrumentation that will supplement the astrometric information. But data processing was not yet completed for systems with separations below ~ 2 arcsec in early Gaia information releases. Processing issues have been related to the detector windowing. Hopefully, these matters will be resolved as experience grows, allowing the mission to succeed with its planned objectives.

An early example of the application of Gaia DR2 data to a specific issue in binary star astrophysics was with the young and massive close system HX Vel, that appears fairly centrally within the galactic cluster IC 2395. The cluster is located, according to the WEBDA internet resource of the University of Vienna, about 705 pc from the Sun, at galactic co-ordinates $\sim 266.6°$ long. and $-3.61°$ lat., having an estimated age around 15 Myr. Gaia parallaxes of several members of IC 2395 were in the range 1.32 — 1.41 mas, ie. at distances ~710 — 760 pc, consistent with the WEBDA information, while the DR2 parallax of HX Vel itself corresponds to a distance of 754 pc that would clearly agree well with cluster membership.

By contrast, the HIPPARCOS parallax put HX Vel at a distance of about twice that given for the cluster, but the positional accuracy of HIPPARCOS, at 700 pc distance, was sufficiently low that its parallax estimate is still reasonably consistent with Gaia's improved measurements. Complications due to the multiple star context of the system would have probably affected the HIPPARCOS parallax.

On the other hand, the distance of HX Vel indicated by its apparent brightness, and separate inferences about its inherent luminosity, turns out to be rather low ($\lesssim 650$ pc). The situation might be ameliorated if certain of the adopted stellar model properties were changed, for example, by increasing the masses of the component stars. But the distance modulus cannot be increased enough to permit cluster membership without exceeding the expected uncertainties of the masses. Further studies of comparable photometric distance moduli have suggested clusters seemingly extended along the line of sight, compared with the plane of the sky. The explanation must be that the distance moduli of cluster stars are often systematically inaccurate. This might be due to the crowded nature of the field affecting brightness measurements, or the role of intervening, possibly inhomogeneous, absorptive material. In the case of IC 2395, the situation becomes more intriguing by the cluster's relationship to the much larger Vela OB1 association. It has been surmised that gravitational interactions with other massive inhomogeneities in the Galaxy have disrupted the association into separate substructures (Vela OB1 A, B and C) – IC 2395 more or less coinciding with Vela OB1 C.

Gaia's capacity to deliver reliable absolute information will thus provide important constraints in relating a wide range of observational to theoretical subjects in astrophysics.

5.6   Bibliography

5.1 W. Herschel's third list of 145 double stars was published in the *Mem. Royal Astron. Soc.*, Vol. 1, p166, 1821. A number of well-respected catalogues and compendia of information on double stars appeared between the Herschels' and their modern electronically available descendants, including those of S W Burnham (1906), F. W. Dyson (1921), R G Aitken (1932), and W H van den Bos (1957). The Washington Double Star Catalogue is referenced to B. D. Mason, G. L. Wycoff, W. I. Hartkopf, G. G. Douglass & C. E. Worley, *Astron. J.*, Vol. 122, p3466, 2001. The binary star data referred to in connection with Fig 5.1 is that of O. Yu. Malkov, V. S. Tamazian, J. A. Docobo, D. A. Chulkov; *Astron. Astrophys.*, 546, 69, 2012 (Table 2). This table is available in electronic form at the CDS via http://cdsarc.u-strasbg.fr/viz-bin/qcat?J/A+A/546/A69. The general database of O. Malkov, A. Karchevsky, P. Kaygorodov, D. Kovaleva & N. Skvortsov appearing as *arXiv*; 1811.04100 (2018), affords an interesting perspective on the different classification arrangements of binary stars. Applications of such extensive data to cosmogonic problems of binary stars were discussed by D. Kovaleva, O. Malkov, L. Yungel'son & D. Chulkov; in *Baltic Astron.*, Vol. 25, p419, 2016.

C. W. Allen's (1974) list of the 100 nearest stars originally given in *Astrophysical Quantities*, Athlone Press, London; has been republished, under the supervision of G. F. Gilmore and M. Zeilik in *Allen's Astrophysical Quantities, 4th Edition*, ed. A. N. Cox, Springer, 1999, p471. O. J. Eggen's compilation of 333 visual binaries was published in *Astron. J.*, Vol. 61, p405, 1956. G. P. Kuiper's list appeared in the *Astrophys. J.*, Vol. 95, p210, 1942; P. van de Kamp's was in *Publ. Astron. Soc. Pacific*, Vol. 81, p5, 1969.

The concept of a power-law distribution in the masses of stars stars was introduced by E. Salpeter in the *Astrophys. J.*, Vol. 121, p161, 1955, and has been developed in many studies since. A commonly used form is that of P. Kroupa *Mon. Not. Royal Astron. Soc.*, Vol. 322, p231, 2001, and its apparent widespread uniformity discussed bv the same author in *Science*, Vol. 295, p82, 2002.

A. Labeyrie's paper on high time-resolution imaging appeared in *Astron. Astrophys.*, Vol. 6, p85, 1970. The subject is reviewed in Chapter 3 of R. W. Hilditch's *An Introduction to Close Binary Stars*, CUP, 2001, see also the same author's general review in *ASP Conf. Ser.*, Vol. 318 (eds. R. W. Hilditch, H. Hensberge & K. Pavlovski), p3, 2004. W. I. Hartkopf presented a summary article on speckle interferometry in *Complementary Approaches to Double and Multiple Star Research; Astron. Soc. Pacific Conf. Ser.*, Vol. 32, IAU Colloquium 135, eds. H. A. McAlister & W. I. Hartkopf, p459, 1992. Principles of Fourier spectral analysis applied to astronomical data-collection are set out in R. N. Bracewell's *The Fourier Transform and its Applications*, McGraw-Hill Book Co., 1978. The SUSI optical interferometer was reviewed by J. Davis in *Complementary Approaches to Double and Multiple Star Research; op. cit.* p521), see also R. Hanbury-Brown, *The Intensity Interferometer: Its Application to Astronomy*, Taylor & Francis, 1974. Its application specifically to binary stars was discussed by A. Kelz in *The Origins, Evolutions, and Destinies of Binary Stars in Clusters*, eds. E. F. Milone & J-C. Mermilliod, *Astron. Soc. Pacific Conf. Ser.*, Vol. 90, p51, 1996.

5.2 E. Öpik's lengthy study on the properties of visual binaries is in the *Publ. L'Obs. Astron. Univ. Tartu*, Vol. 25(6), pp1-167, 1924. Their distribution was further considered by P. P. Eggleton, M. J. Fitchett & C. A. Tout in the *Astrophys. J.*, Vol. 347, p998, 1989.

Low mass spectral binaries, such as that shown in Fig 5.3 were discussed by D. C. Bardalez Gagliuffi et al. (10 authors) in the *Astrophys. J.*, Vol. 794, p143, 2014. A more recent review, concentrating on astrometric techniques used to confirm the binary nature of low mass spectral binaries is that of J. Sahlmann et al. (10 authors) in *Mon. Not. Royal Astron. Soc.*, Vol. 495, p1136, 2020.

J. Dommanget's invited paper at the Bamberg conference to honour the 200th anniversary of F. W. Bessel (*Astrophys. Space Sci.*, Vol. 110, p47, 1985) looked forward to the role that the HIPPARCOS satellite would play in advancing knowledge of binary systems, particularly for visual binaries of shorter period and with determinable parallax. The same author's paper at IAU Colloquium 80 on Double Stars: Physical Properties and Generic Relations in the previous year (*Astrophys. Space Sci.*, Vol. 99, p23, 1984) looked back over historic observations of visual binaries and their role in determining stellar masses, as well as the questions they raise in relation to the origins of stars in general. A good review of the HIPPARCOS satellite: its functioning, aims and organization; was given by J. Kovalevsky in *Astrometric techniques: IAU Symp. 109*, eds. H. K. Eichhorn & R. J. Leacock, p581, 1986. Its application to double star astrometry was noted in the article of L. Lindegren et al (15 authors) in *Astron. Astrophys.*, Vol. 323, L53, 1997. J. T. Armstrong's contribution on sub-mas optical astrometry and binary stars to *IAU Symp. 166* (The astronomical and astrophysical objectives of sub-mas optical interferometry), eds. E. Hog, P. Kenneth Seidelmann, CUP, p193, 1995; are relevant to this context. These aims were further discussed in the context of the Navy Optical Interferometer by J. T. Armstrong et al. (15 authors) *Astrophys J.*, Vol. 496, p550, 1998.

The AB Dor system was reviewed by R. Azulay et al. (11 authors) in *arXiv*:1504.02766 [astro-ph.SR], 2015, including the use of long baseline interferometry in the determination of the orbits within this multiple system, and the significance of the derived absolute parameters of the component stars to the theory of Pre-Main Sequence dwarfs. A review referring to multiwavelength observations and physical interpretations of the emission mechanisms from AB Dor was published by O. B. Slee; N. Erkan; M. Johnston-Hollitt & E. Budding, *Publ. Astron. Soc. Australia*, Vol. 31, p21, 2014.

5.3 The paper of J. N. Bahcall & R. M. Soneira on the distribution of stars over the sky is in the *Astrophys. J. Suppl. Ser.*, Vol. 47, p357, 1981; that of D. Scott & C. Tout was in the *Mon. Not. Royal Astron. Soc.*, Vol. 241, p109, 1989.

5.4 The parametrization of visual binary orbits (Section 5.4.1) is presented in a parallel way to that of W. M. Smart's *Text-Book on Spherical Astronomy*, CUP, 1960, Chapter 14, with some differences of approach, see also W. H. van den Bos, in *Astronomical Techniques*, ed. W. A. Hiltner, Univ. Chicago Press, Chapter 22, p537, 1962. In view of ambiguities, it would be constructive if investigations related their procedure to one accepted set of parameter meanings, or, failing that, spelled out, or made clear, their adopted formulae. Smart's *Text-Book* (op. cit.) gives a detailed account of the method of H. J. Zwiers, originally published in the *Astron. Nach.* Vol. 139, p369, 1896. Optimization of the parameter-set is often effected, in this type of problem, via the Levenberg-Marquardt technique, or a variant of it (cf. D. Marquardt, *SIAM J. Appl. Math.*, Vol. 11 (2), p431, 1963; K. Levenberg, *Quart. J. Appl. Math.*, Vol. 2, p164, 1944).

Ideas surrounding the formulation and application of 'dynamical parallaxes' were discussed by F. H. Seares in the *Astrophys. J.*, Vol. 55, p165, 1922; but a concise summary can be found in W. M. Smart's book *op. cit.*, p356. The use of an iterative procedure to improve the mass estimates using spectroscopic and luminosity data was discussed by H. N. Russell, W. S. Adams & A. H. Joy in *Publ. Astron. Soc. Pacific*, Vol. 35, p189-193, 1923. The application of dynamical parallaxes to the Hyades cluster was examined by W. D. Heintz in *The Observatory*, Vol. 89, p147, 1969.

F. B. Wood called attention to the 'puzzling' system R CMa in the *Contr. Princeton Univ. Obs.*, Vol. 22, p31, and it has been a subject of much subsequent discussion. The astrometric data considered in Section (5.4.4) was collected by I. Ribas, F. Arenou and E. F. Guinan in the *Astron. J.*, Vol. 123, p2033, 2002. The discussion here follows the lines of E. Budding & R. Butland's 2011 paper in the *Mon. Not. Royal Astron. Soc.*, Vol. 418, p1764. The light travel time effect (LTTE) was spelled out by J. B. Irwin in the *Astrophys. J.*, Vol. 116, p211, 1952 (see also J. B. Irwin, *Astron. J.*, Vol. 64, p149, 1959). Irwin cites J. Woltjer Jr. (*B. A. N.*, Vol. 1, p93, 1922) as having introduced the idea. Failure to secure spectroscopic confirmation of a third body in the R CMa system was reported by H. Lehmann, V. Tsymbal, F. Pertermann, A. Tkachenko, D, E. Mkrtichian & N. A-thano, in *Astron. Astrophys.*, Vol. 615, A131, 2018. The compilation of O – C data referred to is that of J. M. Kreiner, C-H. Kim, & I-S. Nha, who published *An Atlas of O–C Diagrams of Eclipsing Binary Stars*, ISBN 83-7271-85-6 in 2000. C. R. Chambliss in *Publ. Astron. Soc. Pacific*, Vol. 104, p663, 1992, surveyed 80 bright binaries for multiplicity. He noted that the most frequently encountered type of eclipsing binary in multiple, especially triple, systems is that with two young, early-type stars. EW systems are also not uncommonly found in multiple stars.

5.5 Source information on the HIPPARCOS satellite, its design and results, are often associated with the ESA publication: *The HIPPARCOS and Tycho Catalogues*, ESA SP-1200, ESA Publ. Noordwijk, 1997; see also M. A. C. Perryman et al. (20 authors) *Astro. Astrophys.*, Vol. 500, p501, 1997. *HIPPARCOS, The New Reduction of the Raw Data*, by F. van Leeuwen, *Astron. Astrophys. Lib.*, ed. S. N. Shore, Springer, 2007, is also generally cited in this context, and that provides a background for the structure and use of Eqn (5.29).

HIPPARCOS parallaxes can be used for the determination of stellar masses from visual and astrometric binaries, but the implication is then that these binaries have accurate orbital parameters. S. Söderhjelm provided such parameters for 205 systems, combining

HIPPARCOS astrometry with ground-based observations in a comprehensive review in *Astron. Astrophys.*, Vol. 341, p121, 1999. Procedures to derive masses from HIPPARCOS observations of double stars were also discussed by C. Martin, F. Mignard, & M. Froeschle in *Astron. Astrophys. Suppl. Ser.*, Vol. 122, p571, 1997.

T. F. Worek's model for the triple system δ Lib was given in the *Publ. Astron. Soc. Pacific* Vol. 113, p964, 2001. The discussion of the system presented in Section 5.5.1 follows that of E. Budding, V. Bakış, A. Erdem, O. Demircan, L. Iliev, I. Iliev & O. B. Slee in *Astrophys. Space Sci.*, Vol. 296, p371, 2005. J. Tomkin's, spectroscopic detection of the secondary component in the system was published in the *Astron. J.*, Vol. 221, p608, 1978. A review of close binary stars that may have wide orbit companions was presented by P. Zasche, M. Wolf, W. I. Hartkopf, P. Svoboda, R. Uhlar, A. Liakos and K. Gazeas in *Astron. J.*, Vol. 138, p664, 2009. The review included a catalogue of 44 examples for which more than 10 astrometric measurements showing variations of position angles of more than 10° confirmed their multiple star status. δ Lib was not in this confirmed list but it was included in a list of 74 additional systems of interest, that did not quite meet the selection criteria.

ESA's successor to the HIPPARCOS mission, Gaia, is on a greatly expanded scale, issuing large amounts of data in a programme of regular releases. The aims of the mission were given by J. de Bruijne, R. Kohley & T. Prusti in *Proc. SPIE* 7731, (2010). Up-to-date information on the mission can be sought from relevant websites, such as ESA's https://sci.esa.int/web/gaia/-/28820-summary, or, specifically on binaries at -/31441-binary stars. Gaia's evidence on δ Lib remains to be reconciled with the earlier studies discussed in the foregoing. The application to HX Vel was raised by M. G. Blackford, A. Erdem, D. Sürgit, B. Özkardeş, E. Budding, R. Butland & O. Demircan, in *Mon. Not. Royal Astron. Soc.*, Vol. 487, p161, 2019.

6

Spectroscopic Binaries

6.1 Introduction

We have seen in the previous chapter how the interpretation of data associated with visual binaries and astrometric techniques have enhanced knowledge of stellar physics. In this chapter we concentrate on the application of the parallel, but rather separate, techniques of stellar spectroscopy to stars in relatively close proximity. The role of spectral lines is prominent in this context. Although this is a big subject of study in its own right, well beyond the compass of this book, it will be taken as understood that the distribution of radiation with wavelength in the light coming from stars consists of a smoothly varying *continuum*, interrupted here and there by sudden and short variations of flux – *lines* – that can be above or below the continuum level, but usually the latter for our context. In some ('cool') stars, groups of lines are so plentiful and close together that they appear as bands: formations associated with the properties of atoms bonded together as molecules, rather than single atoms or ions of a particular element. There are also places in the spectrum where the entire continuum changes suddenly: spectral discontinuities, the most familiar of which is probably the *Balmer discontinuity* at 364.6 nm, in the near ultra-violet. But it is the properties of spectral lines – in particular their positions and shapes – that are most relevant to the themes of this book.

The spectral features that are observed: absorption or emission lines, or continuum discontinuities; are critically dependent on the composition, local temperature, density and gravity conditions in the regions in which the source photons originate and are transported. Spectral data thus have an inherently large information content that has stimulated a vast literature concerning the physical properties of stars and other astronomical light sources. We apply some of this information in assessing the nature of the stars in the binary or multiple systems considered. Usefully accurate measures of the mass, surface temperature and size of component stars are often deduced as a result; likewise their composition and age.

We will be particularly concerned with the motions of components deduced from Doppler shifts. The role of the Doppler effect in shifting apparent wavelengths of spectral lines in dependence on the relative velocity of the source follows the general formula

$$\frac{\lambda_o}{\lambda_s} = \left(1 + \frac{v}{c}\right)\gamma \, , \tag{6.1}$$

where λ_o, λ_s are the observed and source wavelengths; v is the velocity of the source, c that of light, and γ is the Lorentz factor $\gamma = 1/\sqrt{(1 - v^2/c^2)}$ arising from the dilation of time

DOI: 10.1201/b22228-6

to the observer*. Directly or indirectly, the Doppler effect offers the key to understanding many of the source's properties.

It is on this basis that periodic variations in radial velocities of the components of binary systems are interpreted in terms of orbital and other motions. Rotational velocities are also of high significance to the physics of close stellar systems. From the viewpoint that the observed spectrum is accumulated from the contributions of local regions that are small compared to the overall dimensions of the star, we speak of surface properties, such as a local surface velocity that is linked to a bodily rotation.

Let us note that optical Doppler displacement, in the context of binary stars, is a fairly small effect in terms of practical measurement procedures. Thus, at a wavelength of 5000Å a shift of 0.02%, i.e. 1 angstrom, already means 60 km s^{-1}, which is comparable to many binary orbit amplitudes. The determination of stellar radial velocities to an accuracy of 1 km s^{-1}, in general, has only been routinely achieved in the last few decades. This quality of data calls for stringent observational procedures as well as modern detection and data-processing technology.

A key parameter is the *spectral resolution* $(\lambda/\Delta\lambda)$ of the spectrograph used to collect the data. The Doppler velocity effect, if only one feature at the limit of resolution was involved, might suggest a required resolution of order 10^5. While such high resolutions may be found in specialized facilities of the present day, they are not necessarily a prerequisite for binary system research. There are often numerous spectral features, or components of features, capable of contributing to an assessment of the net shift, while these features may be affected by other issues such as noise in the data compared to the basic signal of the spectrum, the blending of lines, mechanical and optical performance of components, and the smearing out of features due to changes of velocity during the process of recording the signal. The observer normally has to compromise in order to deal with data-collection in an appropriate way. Important though these matters are, we will be assuming in what follows that the practicalities of good data-collection were already solved.

A similar stance concerns how radial velocity (rv) values are extracted from an appropriately wavelength and flux calibrated spectral image. Although for certain types of star Eqn (6.1) may be applied, more or less as it stands, to the modelling of individual lines, a widely used approach is to cross-correlate an observation spectrum with a suitable template image. This image would have similar properties to the spectrum of the observed star, and a known rv. The cross-correlation function (ccf) is formed, in principle, by multiplying the run of the two spectral flux values together, in dependence on a regular sequence of accurately known relative displacements in the direction of dispersion. The ccf will maximize when the mean separations of corresponding spectral features minimizes, allowing the velocity displacement of the observed spectrum with respect to the template to be determined. The application of this idea to real rv evaluation involves considerably more detailed operations than indicated simply by this principle, but our course again adopts that rv values can be provided that allow us to discuss their physical interpretation. Further information on the data side of this subject is given in the relevant sources mentioned in Section 6.6.

Several thousand spectroscopic binaries have been identified so far with orbital and other parameters catalogued. New satellite programs are likely to extend this tally by orders of magnitude. Fig 6.1 shows the distribution of radial velocity amplitudes with period drawn from a short selection from the online (SIMBAD/Vizier) *Catalogue of Spectroscopic*

---

*The Lorentz factor is effectively unity for most, if not all, situations involving observations of classical close binary stars.

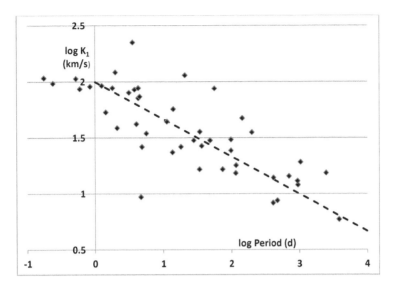

Figure 6.1 A sample of 50 spectroscopic binary orbits from the compilation of D. Pourbaix et al. (2020) showing the trend of primary radial velocity amplitude K_1 versus period P.

Binaries, whose trend $(\log K \sim -\frac{1}{3}\log P)$ again confirms, in a general way, the essential validity of Kepler's third law (cf. Fig 5.1).

Of course, measured shifts of spectral lines will include the effects of motion of the platform from which the observations are made: these have to be dealt with as part of the reduction process. This point requires closer attention with the advancing precision of data. It was shown (Section 1.3.2.5) that the velocity of an object moving in a Keplerian orbit with semi-major axis a, period P and eccentricity e can be decomposed into two constant components $w_\theta = 2\pi a/(P\sqrt{1-e^2})$ perpendicular to the radius vector, and $w_y = 2\pi ae/(P\sqrt{1-e^2})$ perpendicular to the major axis. By convention, radial velocities are taken as positive for recession. A star receding from the Sun with velocity V_0, would then have an apparent radial velocity V_1 as a result of these added components of the terrestrial motion, where

$$V_1 = V_0 - w_\theta \cos\beta \sin(\lambda_\odot - \lambda_\star) + w_y \cos\beta \sin(\lambda_\omega - \lambda_\star) \ , \tag{6.2}$$

with $\lambda_\odot$, $\lambda_\star$ and λ_ω as the longitudes of the Sun, star and perihelion respectively. β is the star's ecliptic latitude. The second and third terms on the right in Eqn (6.2) would have to be added back with the opposite sign to correct the measurements for the radial velocity of the Earth with respect to the Sun, an operation known as the *heliocentric correction* for the radial velocity. Numerical values are given[*] for a, P and e as 1.4959787066×10^8 km, 3.15581495×10^7 sec and 0.01771022, so that $v_\theta = 29.78119$ and $w_y = 0.52743$ km s^{-1}. These quantities are here specified to an accuracy greater that what is generally available in stellar spectroscopy, which is not yet more precise than ~ 0.001 km s^{-1}. However, plans to achieve such precision are in progress, for example in exoplanet research.

A heliocentric correction for the time of the observation also occurs for the observer's moving platform. This refers to the light travel time over the Earth's distance from the Sun projected in the direction of the star, s/c, say, where $s = \boldsymbol{R}.\boldsymbol{S}$ the scalar product of the solar

---

[*]cf. *Astrophysical Quantities* (4th ed., A. N. Cox Editor, Springer, 1999) for these and similar parameters in this section.

distance vector $\boldsymbol{R}$, and the unit vector in the direction of the star $\boldsymbol{S}$, and c is the velocity of light. The product is directly formed by reference to the equatorial coordinate system, with $S_x = \cos\alpha\cos\delta$, $S_y = \sin\alpha\cos\delta$, $S_z = \sin\delta$; and $R_x = R\cos\lambda_\odot$, $R_y = R\sin\lambda_\odot\cos\epsilon$, $R_z = R\sin\lambda_\odot\sin\epsilon$; ϵ being the obliquity of the ecliptic plane. The time correction $\Delta\tau$, in days, is then

$$\Delta\tau = -0.0057755\left[R\cos\lambda_\odot\cos\alpha\cos\delta + R\sin\lambda_\odot(\cos\epsilon\sin\alpha\cos\delta + \sin\epsilon\sin\delta)\right] . \qquad (6.3)$$

For a precision of better than $1\ \mathrm{km\ s^{-1}}$ (shifts of less than 0.02 Å at 6000 Å), the effect of the Earth's rotation should be taken into account, and a similar calculation to the foregoing shows that

$$V_2 = V_1 - v_{\mathrm{rot}}\cos\phi\cos\delta\sin H , \qquad (6.4)$$

where the equatorial rotation velocity $v_{\mathrm{rot}} = 0.46510\ \mathrm{km\ s^{-1}}$, and ϕ is the latitude of the observer; δ and H are the latitude and hour angle of the observed star.

Additional corrections for the motion of the Earth about the centre of the Earth-Moon system (a term of order 13 m/sec) and the motion of the Sun about the *barycentre* of the solar system (Jupiter's effect on the Sun involves a motion of about 12 m/sec on a 11.9 y period, that of Saturn ∼3 m/sec on a 29.5 y period) are required for high accuracy rv determination. Other effects having amplitudes $\lesssim 1$ m/sec would be taken into account for work of the highest precision. Such corrections to a measured rv would normally be included in initial data-reduction programs using formulae of essentially similar form to the foregoing, where quantities such as $\lambda_\odot$, $\lambda_\star$, R, ϵ, α, δ, ϕ, H etc. are logged automatically by the telescope's controlling program.

The foregoing principles for dealing with the motion of the observer as a prelude to analysis apply also at the source, notably in situations where a regular series of repetitive phenomena occur from a source that has some extra accelerative behaviour. This may typically arise because the source is a close eclipsing binary in orbit with a wide companion about the whole system's barycentre. In Section 5.4.4, the third of Eqn 5.20 – essentially related to Eqn 6.3 – was applied as a fitting function to determine the light travel time appearing in the O – C data in terms of the wide system's parameters. In fact, the phases of measurements made on a component in a single binary orbit should take into account the LTT across the orbit. If the primary eclipse is a transit, for example, the secondary occultation would be observed slightly later than phase 0.5 due to the LTT. Generally speaking, however, this difference amounts to just a few seconds in a period that could be typically of order 10^5 sec or greater, and, until quite recently, below the level of available accuracy of light curve specification. Alternatively, one or other of the component stars may exhibit regular pulsations. Radio pulsars in binary systems are particularly fine examples of precise regularity in this connection. In that case, the shifts in frequency of observed pulses can be analysed not only using the LTT effect but also including its derivative, the Doppler effect.

6.2   Spectroscopic Binary Radial Velocity Curves

Radial velocity observations for a spectroscopic binary system showing cyclic variations, being phased according to an assigned period and plotted to show the form of the variation constitute a *radial velocity curve*. This data-set, from an analysis point of view, corresponds to the positionally detected orbit of the previous chapter, or the light curve of the next. This correspondence becomes particularly significant for close pairs, when spectroscopic and other effects may be examined in parallel and treated in combination.

If a pair of stars are sufficiently separated that the relative distortions of their surface light distribution due to proximity effects can be neglected, then their motions are taken to

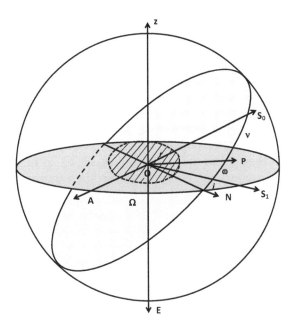

Figure 6.2 Geometry of an orbiting component of a spectroscopic binary, the origin O now set at the system's centre of mass (cf. Fig 1.10). The nodal angle AN, marked as Ω, has no bearing on the radial velocity. The orbital plane NP, where P denotes the periastron position, is inclined at angle i to the local sky tangent plane. The longitude of the periastron ω corresponds to the arc NP. The central ellipse marks the orbit of a stellar component, located in the direction OS_0 at radius r and true anomaly ν. Spectroscopically observed Doppler shifts allow the motion in z-direction to be derived.

be those of mass points. These motions fit into the framework of the Keplerian formulae in Chapter 1, with a parallel in the zero-order model of the photometric data analysis. Spectral line profiles, in such idealized circumstances, are generally found to be symmetrical: their displacements are those of the *centre of light* of the source, and coincide with its *centre of mass*. Proximity effects alter this: there are then significant tidal distortions of the surfaces or reflection of light from the companion to separate the light and mass centres.

Stars appearing as single light sources optically, but formed by pairs whose orbital Doppler shifts are distinctly measurable make up the class of *spectroscopic binaries*. Most of these sources are sufficiently separated to be essentially free of proximity effects. In practice, this class overlaps with that of close, often eclipsing, binaries, though the terminology has somewhat distinct usages related to the observational history of particular stars. Spectroscopic binaries are further divided into single or double lined systems (SB1 or SB2), referring to the detection of one or both sets of spectral lines. These designations can extend to SB3 or higher if more than two spectra are visible.

6.2.1 Orbit Parametrization

The modelling of rv curves is again open to ambiguities of prescription, partly related to parameter designations, but generally adopted procedures can be followed by referring to Fig 6.2. The conventional Eulerian arrangement has the nodal angle Ω, measured from a fixed direction (usually north), or reference point on the sky plane, and an inclination

i – a positive angular turn at the 'ascending' node as shown in Fig 6.2. At this ascending node the orbital motion of the star is outward, or 'into', the sky. For spectroscopic binaries the inclination between the binary's orbital plane and the observer's local sky plane is often relatively high, facilitating the clarity of rv variations.

In the arrangement of Fig 6.2, the observer looks in the direction of increasing z, to observe the system as shown in Fig 6.2 with the secondary approaching its upper culmination. Spectrometric data does not directly involve the nodal angle Ω, but, for definiteness, let us set it to 90°. The star under consideration, here the secondary in its relative orbit, can be visualized as moving from the node (at the west in this case), deeper into the sky travelling south in a clockwise motion. After passing through the 'superior conjunction' it reaches the easterly node. From that point, the motion is inward, i.e. in front of the primary, and northward to the inferior conjunction that is often used as a fiducial reference point.

If the binary were to show eclipses, the secondary would pass, east to west, (transit) with its centre across the northerly (conventionally 'upper') part of the primary's disk, the inclination here being less than 90°. If the inclination were greater than 90° the transit would move across the southerly part of the primary disk and the sense of the rotation on the sky would be anticlockwise: a retrograde orbit as seen from below. If the ascending node is switched from 90° to 270° with the inclination $i < 90°$ as before, the transit would occur across the southern part of the primary's disk from west to east. If the inclination is replaced with its supplement the transit is still west to east, but the transiting secondary's centre point is northward of the primary's. All four of these possibilities are equivalent as far as the rv evidence alone is concerned, which does not depend on the nodal angle and is symmetrical about $i = 90°$.

Positive displacement z beyond the sky-plane satisfies

$$z = r \sin(\omega + \nu) \sin i \; , \tag{6.5}$$

where r is the radius from the origin at the system centre of gravity to the star whose true anomaly is ν. That is measured from the periastron position, located at angle ω along the orbital plane from the ascending node at N.[*] The radial velocity is then given by

$$V = \frac{\mathrm{d}z}{\mathrm{d}t} + V_\gamma \; , \tag{6.6}$$

where V_γ, the 'γ-velocity', is the radial velocity of the centre of gravity of the system that is added to the periodic motion in forming the rv curve.

Eqn (1.25) gave the value of r with an indirect dependence on time as $r = a(1-e^2)/(1+e\cos\nu)$. We had also from Section (1.3.2.2) the constant angular momentum $h = r^2 \, \mathrm{d}\nu/\,\mathrm{d}t = na^2\sqrt{1-e^2}$, where the mean angular motion $n = 2\pi/P = \sqrt{G(M_1 + M_2)/a^3}$. Differentiating (6.5), and with a little rearrangement, using the formulae of Keplerian motion, we find

$$\frac{\mathrm{d}z}{\mathrm{d}t} = \frac{na\sin i}{\sqrt{1 - e^2}}[\cos(\omega + \nu) + e\cos\omega] \; . \tag{6.7}$$

In the case of zero eccentricity, this produces a perfect sinusoid, which, if compared with photometric data for the same system, would be phased from the inferior conjunction when the periodic term in the radial velocity is zero.

---

[*]A. H. Batten (1973) pointed out the possibility of confusion between the longitude ω, and a comparable parameter ϖ. The former is measured in the orbital plane from the node and occurs more directly in observational contexts. The latter, also referred to as a periastron (or perihelion) longitude, appears in classical discussions of orbital dynamics. It includes the angle ω added to a separate angular component measured from a fiducial point in a fixed frame of reference.

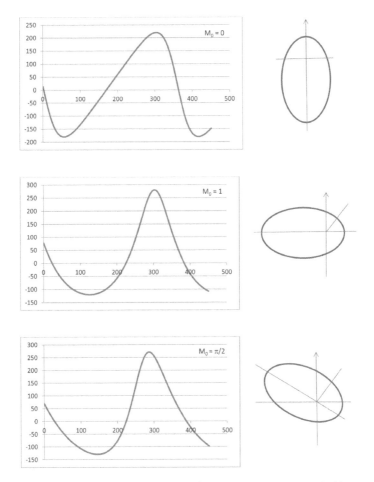

Figure 6.3 Radial velocity curves for the primary of an eccentric spectroscopic binary at different orientations of the major axis. The primary orbits on the right indicate the line of sight with an arrow and the principle axis of the ellipse with a straight line. The plane of the sky, through the centre of mass and perpendicular to the page, is shown as a horizontal line. Another line is sometimes used to mark the mean anomaly at phase zero M_0, measured anticlockwise from the periastron (see text for more details).

When eccentricity is present the rv curve is still a simple cosine function in the true anomaly ν, but the curve plotted against time becomes distorted, due to the rather complex dependence of the true anomaly on time. The connection of ν to time is usually evaluated through use of the intermediate variable M, the mean anomaly, which performs a uniform increase of 0 to 2π in the same period as the non-uniform increase of ν (see Section 1.3). This needn't detain us now other than to note that the form of the distortion, for a given value of e, will depend on the value of the parameter that fixes the orientation of the major axis of the ellipse to the line of sight. This is, ostensibly, the longitude of periastron ω, but more immediate to the relationship with time is the quantity M_0; the value of the *mean anomaly* at the zero point of the arranged orbital phases. It is implicit in the lead up to (1.8) that $\omega = \omega(e, M_0)$, i.e. ω is fixed for given values of e and M_0. From Fig 6.2 it can be seen that at the inferior conjunction (lowest point of the great circle in the plane of the orbit) that $\nu(e, M_0) = 270° - \omega$. This corresponds to the relative position of the secondary. But binary star spectroscopic data is usually better defined for the primary; in fact, sometimes only for the primary. In that case, e and M_0 having been determined through optimization

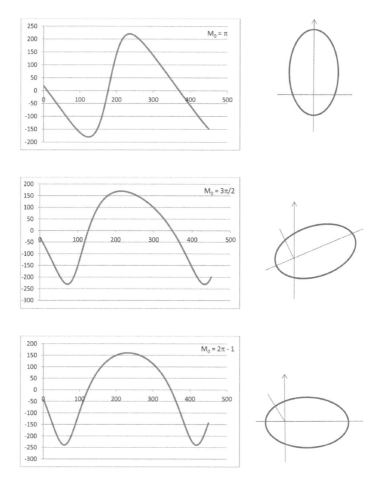

Figure 6.4   Radial velocity curves for the primary of an eccentric spectroscopic binary at different orientations of the major axis (see text for clarification).

as in (5.22), the orbit orientation is then identified through

$$\omega = 90° - \nu(e, M_0) \ . \tag{6.8}$$

Representations of eccentric Keplerian rv curves are shown in Figs 6.3 and 6.4. The relationship of M_0 to the observed distortion of the rv curve is not so obvious, but can be interpreted from the sketched orientations of the orbits in these diagrams. The line of sight is marked by a vertical upward arrow. The plane of the sky, perpendicular to the page, is indicated by the horizontal line. The inclination is set at 90°. In the uppermost diagram, in Fig 6.3 the orbit of the primary is shown with the mean anomaly at phase zero $M_0 = 0$. The true anomaly is also zero in this instance, the primary being 'beyond' the sky plane at the periastron position. The longitude of the periastron ω, measured from the ascending node on the right, is 90 deg. The eccentricity used in the rv calculations was set at $e = 0.3$, but this is exaggerated in the orbit sketches for illustrative purposes. A typical γ-velocity of $+20$ km s^{-1} has been added for completeness.

In the second figure, M_0 has been increased to a value of 57.3 deg (1 radian), an angle that would be lagging behind the true anomaly at phase zero. The value of ω, turning in the opposite sense to M_0, and fixed by the constraint $\omega = 90° - \nu(e, M_0)$ has dropped toward zero, so the major axis of the ellipse becomes close to the plane of the sky. There is still a recessional velocity at phase zero, so the rv there is now greater than 50 km s^{-1}. The Keplerian motion of the star ensures that it moves through its maximum recession rather quickly, while taking much more time to pass through the region of the curve with low approach speed.

The third figure shows M_0 at 90 deg, but the star itself is now ~120 deg from the periastron position. The value of ω has fallen to ~−30 (equivalently 330) deg. The recession maximum has moved down a little in phase, measured anticlockwise from the vertical arrow, but the curve, on the whole, is similar to the one above it.

When M_0 reaches 180 deg, the rv curve (top of Fig 6.4) has become a mirror image of its form at phase zero, and the sequence follows an inverse pattern to the one in Fig 6.3, ω finishing up at a little less than 180 deg in the lowest curve of Fig 6.4.

For many catalogued systems we are given a value of ω, but, as noted above, to compute the orbit over a range of phases we need to specify M_0. In other words, it is required to know what is the value of the mean anomaly M_0 for the given value of the true anomaly $\nu = 90° - \omega$. A relevant formula is[*]

$$M = \nu - 2[\beta(1 + e')\sin \nu - \beta^2(\frac{1}{2} + e')\sin 2\nu + \beta^3(\frac{1}{3} + e')\sin 3\nu \ ... \ ] \ , \qquad (6.9)$$

where $\beta = (1 - e')/e$ and $e' = \sqrt{1 - e^2}$. To the lowest order in e, this reduces to the easily verified $M = \nu - 2e \sin \nu$, i.e. $M_0 = 90° - \omega - 2e \cos \omega$. This will allow a quick assessment of the connection between M_0 and ω, but an iterative machine calculation of (1.8) may be as effective in practice.

For a given rv data-set, Figs 6.3-4, or similar, may help to visualize the rv-curve's shape in dependence on the orientation of the stellar orbits in space, though it should be kept in mind that the arrangements shown in Figs 6.3-4 correspond to the motion of the primary star with respect to the system centre of gravity. The motion of the secondary will be a reflection of these curves about the line $V = V_\gamma$ and scaled according to the inverse mass ratio. Also, the presumption of Keplerian motion is still a first approximation for the orbits of real binary stars which do not remain fixed in space. Notably, the major axis of the ellipse, or apsidal line, is subject to the secular progression, discussed in Chapter 3, so the rv curve of a given binary may appear differently at different epochs.

Apsidal motion is not the only factor complicating the appearance of rv-curves. Already in the early years of the 20th century J. M. Barr had pointed out an asymmetry in the observed distribution of longitudes of periastron (ω) of spectroscopic binaries. This has been confirmed in subsequent studies with greater sample sizes. A. H. Batten's selection from well-studied examples in his *Catalogue of Spectroscopic Binary Orbits* is reproduced in Fig 6.5, where the 'Barr effect' preference for the first quadrant ($0 \lesssim \omega \lesssim 90°$) is evident.

There would be no *a priori* reason to suppose any systematic preference for the orientation of the major axis of an eccentric binary, moving as two mass points in a Keplerian orbit, to the line of sight. But the possibility of other factors affecting apparent Doppler shifts is not difficult to admit. One of the key ones relates to the 'spurious eccentricity' noted by E. F. Carpenter (1930) in the rv-curve of the classical Algol binary U Cep. A relatively large proportion of close and eclipsing binary systems are Algols, which would tend to bias statistical results towards their physical properties. There is photometric evidence, as well as sound theoretical reasons, to believe that Algols should have effectively circular orbits. Part of the accepted modelling of Algol systems is that they have evolved into a process of mass-transfer, between and around the components.

---

[*]See D. Brouwer and G. M. Clemence' *Methods of Celestial Mechanics*, Academic Press, 1961; p65.

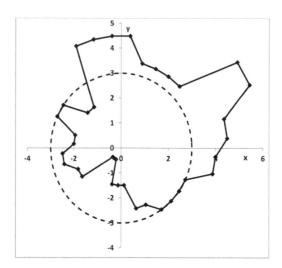

Figure 6.5   The smoothed distribution of the orientations of 110 eccentric orbit axes in 10 degree intervals, according to A. H. Batten in *Binary and Multiple Systems of Stars*, Pergamon Press, 1973. The longitude of periastron parameter ω proceeds in the conventional sense in this diagram from zero along the x-axis to 90° in the direction of the y-axis. The 'Barr effect' relates to the distribution's preponderance in the first quadrant and deficit in the third quadrant.

It is such transferring and circumbinary gas in proximity to the stellar pair that affects the net shape of line profiles and can mislead interpretation on rv-curve parametrization. Mass-transferring Algols would then have their radial velocity curves, which should be symmetrically sinusoidal, systematically distorted so as to assume a form like the eccentric example shown at the top of Fig 6.3, or rather something between the first and second panels in Fig 6.3. This corresponds to the low longitudes of periastron ($0 \lesssim \omega \lesssim 90°$) shown in Fig 6.5. The distribution could be accounted for if a net recessive component was added throughout much of the phase range of a mass-transferring Algol's rv curve. The example presented by Batten using spectrograms of U Cep shows that this can be explained by the addition of an emission feature with an extended blue component. This picture can be reconciled with the mechanics of the process, where circulating accreted matter could appear more extensive on the approach side, from the observer's point of view.

Such circulating matter would affect the profiles of some lines more than others: the hydrogen lines appear sensitive in this respect. In the case of U Cep, which appears to be relatively active in mass transfer, the strong K line of ionized calcium shows an oscillatory behaviour, in phase with the orbital motion but with only about a sixth of its amplitude. These phenomena have led to the idea of an additional cloud-like accumulation of plasma well beyond the near-photospheric layers normally associated with spectrum formation. Observed spectral effects can be accounted for by integrating contributory line-forming processes through this structure.

6.2.2   Preliminary Assessment

An instructive approach to the analysis of a normal binary's rv data, as classically set out by R. Lehmann-Filhés, treats the parametrization as a direct problem. This starts by considering the maximum and minimum on a curve such as those shown in Fig 6.3-4 and relating these to Eqn (6.7). These positions correspond to the extrema (+1 and –1) of the variable cosine term, i.e. when $\nu + \omega = 0$ or π. We fix attention on just one of the stars in

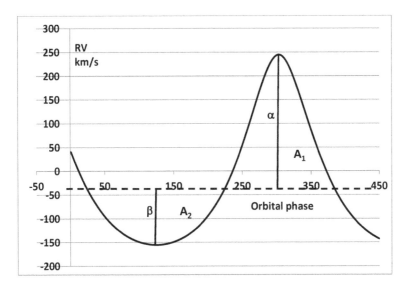

Figure 6.6 The form of the radial velocity curve gives indications on parameter values. The value of the γ-velocity, for example (dashed line), balances the area above the line with that below. The areas of the subsections A_1, A_2, could then be used in estimating other parameter values (see text).

the binary system, though data may be available for both components. For clarity, let us set the suffix 1 to the star in question, taken to be the system's primary.

Writing the velocity coefficient of the cosine terms in (6.7), i.e. $na\sin i/\sqrt{1-e^2}$, as K_1, we have

$$2K_1 = \alpha_1 + \beta_1 \; , \tag{6.10}$$

say, where α_1 and β_1 are the measured upper and lower excursions of the curve from the line $V_1 = V_\gamma$. The location of the γ-velocity on the rv axis is as yet unspecified, but if it were known the constant term in the square brackets on the right of (6.7) would be given as

$$e\cos\omega = \frac{\alpha_1 - \beta_1}{\alpha_1 + \beta_1} \; . \tag{6.11}$$

In this way, $K_1 = na_1\sin i/\sqrt{1-e^2}$ and $e\cos\omega$ would be directly determined.

V_γ can be found, in principle, by noting that the area of the curve ($\propto \int \dot{z}\,\mathrm{d}t$) above the line $V = V_\gamma$ should be the same as that below it, since the shift in z_1 from closest to most distant points (when $\dot{z}_1 = 0$) is the same whether travelled on the slow or fast part of the orbit, while the total excursion in z over the complete orbit is zero. Early versions of this method involved areal measurement, either by a planimeter or a numerical quadrature. For an initial parameter estimation, however, eye-location of the line $V = V_\gamma$ suffices.

The procedure can be extended to separate the parameters e and ω. Take the line $V = V_\gamma$ as the abscissal axis of the rv ($\dot{z}_1$) variation. It can then be shown that

$$e\sin\omega = \frac{2\sqrt{\alpha_1\beta_1}}{\alpha_1 + \beta_1}\left(\frac{A_2 - A_1}{A_2 + A_1}\right) \; , \tag{6.12}$$

where A_1 is the area below the rv curve, between the maximum and the point where the descending curve crosses the $\dot{z}_1$ axis, and A_2 is the complementary area under the ascending branch of the curve from the initial zero of $\dot{z}_1$ to the maximum — see Fig 6.6.[*]

---

[*]Measurements made on this sketch confirm its similarity to the second curve in Fig 6.3, with a moderate

With ω thus determined, the periastron, from which ν is defined, can be located, noting that $\nu + \omega = 0°$ at the rv maximum or $180°$ at the minimum. The value of ν at any particular time was specified by Eqn (1.8): an equation involving the mean anomaly M. The inverse of this relationship (6.9) yields the mean anomaly when the true anomaly is $-\omega$, $M(e, -\omega)$, say. But the phase of this maximum can be read from the rv curve. We thus connect the sought epoch of periastron passage t_0 to the assigned reference epoch HJD_0, used in phasing the rv curve, by

$$t_0 = \mathrm{HJD}_0 - PM(e, -\omega)/2\pi \ . \tag{6.13}$$

Since K_1 is defined as $2\pi a_1 \sin i/(\sqrt{1-e^2}P)$, with P and e known, $a_1 \sin i$ becomes determined. In a similar way, if the secondary's radial velocity variations can also be measured, we will have $K_2 = 2\pi a_2 \sin i/\sqrt{1-e^2}P$, so that the mean orbital separation $a = a_1 + a_2$ appears in this formula obtained by adding $K_1 + K_2$. This is of special interest, since Kepler's third law relates a to the sum of the masses $m_1 + m_2$ and the period P, while the ratio of the masses m_2/m_1 is K_1/K_2, thus suggesting a way to evaluate the masses of the two stars. However, the factor $\sin i$ in the definition of the K values cannot be eliminated from the kind of data and methods considered so far.

6.2.3   Stellar Masses

In solar units (solar mass, astronomical unit of distance, and year of time), Kepler's third law assumes the simple form

$$a^3 = (M_1 + M_2)P^2 \ . \tag{6.14}$$

In the same units, the mean velocity in the relative orbit is

$$na = 2\pi a/P = (K_1 + K_2)\sqrt{1-e^2}/\sin i \ . \tag{6.15}$$

Velocity amplitudes are usually given in km s^{-1}. To convert these to AU y^{-1} they should be divided by the appropriate distance/time ratio, i.e. 4.7404 km s^{-1}. Combining (6.14) and (6.15), we can then write

$$M_1 + M_2 = CP(K_1 + K_2)^3(1 - e^2)^{3/2}/\sin^3 i \tag{6.16}$$

where C incorporates the appropriate unit conversions for P (usually given in days) and the K-values, as well as the factor $8\pi^3$ in the denominator.[*]

If only one component of the binary's spectrum can be identified, the amplitude of the single radial velocity curve can still be used with the observed period and eccentricity values to construct a *mass function*. Although this will not furnish both masses directly, it can give insights into their likely values. We thus rewrite (6.16) as

$$M_1 + M_2 = CPK_1^3 \left(\frac{M_2 + M_1}{M_2}\right)^3 (1 - e^2)^{3/2}/\sin^3 i \tag{6.17}$$

Bringing the unknowns over to the left side, we will have the mass function $f(M)$ as

$$f(M) = \frac{M_2^3 \sin^3 i}{(M_1 + M_2)^2} = CPK_1^3(1 - e^2)^{3/2} \ . \tag{6.18}$$

---

eccentricity ($\sim$0.4) and low periastron longitude ($\omega \sim -20°$).

[*]The result is often cited as 1.038×10^{-7} in older literature. Recent data yields the number $C = 1.036149 \times 10^{-7}$.

This mass function has had particular relevance for the classical Algol group of binary systems, where only the primary component's spectrum readily stands out. Since the 1980s, however, improved spectrographic methods have allowed direct identifications of the secondary star's features in an increasing number of Algols. In the case of the R CMa-type Algols, where there are unusually low mass components, the point takes on an added significance that is discussed in Chapter 10. This is also becoming increasingly relevant to the growing data collections of exoplanet spectroscopy. A connected point, relating to the evolution of both close binary systems and planetary systems, concerns the special role that bodies of low mass can play in angular momentum redistribution.

A similar calculation works with the light travel time effect (encountered in 5.4.4), when a close eclipsing pair may have a distant companion, whose mass relative to that of the close binary is q, with which it shares a wide orbit around a common centre of gravity. In that case, the amplitude of the periodic timing differences for the eclipses of the close pair τ_{lt}, say, come from the varying distances projected in the line of sight. The binary orbit axis would be inclined to this line at an angle i, say. The corresponding form for the orbit of a wide binary of total mass M ($\odot$) and with period P becomes

$$\tau_{lt} = P^{2/3} M_1^{1/3} \left(q/(1+q)^{2/3} \right) \sin i \sqrt{1 - e^2 \cos^2 \omega}/c \ . \tag{6.19}$$

If working in solar units, the constant c becomes the number of AU travelled at the speed of light in a year $= 6.32421 \times 10^4$. The time unit generally used for both τ and P would be days, however, in which case the appropriate denominator becomes 8.84723×10^3.

6.3   Rotational Broadening

The introduction to this chapter referred to the changing apparent wavelengths of spectral lines as the basis for gathering rv information, but neglected the available information in the line about the circumstances of its formation. In what follows we develop a simple approach to this topic, with a mind to rotation-related effects in binary stars.

In Figure 6.7, a star's axis OP of a uniform rotation with angular velocity ω is set in the y, z plane of a 3-dimensional Cartesian frame, such that the z-axis coincides with the line of sight. We scale the axes to the mean radius $\bar{R}$ of the star. The mean equatorial speed of rotation of the star we set to be V_{rot}, which projects into $V_{\text{rot}} \sin i$ in the line of sight at the points where the stellar disk intersects the x-axis.

The line's intensity $I(\lambda)$ relative to the continuum level, that, for convenience, we normalize to unity, will depend on the surface distribution of light from disk segments at varying x through the line's Doppler spreading. This is determined by the model's unique dependence of the projected velocity $V_{\text{rot}} \sin i$ on x. The practical detection of this rotational effect requires that $V_{\text{rot}} \sin i$ be great enough to spread the line over a wavelength range significantly greater than the inherent broadening of the line, whether that is due to kinetic effects in the medium in which it was formed or optical effects in the spectrograph that limit its resolution, yet not so wide that shallowness prevents a clear identification of the line against the continuum.

If rotational broadening can be thus separated out, the photosphere's light distribution about the y-axis, will be reflected in the intensity distribution in the broadened spectral line centred at wavelength λ_0. The relative intensity in the Doppler-shifted wavelength region λ to $\lambda + \delta\lambda$ is associated with a strip of the star's photosphere having a radial velocity v to $v + \delta v$, say, relative to the mean wavelength λ_0, where (6.1), to a sufficient approximation, can be written as

$$\frac{\lambda - \lambda_0}{\lambda_0} = \frac{v}{c} \ . \tag{6.20}$$

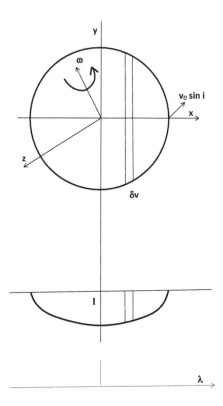

Figure 6.7   A uniformly rotating stellar model has angular velocity ω about the axis OP inclined at angle i to the line of sight. A strip of the projected disk, characterized by its radial velocity, corresponds to a particular displacement of the broadened line profile $I(\lambda)$, as shown.

The radial velocity v, at the point x, y, z is given by the z component of the vector product of the angular velocity $\boldsymbol{\omega}$ and radius vector $\boldsymbol{R}$, i.e.

$$v = (\boldsymbol{\omega} \times \boldsymbol{R})_z, \tag{6.21}$$

where we can expect that $\boldsymbol{\omega} = \omega_0 f(\zeta^2)\hat{\boldsymbol{\zeta}}$ in general, $\hat{\boldsymbol{\zeta}}$ representing the direction of the polar axis OP in the y, z plane. A convenient, and generally adopted, first approximation is to regard the function $f(\zeta^2)$ as a constant. In a similar way, we can scale units so that $x^2 + y^2 + z^2 = 1$ provides a usable approximation to a photospheric surface of mean radius $\bar{R}$. On this basis we approach the main effect of rotation on the line profile. We then have

$$\frac{\lambda - \lambda_0}{\lambda_0} = \frac{x\omega \sin i}{c}, \tag{6.22}$$

where the velocity of light c is now in units of the stellar mean radius $\bar{R}$ per second. Strips of the surface where $v = \text{const.}$ correspond to the parallel strips $x = \text{const.}$

The dependence of I on λ can be converted, through Eqn (6.22), into one on x, so that we write, for the relative intensity of the line at some typical value x' along the profile, the convolution

$$I(x') = 2 \int_{-1}^{1} \int_{0}^{\sqrt{1-x^2}} F_x(y)dy \; r(x' - x)dx, \tag{6.23}$$

where $r(x' - x)$ expresses the remainder (i.e. emergent and assumed symmetric) component of an appropriately scaled flux F_x, at the argument $(x' - x)$, associated with the inherent line formation process. The equivalent width w of this feature is

$$w = \int_{-\infty}^{\infty} (1 - I(x')) \, \mathrm{d}x', \tag{6.24}$$

or more fully

$$w = \frac{\int_{-\infty}^{\infty} \int_{-1}^{1} \int_{0}^{\sqrt{1-x^2}} F(x, y) \, \mathrm{d}y \, [1 - r(x' - x)] \, \mathrm{d}x \, \mathrm{d}x'}{\int_{-1}^{1} \int_{0}^{\sqrt{1-x^2}} F(x, y) \, \mathrm{d}y \, \mathrm{d}x}. \tag{6.25}$$

As indicated above, the separating out of the rotational broadening implies that its Doppler shift predominates in the observed displacement of contributions to the line across the relevant wavelength range. This broadening is appreciably greater than the intrinsic broadening given by the function $r(x')$. It is from this predominating effect, associated with bulk motion of the source material, that the sought information is resolved.

The discussion is simplified if the width of the broadened line where it joins the continuum, $\Delta\lambda_c \gtrsim 2w$. We can then assume that the feature is *optically thin*, i.e. the flux at a particular wavelength from a given area of surface is in direct proportion to the local surface area projected in the line of sight. For the He I lines in fast rotating early type stars equivalent widths can be as great as 0.6 Å, so that line widths at base should be broader than ~ 1 Å, or projected equatorial rotation velocities of at least ~ 30 km s^{-1}. In practice, $\Delta\lambda_c$ may be several times greater than this lower limit, with projected rotational velocities of around 100 km s^{-1} in the well-inclined systems of typical target stars. This is an order of magnitude greater than thermal velocities of photospheric helium atoms, which compare with modern instrumental resolutions of $\sim$0.1Å . The central depths of rotationally broadened He I lines, taking into account binarity of the source, would then turn out to be only several percent of the continuum, calling for a high signal to noise ratio in the spectral image.

In these circumstances, the convolution in (6.23) of the broad and smoothly varying source function with the narrow transmission of the non-Doppler-shifted kernel tends to the product of the source with the delta-function $\delta(x')$, i.e. we can write $r(x') = 1 - d_0' \, \delta(x')$, where d_0' measures a line depth equivalent to that of the inherent absorption but scaled to the effective resolution width Δx_r, say.

We can now rewrite (6.23) for the line profile

$$I(x') = 2(1 - d_0') \int_{0}^{\sqrt{1-x'^2}} F_{x'}(y) \, \mathrm{d}y \tag{6.26}$$

The form of the function F can be spelled out following the standard normalized linear limb-darkening law prescription. In that case, we have a constant *coefficient of limb-darkening* (u), on the assumption that the localised brightness over the stellar disk declines linearly towards the limb in dependence on $u \cos\theta$, where θ is the angle between the line of sight and the local outward normal on the assumed spherical surface of the star (see Section 7.3.2). We then have

$$F = \frac{3F_0}{(3 - u)} \left[(1 - u) + uz \right], \tag{6.27}$$

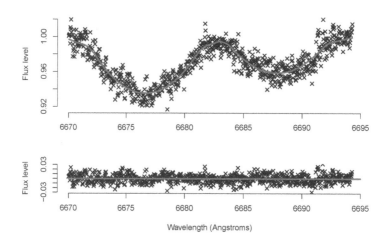

Figure 6.8 A profile model for the HeI $\lambda 6678$ lines of the short period close binary V Puppis at 1st elongation (primary approaching), modelled with a fitting function that convolves a narrow gaussian with the rotational broadening according to (6.23). The 'dish-shaped' component (6.28) is preponderant in the observed profile. Residuals from the fitting are shown beneath.

where $z = \sqrt{1 - x^2 - y^2}$ for the adopted approximation, and normalization of the continuum flux would imply $F_0 = 1/\pi$. For the rotation-broadened profile the result is

$$
\begin{aligned}
I(x) &= 1 - \frac{3d_0}{3 - u}\left((1 - u)\int_0^{\sqrt{1 - x^2}} \mathrm{d}y + u\int_0^{\sqrt{1 - x^2}} \sqrt{1 - x^2 - y^2}\,\mathrm{d}y\right) \\
&= 1 - \frac{3d_0}{3 - u}\left((1 - u)\sqrt{1 - x^2} + \frac{\pi}{4}u(1 - x^2)\right) \quad (-1 < x < 1),
\end{aligned} \tag{6.28}
$$

i.e. the conventional 'dish-shaped' profile of a line broadened by rotation. Applications of this approach to real spectral line effects are shown in Figs 6.8 & 9.

In the case of an undarkened disk ($u = 0$), (6.28) specifies an elliptical profile whose co-eccentricity is d_0, i.e. the shape of the star is linearly scaled to the profile's central relative depth. This can be regarded as an adjustable unknown in a profile modelling problem (cf. Figs 6.8-9). The area of the original kernel $d_0'\,\delta x$ does not transform linearly into that of the semi-ellipse $\pi d_0/2$, because the narrow width Δx_r of the resolved line in the non-rotating limit entails that the feature would not then be optically thin according to the foregoing discussion. For the fully darkened case, the limb transforms into a parabolic shape, from which the scaling can be determined by dividing by $3\pi d_0\sqrt{1 - x^2}/8$. If the limb darkening coefficient were known, an appropriately weighted divisor could recover the stellar perimeter's departure from circularity from the line profile.

While the separate forms of the functions F and r in (6.23) – the latter usually a Gaussian – can be spelled out explicitly, this is not necessarily true for the convolution integrals that replace $\sqrt{1-x^2}$ and $(1-x^2)$ in the more general case of (6.28). But these integrals can be simply derived by a direct numerical quadrature, with the photospheric disk divided into m strips parallel to the y-axis, where m compares with the ratio of rotational over instrumental broadening scales. For results such as those in Figs 6.8-9, m was typically ~ 20.

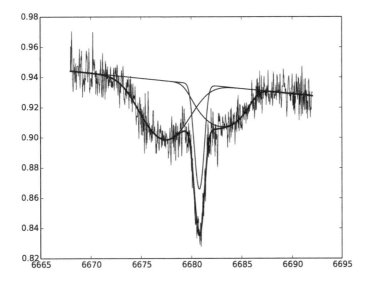

Figure 6.9 Dish-shaped convolved with gaussian components are here combined to model the radial velocity shifts of the blended HeI $\lambda6678$ feature in the triple system V454 Car (after Butland et al., 2019).

6.4 The Effect of Eclipses

Quite early in the spectroscopic studies of close binary systems it was discovered that there is a definite eclipse-related effect in the radial velocity curve. In a comparable way to the mapping of the source through the rotational effect in line profiles, there is a mapping of the photometric eclipses by a Z-shaped variation in the quasi-sinusoidal trend of the radial velocity curve. An example is shown in Fig 6.10, that stimulated its producer, Dr R M Petrie of the Dominion Astrophysical Observatory in Victoria, B.C., Canada, to produce a theoretical explanation. This explanation, and the next subsection dealing with the effects of proximity of the components, follow a parallel course to their counterparts in the photometric discussion, pursued more fully in the next chapter. We will restrict ourselves to preliminaries here.

An important principle, relating to this discussion, follows from the foregoing idea of surfaces of constant radial velocity forming strips parallel to the projected rotation axis. This arises when we regard the effect of eclipses on the observed radial velocity as due to the removal of part of the visible surface that would otherwise contribute to the net Doppler shift.

The situation shown in Fig 6.11 has the instantaneous line of centres projected on to the sky forming the x-axis, with the z-axis along the line of sight. The orbital movement of the eclipsing star relative to the eclipsed one is indicated by the dashed curve in the plane perpendicular to the rotation axis, that projects onto the sky as y'. The angle ψ, say, between the y and y' axes will vary with the orbital movement in time. The lines $x' = $ const. are given in the $x - y$ plane by the linear transformation

$$x' = x \cos \psi + y \sin \psi \ . \tag{6.29}$$

The related direction cosine, n_1 (see Eqn 1.12, and Fig 1.11), between the y and rotation axes at orbital phase ϕ and inclination i is given as $n_1 = -\sin\phi \, \sin i / \sqrt{1 - \cos^2\phi \, \sin^2 i}$.

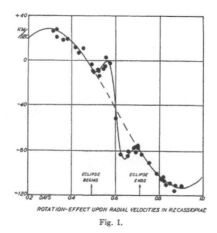

Fig. I.

Figure 6.10 The eclipse effect on the radial velocity curve of RZ Cas as shown in the original discussion of R. M. Petrie in *J. Royal Astron. Soc. Canada*, Vol. 32, p257, 1938.

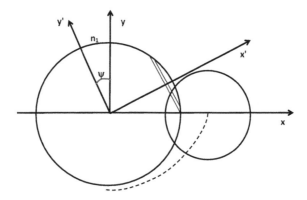

Figure 6.11 The x, y, z frame is centred on the centre of the eclipsed star. Orbital motion is indicated by the dashed line (with exaggerated curvature). The line of sight is perpendicular to and into the page. The x-axis is in the instantaneously projected direction of the eclipsing star's centre. The y'-axis is the projection in the sky plane of the eclipsed star's rotation axis. This is at an angle ψ to the y-axis. Strips of constant $V_{\rm rot}$ on the eclipsed star are parallel to the y'-axis as shown.

The effect of eclipses on the observed radial velocity is taken to be due to the effective removal of part of the visible surface that would contribute to the net Doppler shift. The radial velocity decrement Δv from the mean trend of the earthward hemisphere is regarded as the sum of each local radial velocity weighted element of surface $\delta a'$ that has been eclipsed out of what would otherwise be the full radial velocity contribution. This representative velocity calculation should take account of the varying amount of residual surface, so that, using the notation a' and A to represent areas of the (variable) eclipsed region and the star being eclipsed, respectively, we have:

$$\Delta v = \frac{\int_{a'} v_z \, \mathrm{d}a'}{\int_{A-a'} \mathrm{d}A} \qquad (6.30)$$

The areal integrals in (6.30) can be represented by a systematic notation of the form α_n^m that will be more precisely defined and developed in the next chapter.

Using (6.29), and the eclipse integrals α_n^m, and noting that $\cos\psi\sin i = -n_1$, we can express the eclipse effect* in Fig 6.10 in terms of the α_n^m-integrals as

$$\frac{\Delta v}{V_{\rm rot}} = \frac{-n_1[(1-u)\alpha_0^1 + u\alpha_1^1]}{(1-u)(1-\alpha_0^0) + u(\frac{2}{3} - \alpha_0^0)} \ , \tag{6.31}$$

where u is again the coefficient of limb-darkening. The velocity shift Δv – the numerator in the *rotation factor* given by (6.31) that gives the scale of the eclipse effect in terms of the equatorial rotation velocity $V_{\rm rot}$ – would generally be added to the net radial velocity variation during eclipse phases of the star in question.

Interestingly, the notation of (6.31) can be generalized to an unaligned rotation axis displaced by obliquity and precession angles ϵ, β to the orbital plane (cf. Fig 1.12) if we resolve the angular velocity vector $\boldsymbol{\omega}$, of magnitude ω_1 into components $\omega_\xi = \omega_1 \sin\epsilon \sin\beta$, $\omega_\eta = \omega_1 \sin\epsilon \cos\beta$, $\omega_\zeta = \omega_1 \cos\epsilon$. The product $V_{\rm rot} n_1 = R_1\omega_1 n_1$ in (6.31), rearranged as an expression for Δv, then becomes replaced by $R_1(m_1\omega_\eta + n_1\omega_\zeta)$. On this basis, the result of the eclipse of the rotating star may be interpreted as the sum of the effects of the three components ω_ξ, ω_η, and ω_ζ. The third of these corresponds to the normal aligned case and it produces the main Z-shaped variation with central zero. The second makes for an upward or downward shift of the 'Z' depending on the sign of $\cos i$, resulting in an asymmetric net effect. The third component ω_ζ produces no shift of the rv curve, but it reduces the relative scale of the other two. This modified version of (6.31) is added to the rv fitting function for the uneclipsed phases so as to allow optimal parametrization according to the procedure expressed by (5.22). This generalized Rossiter effect will then, in principle, include ϵ and β, so that a direction of the rotation axis corresponding to the observed rv-eclipse can be estimated.

This conventional explanation of the Rossiter effect adopts that an eclipsing body disturbs a spectral line through a convolution of the local Doppler shift with other processes, particularly instrumental broadening. The measured movement of the spectral feature is then that of its 'light centre'. This situation is essentially different to that considered in (6.28), where rotational motion dominates the profile and can be effectively separated from the inherent broadening. If, in a very high resolution system, a line profile broadened by a strong rotation were to be examined as its stellar source was undergoing eclipse the line centre wavelength would not necessarily be disturbed. The eclipsing body would simply register as a feature on the generally weakening profile. As mentioned in Section 6.3, the broadened profile is a mapping of the net surface flux distribution with longitude. When eclipses occur, instead of the explicit integral for the profile given in (6.28) the contribution of a given strip to the quadrature depends on how much of the strip is eclipsed. The location of the eclipsing star's arc in relation to each strip is determined from the given set of parameters for the eclipsing binary model at a given phase. The quadrature then proceeds using these determined limits to add in the contribution from the uneclipsed part of each strip. The same reasoning applies to local inhomogeneities in the flux arising from effects other than eclipses, e.g. surface maculation.* The results of rv determinations in close binaries from their spectral data thus depend on the processing method used to determine wavelength shifts, that relate also to the available resolution.

6.5 Proximity Effects

The idea expressed in the transition from (6.30) to (6.31) is that an effect on the radial velocity integrated over a given area of the surface has a parallel in a corresponding

---

*Also known as the Schlesinger, or Rossiter, or Rossiter-McLoughlin effect.
*This is the basis of the 'Doppler imaging' technique (cf. Vogt & Penrod, 1983).

photospheric areal integration, and the transformation can be effected simply by using corresponding α-integrals with the x component (upper suffix) increased by 1. This allows rapid progress in dealing with the usually relatively small effects of surficial distortion on the apparent net radial velocity of a star in a close binary system. The parallel photometric effects are presented in detail in Section 7.3.7.

It can be shown that the tidal distortion introduces a shift in the centre of light from the centre of mass that is, at most, of order $qr_1^4 \times v_{\text{orb}}$, where q is the mass ratio and r_1 is the radius of the star in question as a fraction of the separation of the two stars. This would amount to typically $\sim 10^{-4} \times v_{\text{orb}}$, and so, until relatively recently, less than the accuracy of available radial velocity measurements of v_{orb}. By analogy with photometric proximity effects, a similar result could be expected for the effects of reflection in many cases, although not for the large cool secondaries of classical Algols. These systems, with their highly contrasting masses and temperatures yet comparable sizes, may shift the light centre from the body centre of gravity at elongation by a few percent. This will also depend on which particular lines are measured, as well as the light distribution over the pear-shaped surface.

In summary, proximity effects on radial velocity determination would, in general, be slight. But there are cases where line-shifting processes become physically complex and theory is still incomplete. These matters call for more detailed analysis in the very high accuracy context of present-day radial velocity data.

6.6   Bibliography

6.1 The basic idea of a hot, relatively dense, source of continuum radiation overlain by a cooler atmosphere that can selectively absorb energy to produce discrete spectral lines, with particular dependence on the atmospheric presence of corresponding elements was famously set out by G. Kirchhoff in the *Monatsberichte Berliner Akad.*, Okt. 1859, p783. This work, on its appearance, aroused controversy, particularly regarding the issue of priority in the recognition of the physical processes at play. The matter is well reviewed in J. Hearnshaw's *The Analysis of Starlight*, CUP, 1986, Chapter 3. C. Doppler's paper on the effect that bears his name was cited in Section 1.4. The relativistic correction, included in (6.1) was introduced by A. Einstein in the *Annalen der Physik*, Vol. 17, p891, 1905, accessible as *The Principle of Relativity*, Dover, NY, 1923, pp 35–65. The concept of rotational broadening of spectral lines can be traced to W. de W. Abney's note in the *Mon. Not. Royal Astron. Soc.*, Vol. 37, p278, 1877. A modern text comprehensively reviewing observational spectroscopy and its theoretical modelling is *The Observation and Analysis of Stellar Photospheres*, of D. F. Gray, CUP, 2005.

The *Ninth Catalogue of Spectroscopic Binary Orbits* of D. Pourbaix, A. A. Tokovinin, A. H. Batten, F. C. Fekel, W. I. Hartkopf, H. Levato, N. I. Morell, G. Torres, and S. Udry, published in *Astron. Astrophys.*, Vol. 424, p727, 2004; was the latest in a series of such catalogues gathered over the last several decades. The catalogue is maintained in a regularly updated online version at http://sb9.astro.ulb.ac.be/ (latest version, Pourbaix, 2020). Nowadays, data from such catalogues are generally accessed through the *Centre de Données Astronomiques de Strasbourg* (CDS) via their Vizier cataloguing service. In the present case, this would be found at https://vizier.u-strasbg.fr/viz-bin/VizieR?-source=B/sb9 .

6.2 An excellent review of the practicalities of spectroscopic data-collection and radial velocity determination appears in R. W. Hilditch's *Introduction to Close Binary Stars*, CUP, 2001, (Chapter 3). The important introduction of the cross-correlation principle to rv determination by R. F. Griffin in *Astrophys. J.*, Vol. !48, p465, 1967, (foreseen by P. B. Fellgett in *Optica Acta*, Vol. 2, p9, 1953), should be noted. Griffin, having introduced the technique, has gone on to produce over 260 papers of follow-up studies of spectroscopic binaries (for a recent example, see *The Observatory*, Vol. 139, p111, 2019). Griffin's rv spectrometer has inspired similar instruments worldwide. A well-known example is the CORAVEL Cassegrain spectrophotometer that compares the spectrum of a target object with an optimally selected mask in the focal surface. Apart from rvs, stellar rotation and metallicity can be derived simultaneously. CORAVEL was described by A. Baranne, M. Mayor & J. L.. Poncet in *Vistas Astron.*, Vol. 23, p279, 1979.

Later developments saw the adaptation of the cross-correlation principle to software applications, e.g. the VCROSS program of G. Hill *Publ. Dom. Astrophys. Obs. Victoria*, Vol. 16, p59, 1982. It seems intuitively clear that cross-correlation would work better for cooler type stars ($\sim$ F type and later), that are characterized by a large number of discrete absorption features. Early type stars show few such features in the optical range, but have more to offer in the UV. This gave rise to the collection of SB rv curves of O and B type stars, using UV data from the International Ultra-violet Explorer (IUE) satellite, by D. J. Stickland & C. Lloyd in *The Observatory* journal. A summary, concentrating on the systems that still have relatively few observations, was published in paper 31 of the series in *The Observatory* Vol., 121, p1, 2001.

Despite the great success of cross-correlation methods, care is still required in taking into account potential problems due to an unsuitable choice of masking or template spectrum, and a non-optimal choice of spectral range, for example with insufficient lines of the right type or intrusions from the telluric features from the Earth's atmosphere, as well as effects of line-blending, or local non-uniformities due to physical effects at the source such as maculation or circumstellar gaseous streams. It has become clear that some process of separating, or 'disentangling', the individual intrinsic spectra of the stars of interest from the disturbing effects of inter-component or extraneous blending is desirable in order to extract the required information from the data. Various approaches bearing on this problem have appeared, including the broadening function technique of S. M. Rucinski, *Astron. J.*, Vol. 104, p1968, 1992; singular value decomposition of the component spectra, as done by K. P. Simon & E. Sturm, *Astron. Astrophys.*, Vol. 281, p286, 1994; and the Fourier decomposition KOREL program of P. Hadrava, *Astron. Astrophys. Suppl. Ser.*, Vol. 114, p393, 1995. Of course, data of uniformly high quality are a pre-requisite to greater information yield, whatever procedure is used to construct rv curves.

Formulae for the basic corrections arising from the movements of the earth-based observer are given in W. M. Smart's *Text-Book on Spherical Astronomy*, CUP, 1977, Chapter 14. More detailed discussions are from P. Stumpff *Astron. Astrophys.*, Vol. 78, p229, 1979, and J. T. Wright & J. D. Eastman, *Publ. Astron. Soc., Pacific*, Vol. 126, p838, 2014. On-line calculations are provided by such resources as http://astroutils.astronomy.ohio-state.edu/exofast/barycorr.html. A number of alternative expansions in elliptic motion, including formula (6.9), can be found in D. Brouwer & G. M. Clemence's *Methods of Celestial Mechanics*, Academic Press, 1961, Chapter 2. Radial velocity measurement precision reaching to cm s^{-1} are discussed by C. H. Li et al. (9 authors) in *Nature*, 2008, Apr 3;452(7187):610-2.

The early rv-curve interpretation recapitulated in Section 6.2.1 was published by R. Lehmann-Filhés in the *Astron. Nachr.*, Vol. 136, p17, 1894. It was reviewed by H. D. Curtis in *Publ. Astron. Soc. Pacific*, Vol. 20, p133, 1908, and in R. G. Aitken's *The Binary Stars*, McGraw-Hill, 1935, Chapter 6, along with several other techniques. Lehmann-Filhés is also remembered for what has become known as the 'differential corrections' method. This allows improvements to be made from an initial parameter-set solution through writing out the the effect of small changes in the parameters on the basic equations of condition using a linear Taylor expansion and relating this to the corresponding residuals. The number of parameters involved is generally much less than the number of observations so the over-determinate set of equations of condition is dealt with by least squares minimization (Section 7.1.1). This approach appeared in the *Publ. Allegheny Obs.*, Vol. 1, p33, 1908, though a clear and accessible explanation is given in R. W. Hilditch's *Introduction to Close Binary Systems* (op. cit.), Chapter 3.

J. M. Barr first pointed out the effect that bears his name, and is shown in Fig 6.5, in 1908 in the *J. Royal Astron. Soc. Canada*, Vol. 2, p70. It is discussed in Chapter 8 of A. H. Batten's *Binary and Multiple Systems of Stars*, Pergamon Press, 1973, where it is related to the presence of circumstellar matter. For further details of the distribution of apparent orientations of spectroscopic binary orbits see A. H. Batten & M. W. Ovenden's paper in *Publ. Astron. Soc. Pacific*, Vol. 80, p85, 1968. The effect is also recalled in D. M. Popper's (1980) review of the parametrization of binary system data. The case of U Cep was examined in detail by R. H. Hardie in *Astrophys. J.*, Vol. 112, p542, 1950, where the 'spurious eccentricity' effect was accounted for in terms of shifts of the hydrogen line cores relative to the more regular motion of the line formation measured at 1/4 depth (see also O. Struve *The Observatory*, Vol. 71, p197, 1951).

The particular role that third bodies in inclined orbits could play in the evolution of binary systems, associated with the Kozai mechanism, even with low mass third bodies, was noted by P. P. Eggleton & L. Kisseleva-Eggleton in *Close Binaries in the 21st Century: New Opportunities and Challenges*, eds. A. Giménez, E. Guinan, P. Niarchos, & S. Rucinski, Springer, 2006, pp73-77.

A critical compilation of accurate stellar masses and radii for 95 detached binaries containing 190 stars (94 eclipsing systems, and α Cen) was published by G. Torres, J. Andersen, and A. Giménez in the *Astron. Astrophys. Rev.*, Vol. 18, pp67–126, 2010. The authors, using updated parameters, found a good consistency with data on host stars of transiting exoplanets.

References to the light travel time effect and Irwin's papers are given in Section 5.6. The subject of pulsars has expanded into a large branch of astrophysics since their first recognition in the 1960s. Several hundred radio pulsars are now known with a few dozen in binary systems usually made up of a condensed object accompanying the observed pulsing neutron star component. Their place in possible evolution scenarios of close binary systems is discussed in Chapter 11. Binary pulsars are reviewed R. H. Manchester's Ellery Lecture of 1993, published in *South. Stars*, Vol. 35, p194, 1994, and subsequently in Chapter 6 of A. Lyne & F. Graham-Smith's *Pulsar Astronomy*, CUP, 2012. The important Hulse-Taylor binary pulsar (PSR B1913+16), announced by R. A. Hulse & H. J. Taylor in *Bull. American Astron. Soc.*, Vol. 6, p453, 1974, led to the accurate determination of neutron star masses as well as paving the way to the development of gravitational wave astronomy. The modulation of the pulsar's signals due to binary effects can be effectively modelled with a ten parameter fitting function relating to the basic Keplerian model and three post-Keplerian effects (apsidal advance rate, gravitational redshift with time dilation, and orbital shrinkage due to gravitational wave emission).

6.3 The discussion of rotational broadening given in Section 6.3 is adapted from that of E. Budding & M. Zeilik in *Astrophys. Space Sci.*, Vol. 222, p181. Formula (6.23) is essentially similar to that derived by G. Shajn & O. Struve, in the *Mon. Not. Royal Astron. Soc.*, Vol. 89, p222, 1929; Eqn (10). The separation of the convolution form into gaussian and dish-shaped components and the consequent limiting forms of absorption line appearances was discussed by S-S. Huang & O. Struve in *Astrophys. J.*, Vol. 116, p463, 1953. The subject has been extensively reviewed by A. Slettebak, *IAU Symp. 111*, eds. D. S. Haynes, L. E. Pasinetti, P. A. G. Davies, p163, 1985; together with other approaches to the determination of stellar rotation values. Fig 6.8 comes from the article of E. Budding, T. Love, M. G. Blackford, T. Banks & M. D. Rhodes, *Mon. Not. Royal Astron. Soc.*, Vol. 502, p6032, 2021. An application of the dish-shaped profile to optimal modelling of spectral line displacements as presented by R. J. Butland et al. (7 authors) in *Mon. Not. Royal Astron. Soc.*, Vol. 482, p2644, 2019; is shown in Fig 6.9.

6.4 Fig 6.10 comes from R. M. Petrie's article in the *J. Royal Astron. Soc. Canada*, Vol. 32, p257, 1938. The essential form of Eqn (6.31) was given by the same author in the *Publ. Dom. Astrophys. Obs.*, Vol. 7, p133, 1938. Eqn (6.31) and its more generalized form referred to in the text appear as (2.8) and (2.10) in Chapter 5 of Kopal's (1959) book (op. cit.). Applications of the Rossiter effect to study the alignment of spin and orbit axes have become widespread in recent years in the context of exoplanet research, where non-alignment is not infrequently observed. A recent example is in M. D. Rhodes, Ç. Püsküllü, E. Budding & T. S. Banks, *Astrophys. Space Sci.* Vol. 365, p77, 2020. For an application in the general stellar context see S. Albrecht, S. Reffert, I. Snellen, A. Quirrenbach & D. S. Mitchell, *Astron. Astrophys.*, Vol. 474, p565, 2007.

The relationship of line-shaping effects to the measured shift representative of the mean motion of the source has been studied by D. J. A. Brown et al. (15 authors) in *Mon. Not. Royal Astron. Soc.*, Vol. 464, p810, 2017; see also T. Hirano et al. (7 authors), in *Astrophys. J.*, Vol. 742, p69, 2011; and G. Boué, M. Montalto, I. Boisse, M. Oshagh, N. C. Santos in *Astron. Astrophys.*, Vol. 550, A53, 2013. Doppler Imaging was introduced in the paper of S. S. Vogt & G. D. Penrod, *Publ. Astron. Soc. Pacific,*, Vol. 95, p565, 1983. A. P. Hatzes & M. Kürster *Astron. Astrophys.*, Vol. 346, p432, 1999, gave an example of its application in their study of the maculation on V824 Ara. The topic arises again in section 10.1.

6.5 Chapter 5 in Z. Kopal's *Close Binary Systems*, Chapman & Hall, 1959, is devoted to the effects of proximity on rv-curves. The analysis of the tidal effect, accounting for the shift between the centre of light and centre of mass by reference to the surface distortion of the stars, is consistent with the parallel treatment of the photometric effects in Chapter 7 of this book. A. H. Batten reviewed the subject in Chapter 5 of his PhD thesis (Univ. Manchester, 1958), dealt with in his paper in *Mon. Not. Royal Astron. Soc.*, Vol. 117, p521, 1957. Batten calculated the biasing effect that the shift of maximum light towards a surface region whose rotation would reduce the apparent scale of orbital radial velocity, and he found it to be significant for the determination of stellar masses at the $\sim$10% level, the high sensitivity of determined mass to the measured velocity amplitude being apparent from Eqn 6.17.

Determining the full effects of reflection, however, requires circumspection, both concerning the formation of the emergent spectrum in a given wavelength range for an illuminated photosphere, as well as how such a spectrum would affect the measurements of selected spectral line positions. A. H. Batten's later review in *Binary and Multiple Systems of Stars*, Pergamon Press, 1973, notes that the properties of the spectral range selected for measurement in practice differs from the bolometric case of the earlier studies. This had been pointed out by M. W. Ovenden in *Mon. Not. Royal Astron. Soc.*, Vol. 126, p77, 1963. Different lines measured for rv variation in spectra of the system 57 Cyg could lead to either a reduction or an enhancement of the apparent rv amplitudes over their true values in dependence on which part of the stellar surface their strengths would maximize. The follow-up study of Ovenden and Napier (1970) found that the different results from the data on 57 Cyg could not be satisfactorily explained by the reflection models adopted in previous treatments. They called for more detailed physical discussion of processes likely to be involved in the heated atmospheres of the component stars, including turbulence.

In fact, more recent observational work on close binary systems, such as those of Popper (1980) and Torres et al. (2010), referred to above, suggest that the level of agreement between modelling and empirical masses is significantly better than the $\sim$10% mentioned by Batten. The full combination of factors thus appears to place the role of the reflection effect on measured rvs to be not as great as originally thought.

7

Photometric Data Analysis

7.1 General Issues

Close binary stars show regular patterns of light variation that can often be explained in the same, or a similar, way to Goodricke's first hypothesis about Algol raised in Chapter 1. It is useful to have this picture in mind as a preliminary to modelling, but the issues now before us concern the optimal fitting of an appropriately formulated model to a set of observations (cf. Section 1.3). The two main issues in this are the physical basis of the modelling and the analytical procedure.

In connection with the former: the observations made by Goodricke, and many other subsequent observers of variable stars, were those of the apparent brightness of a given star in dependence on the time of observation. This remains in present-day stellar photometry, though the techniques of measurement for both the light-flux received and its timing have developed enormously over the last couple of centuries. Some of this development was reviewed in the first chapter, but it is clear that astronomical photometry is a large field in its own right. We shall confine attention mainly to this technology's outcomes for binary and related stellar systems.

With regard to procedure, some of this was encountered already, particularly in the previous two chapters. We develop this topic in what follows.[*] Our starting point is the data sample, i.e. we are given a collection $\{l_k\}$ of discrete observations in a multi-dimensional ($k = 1, ...\nu$) *data space*. These data have a probabilistic character, meaning that any particular observation at time t_i may have some inaccuracy of measurement. So with each reported datum t_i, $l_{k,i}$ there should be also an error assessment $\sigma_{k,i}$. The dimensionality of the data space is thus at least $2\nu + 1$ in general.

Modelling is effected through the use of *fitting functions* $f_k(\{a\}, t)$, that may depend on t through known formulae (e.g. trigonometric functions) involving a set $\{a\}$ of n model-dependent parameters a_j, $j = 1, ...n$. These functions might not have an explicit form, but represent the result of a programmed operation such as a numerical integration, for the given time t: the f formalism can still be applied. Such fitting functions could, in principle, be of a purely empirical nature, such as a Fourier series, or comparable kind of summed expansion.

---

[*]The procedures in question have been incorporated into the program WINFITTER, developed specifically for the kind of optimization problem discussed in this and other chapters. A reference is given in Section 7.6.

Alternatively, they may be derived from a physical model that relates to quantities like the sizes and temperatures of component objects. This second kind of parametrization is what usually pertains to physical research, but the first kind may be useful in understanding the extent of parametrization that a given data-set is able to deliver.

Minimization arises in connection with either approach: physical interpretations are thought more successful if they can account for facts with a minimum of hypothesis; analytical procedures are aimed at minimizing numerical differences between observed and fitted values of dependent variables at given times. While parallels may be drawn, these purposes are separate: a best-fitting function, mathematically, is not necessarily the same as a most-preferred physical model.

We concentrate on linearly separable and self-consistent physical problems. This allows us to drop the k suffix on our data-sets and fitting functions, and assign direct physical meanings to parameters. In a linear problem the ν separate data-sets can be regarded independently and their parameters optimized separately. If the model is self-consistent, the optimal values of a given parameter derived from separate data-sets will agree to within acceptable uncertainties.

From the practical point of view, data-sets involved in binary or multiple star analysis are measurements of changing light-flux levels; the comparable patterns of radial velocity variation as detected from spectroscopic observations; periodic changes in relative positions of stars, or other patterns of behaviour. In all cases, the departures of observations from model calculation can be for either theoretical or observational reasons: it is important to have a knowledge of the likely scale of all such departures when a particular model is used to account for a given data-set.

Optimization brings in the idea of best choice, i.e. there is a selection from alternative candidates based on some quality that can be measured. Let us suppose there is a quantifier of this 'goodness' of the modelling: a goodness-of-fit quantifier $\alpha(\{l\}, \{\sigma\}, f(\{a\}))$, say. Since real measurements contain errors, any particular determination of α is probably not exactly what it would be in the absence of such errors, so a selection process based on the values of α is likely to be somewhat incorrect. Here, the set $\{\sigma\}$, giving the scale of errors in the measurements, allows us to assess corresponding errors in the resulting parameters.

Let us take a particular evaluation of the quantifier α to be the i'th of a series of trials with different parameter values $\{a\}_i$. Since smallness of the difference between an observed and model-predicted set of values of $\{l\}$ reflects the goodness-of-fit using $f(\{a\})_i$, we can reasonably take *minimization* of α as equivalent to the sought optimization.

Consider now an 'estimator' $\beta(\alpha, s)$ that charts the behaviour of α in response to a sequence of steps of variation in the a's. β includes a procedure to determine a new parameter set $\{a\}_{(i+1)}$ aimed at reducing the numerical value of α. If the fitting can not be improved, i.e. the set $\{a\}_i$ is the best, and the β-procedure valid, then further application of β beyond the ith set does not make the differences between $\{a\}_{(i+1)}$ and $\{a\}_i$ significantly different from zero for a well-defined optimum. A convergent approach to the optimal set is then characterized by successive differences $\{a\}_i - \{a\}_{(i-1)} = \Delta\{a\}_i \to 0$, . A numerical (computer-based) optimization operation thus consists of the α-reducing sequence (β_i) making successive $\Delta\{a\}_i$ tend acceptably to zero: acceptably in both number of iterations and significance of the remaining differences after the operation.

In practice, an improvement sequence will terminate at $i = i_{\max}$ (the program's iteration number limit), or when $\Delta\{a\}_i < \epsilon$, where ϵ is a pre-selected small number. The latter condition serves as a deciding criterion, and if the differences $\Delta\{a\}_{i_{\max}}$ are still significant, when assessed with the aid of the error-scales σ, the calculation would be repeated for larger $i_{\max}$, or starting the sequence of $(\beta)_i$ from one that produces a lower α than the original starting set. The paired sequence $\{a\}_i : (\alpha_i)$ makes up a single track exploration of parameter : α hyperspace, showing the behaviour of the quantifier in response to parameter changes near the optimum. There could be many such tracks in a comprehensive optimization program.

7.1.1 Least Squares Principle

In order to bring out essential features of optimization we discuss a simplified linear case of parameter estimation using the least squares principle. Consider a set of m *equations of condition*

$$\mathbf{Y} = \mathbf{XA} + \epsilon \ , \tag{7.1}$$

where ϵ stands for the column of residuals formed by the differences between m measurements of a quantity $\mathbf{Y}$ and the matching linearized data $\mathbf{X.A}$, formed from the m by n model-dependent elements of the array $\mathbf{X}$ and their matrix product with the column vector $\mathbf{A}$ that lists the n sought parameters of the model.

Attention now fixes on the sum of the squares of the residuals formed by writing

$$\epsilon^T \epsilon = \Sigma \epsilon_i^2 \ . \tag{7.2}$$

Using (7.1), and noting that $\mathbf{Y}^T(\mathbf{XA}) = (\mathbf{XA})^T\mathbf{Y}$, (7.2) can be written as

$$
\begin{aligned}
(\mathbf{Y} - \mathbf{XA})^{\mathbf{T}}(\mathbf{Y} - \mathbf{XA}) &= \mathbf{Y^T Y} - \mathbf{Y^T}(\mathbf{XA}) - \\
&- (\mathbf{XA})^{\mathbf{T}}\mathbf{Y} + \mathbf{A^T X^T XA} \\
&= \mathbf{Y^T Y} - 2\mathbf{Y^T}(\mathbf{XA}) + \mathbf{A^T X^T XA} \ .
\end{aligned} \tag{7.3}
$$

The *least squares principle* now asserts that an optimum selection of parameters $\mathbf{A}$ is given when this quadratic form, corresponding to the sum of squared residuals $\Sigma \epsilon_i^2$, is a minimum.

By analogy, the simple quadratic $y = (x + b/2a)^2 + c/a$ takes its lowest value when the squared term forming the first member is zero, i.e. when $x = -b/2a$. This form differs from the general expression

$$y = ax^2 + bx + c \tag{7.4}$$

only by the constant divisor a and the positive constant $(b/2a)^2$. We thus find that the general form (7.4) has a stationary value when $x = -b/2a$, that is a minimum when a is positive.

Returning to the matrix form in (7.3), where we can regard the adjustable parameters $\mathbf{A}$ as corresponding to the variable x, and comparing the coefficients a, b, c with the corresponding data products, we find parallels between (7.3) and the basic quadratic form (7.4). So if the central $n \times n$ real symmetric product $\mathbf{X}^T\mathbf{X} = \mathbf{V}$, say, in the third term on the right of (7.3), consisted only of diagonal elements, $\alpha_1, \alpha_2, \alpha_3 \ldots$, say, these elements would be the positive coefficients of the parameters squared, i.e. A_1^2, A_2^2, A_3^2 etc. The right products $\mathbf{VA}$ would be an n-sized vector, made up of terms $\alpha_i A_i$, along with $-2\mathbf{X}^T\mathbf{Y}$ in the second term, $\beta_1, \beta_2, \beta_3 \ldots$, say, both multiplying the row array $\mathbf{A}^T$. The behaviour of the quadratic expression (7.3) in response to the parameters $\mathbf{A}$ thus depends on the second and third terms, and becomes stationary when $\mathbf{X^T Y} = \mathbf{X^T XA_s}$. In that case,

$$\mathbf{A_s} = (\mathbf{X^T X})^{-1}\mathbf{X^T Y} \ , \tag{7.5}$$

where $\mathbf{A}_s$ refers to the particular vector solving (7.5) with this diagonalized form of the matrix $(\mathbf{X}^T\mathbf{X})^{-1}$.

Formally, the matrix equation (7.5) also holds for the general non-diagonalized counterpart of $\mathbf{V}$, let us call it $\mathbf{U}$, formed by the operation $\mathbf{U} = \mathbf{QVQ}^{-1}$ where $\mathbf{Q}$ is the eigenvector matrix for $\mathbf{U}$. Let us left-multiply (7.5) by $\mathbf{P}^{-1} = \mathbf{Q}$, and right-multiply by $\mathbf{P}$ to find the general linear transform:

$$\mathbf{A} = \mathbf{U}^{-1}\mathbf{B} \ , \tag{7.6}$$

where $\mathbf{A} = \mathbf{P}^{-1} \mathbf{A}_s \mathbf{P}$ and $\mathbf{B} = \mathbf{P}^{-1} (\mathbf{X}^T \mathbf{Y}) \mathbf{P}$. The right side of (7.5) provides the $\Sigma \epsilon_i^2$-minimizing solution for $\mathbf{A}$, expressed from the original data as $(\mathbf{X}^T \mathbf{X})^{-1} \mathbf{X}^T \mathbf{Y}$ for the now general, non-diagonalized, case. The square matrix $\mathbf{U} = (\mathbf{X}^T \mathbf{X})$ is real and symmetric, though now free from the restriction of diagonal form, but it is implied that eigenvalues of $\mathbf{U}$ can be found that are all positive, in order that the solution vector $\mathbf{A}$ have real elements. The form of (7.5) recapitulates (5.22) , of which it is a linearized version.

Although light-curve fitting is a non-linear operation in general, parameter adjustment procedures in the vicinity of the optimum tend towards the linear least-squares problem and the assessment of the probable errors of an adopted parameter set builds closely on the formalism just sketched. In that context, the real symmetric matrix $(\mathbf{X}^T \mathbf{X})^{-1}$ takes the form of the 'determinacy Hessian', derived from numerical differentiation of the fitting function with respect to its adjustable parameters. A determinate problem, i.e. a modelling whose parameters can be assigned distinct real values, entails this matrix be positive definite.

In practice, if the number of parameters in a fitting operation is increased, the lowest eigenvalue of $\mathbf{U}$ tends to decrease its relative magnitude, as the possibility of an approximate linear correlation between members of the increased parameter-set becomes enhanced. This amplifies the elements of the inverse matrix $\mathbf{U}^{-1}$ that scales the errors with which the parameters $\mathbf{A}$ can be specified. Related to this point is the concept of an *information limit*, i.e. that there is a maximum number of modelling parameters that a data-set with specifiable observational error probabilities is able to yield.

7.1.2   Features of Optimization Programs

While $\Delta\{a\}_i < \epsilon$ is often adopted as a satisfactory result for obtaining a practically optimal parameter set $\{a\}$, it is not a sufficient condition. The following situations satisfying this condition can occur: (1) $\Delta\{a\}_i < \epsilon$ could happen, theoretically, for a maximum of α, although this would be unlikely to be mistaken for an optimum in practice, since lower values of α would be visible close to the located extremum (pessimum); (2) the result could can also happen for a 'local' rather than a 'global' minimum, i.e. there is more than one parameter set which seems to give a best selection, but this is within a smaller parameter domain than the entire one available to the parameters $\{a\}$; (3) the quantifier has a 'long valley', wherein successive changes of the a's, that are not necessarily small themselves, produce very small changes in the α's, without reaching a strict minimum. This latter condition is not uncommon in practical optimization problems, and will occur if there is a near correlation between 2 or more of the a's in some region of parameter space.

To illustrate this, suppose there is a positive correlation between a_1 and a_2. This means that a change in a_1 by a certain positive amount δa_1 is equivalent to a change in a_2 by some negative proportion, i.e. $\delta a_2 = -r\delta a_1$. In that case, the effect of $(+)\delta a_1$, if accompanied by $(+)\delta a_2/r$, will be wiped out. An anticorrelation between a_1 and a_2 works the opposite way: this time the effect of a small change of a_1 has the same effect as a small positive change $r\delta a_2$, say. Such correlations may appear anywhere in the effects of parameter variation, however, the same linear correlations need not continue along the whole $\{a_i\} : (\alpha_i)$ track.

We can regard this fitting exercise as a transformation of the information in data space to that in parameter space, and optimization viewed as maximising the number of potentially specifiable parameters. Following the discussion of the previous subsection, a high precision of parameter specification entails that more parameters could be assigned in a more detailed modelling, in general. Also from the least squares principle, maximal parameter resolution is equivalent to minimization of the χ^2 variate, formed in dependence on the n parameters

a_j, $j = 1, ...n$ by

$$\chi^2(a_j) = \sum_{i=1}^{N} [l_o(t_i) - l_c(a_j, t_i)]^2 / \sigma_i^2, \tag{7.7}$$

where σ_i^2 denotes the variance of the (assumed evenly distributed) data at t_i.

The full number of modelling parameters, n, depends on the adopted form of the fitting function. To be consistent with the physics of the model, however, parameters are generally not free to take any arbitrary value, and are sometimes constrained so that one parameter is a definite function of some other(s). Some of the parameters are given their values from separate or previous evidence, or literature sources, than those to be determined from the given data sample. These adopted values may be referred to, in Bayesian terminology, as 'priors', with an implication that they remain unchanged in the best-fitting experiment.

Ideally, the information transformation would be from closed contours of constant probability in data space to closed ellipsoidal contours in an m-dimensional parameter space, having a definite single central point with co-ordinates $a_{j,\text{opt}}$, $j = 1, ...n$. However, as m increases from unity, a certain number will be reached $m_{\max}$, beyond which this no longer holds good, and there is no longer a unique optimizing parameter set. As m is increased, uncertainty factors Δa_j, related to the expected errors in the determinations of the a_j, increase until, when m exceeds $m_{\max}$, such factors, for one or more a_j, pass through a singularity, precluding a single point $a_{j,\text{opt}}$ solution. A finite range of values for the optimal a_j combination, within their permissible ranges, are then equivalent in their ability to minimize χ^2 to within the numerical limits to which it can be specified. In simple terms, a finite data sample can only transform into a finite amount of model information.

Although alternative distinct but separated χ^2 minima can exist in the parameter space, for a subset of $\{a\}$ less than $m_{\max}$, a single best set of parameters is often located for a photometric data problem. This concurs with the classical presentation of a 'solution to the light curve': essentially as a preliminary summary, generally curtailed in the expectation of an ongoing improved data acquisition. There are various techniques to construct the solution sequence $\beta(\alpha, s) \to \alpha_{\min}$, e.g. the genetic algorithm, self-organizing migration, particle swarming and simulated annealing methods. For light curve analysis the Levenberg-Marquardt technique has been found effective and efficient.

A more comprehensive approach looks at the growing quantity, quality and scope of data series influencing modelling outcomes. As these outcomes increase, the range and details of models become correspondingly developed, together with the march of advancing theory. With corresponding improvements to analysis procedures, their pursuit takes us further along Russell's 'royal road' to information on the stars.

7.2 Eclipsing Binary Systems

Eclipsing binaries are a subset of the close binary systems met with in Chapter 1, the separation of the stars being typically an order of magnitude greater than a component's mean radius. The two stars are sometimes separated far enough that this ratio becomes more than a factor 10 (e.g. TV Nor, FO Ori or LW Pup), though the likelihood of observing eclipses must fall with increasing displacement of the pair relative to their sizes.

If the orbital motion of a gravitationally bound pair lies in a plane inclined at a sufficiently small angle to the line of sight (cf. Fig 1.10) then eclipses will occur, producing variations in the light received at the Earth. Such an eclipsing binary gives rise to a succession of alternating photometric minima: 'transits' (the smaller object in front) followed by 'occultations' (the opposite). Eclipses are described as complete when the outline of the smaller disk projects entirely within that of the greater one: a total eclipse if an occultation, or an annular one if a transit. The word partial is used for incomplete eclipses, or it may be applied to the corresponding phases of a light curve.

The radii of the stars in historically recognized, and by now well-known, eclipsing binaries would seldom be extremely dissimilar, because if one star were very much larger than the other, that body would tend to dominate the system's light. In such a situation, both the transits of the smaller body across the larger one, and the occultations, half an orbit later, would have relatively low effects unless the small star were relatively very bright per unit area of its surface.

But this situation has changed dramatically over more recent years with the advent of significantly more accurate detection systems and the greatly advancing interest in, and knowledge of, exoplanets, a large proportion of which have been found photometrically. Consider, for example, the Sun-Earth system, where the planet has a radius of only 1% that of the star: the transit signal is then a 0.01% effect. If we assume, for discussion's sake, that the photon registration has Poissonian properties in its counting, then even a count sample of a steady source of 10^8, in a given integration time, would still show significant noise at the 0.01% level: to obtain a signal/noise ratio of just 10 would require 100 times this signal count number, implying large light flux collectors, very sensitive receivers, and highly stable monitoring conditions. Such detection capabilities would have seemed beyond reach in astronomical photometry through much of the twentieth century, as recalled in the first chapter. But the urge to find Earth-like planets in a cosmic setting has caused such once highly ambitious aims to seem almost routine. Of course, if counting can be collected over long enough time intervals, or the phenomenon in question is repetitive so that many instances of the same effect can be superposed, then the sought effects become more readily manifest.

From what has already been said, it will be realized that, in the alternating cycle of eclipses for a suitably oriented pair of orbitally bound stars, the deeper light minimum is caused by the eclipse of the star having greater brightness per unit area, associated with a higher temperature photosphere. This is normally called the primary minimum, and the star then eclipsed the primary star. Naturally, the other minimum is termed the secondary and corresponds to the eclipse of the cooler secondary star. Just which star is called the primary can be occasionally confusing. In spectroscopic contexts, the primary would normally be regarded as the component putting out more light overall, or perhaps the more massive one — not necessarily the same as the star of greater brightness per unit area — so spectroscopic terminology can differ from that of photometry, at least initially.

The effects of orbital eccentricity may show up in light curves, particularly those of young or unevolved binaries. For the most part, however, many well-studied eclipsing binaries have orbits that have been effectively circularized as a result of friction-like effects in the dynamical interactions of close pairs. Hence, circular phase-symmetric orbits are a practical and convenient simplification for many analyses. A few elementary deductions then follow: (1) the durations of both eclipses are the same; (2) the same photospheric areas are in eclipse at corresponding phases through either light curve minimum; and so (3) the ratio of light lost at such phases yields a corresponding ratio of surface flux averaged over these eclipsed areas, allowing the temperature-dependent surface brightness ratio of the two photospheres to be derived.

When modelling the stellar evolution of a binary system, 'primary' usually refers to the originally more massive star, though it need not remain so. There is compelling evidence that stars lose appreciable mass during the course of their evolution, at a rate that depends on the star's initial mass. In addition to this general effect, mass may be transferred between the components in *interactive* binary evolution. The star originally identified as primary may then become the secondary. Transfer of mass may indeed happen more than once and in more than one direction.

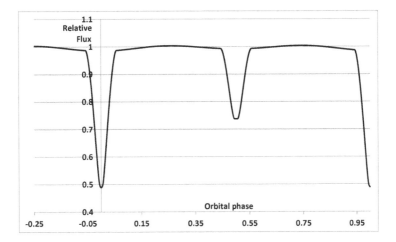

Figure 7.1   Model light curve of EAD type. The configuration shown consists of two Main Sequence type stars revolving in a circular orbit, where the hotter and larger component undergoes an annular eclipse at the primary minimum and the smaller, cooler star is totally eclipsed at the flat-bottomed and less deep secondary minimum. The circular orbit entails that the minima are of equal width. Proximity effects are included in the model, so there is a slight rounding of the light levels between the minima. These would be regarded as constant in a zeroth order (spherical) representation of the system.

The advantage of combinatorial analysis — photometric, spectroscopic and other data types — to maximize parametrization have been noted. Light curves provide a reasonably self-contained branch of the subject, however. Historically, three basic categories in the empirical classification of eclipsing binary light curves were recognized: the 'Algol', 'β Lyrae', and 'W Ursa Majoris' types (Figs 7.1-5). These prototype names turn out to be rather inappropriate, and the short letter designations: EA, EB, EW (etc.) are preferable. This simple classification scheme reflects its early origins in the use of fairly coarse magnitude discriminants. Its usage may decline with the advance of photometric precision, though it still makes for an informative introduction.

EA type light curves have roughly constant light levels – to within $\lesssim 0.1$ mag ($\sim 10\%$ of the mean flux level) – outside of the eclipses that are thus clearly marked; at least the primary one. This light curve type has been subdivided into EAD and EAS groups. The EAD light curves, for orbital periods of a few days, usually result from the mutual eclipses of a pair of normal Main Sequence stars that are of comparable size and separated from each other by at least ~ 5 times their mean stellar radius. Both minima can be discerned, and their depths have a simple relationship to each other if the spectral types, and therefore approximate surface effective temperatures, are known (see Fig 7.1). The primary minimum depth can be shown not to exceed close to one magnitude at optical wavelengths, as a consequence of the stars' Main-Sequence-like properties.

The EAS type light curve is exemplified by Algol ($= \beta$ Per) itself, or the similar system δ Lib shown in Fig 7.2, but perhaps more compellingly by such completely eclipsing systems as U Cep or U Sge. The primary eclipse is deep because of the relatively large size, but cool, photosphere of the 'subgiant' secondary. The relatively low mass of this type of star can be inferred from the low scale of proximity-induced tidal distortion of the bright primary star, so that the light curve shows little of the inter-minimum curvature associated with such effects, at least not at optical wavelengths. They become more noticeable as the more tidally distorted secondary accounts for an increased share of the light variation in the infra-red.

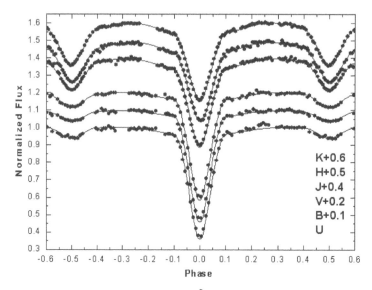

Figure 7.2 Light curves of the classical Algol system δ Lib: one of the brightest of the EAS type systems. The wavelengths range form the near infra-red (top) to ultra violet (bottom), and the increasing relative brightness of the secondary at longer wavelengths is inferred from the succession of relative eclipse depths. The continuous curves show the light variations calculated from the system model.

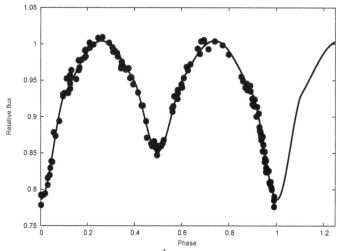

Figure 7.3 Normalized V-band photometry of μ^1 Sco (full circles), and model (continuous curve). The light curve is of the kind generally described as β Lyr type. This is often associated with massive young and unevolved pairs, but in the case of μ^1 Sco, after taking into account the derived mass ratio, the secondary is found to be significantly larger than a Main Sequence star of its mass, and so the physical condition resembles that of the classical Algols.

An EAS light curve is thus distinguished by a deep primary minimum, generally deeper than 1 mag ($\gtrsim 60\%$ of the flux), and the secondary minimum being so shallow as to stand apart from the trend of secondary to primary depth ratios predictable for Main Sequence pairs. Note that it is not *necessary* for an Algol-like system to show an EAS type light curve, even if an EAS light curve is always associated with an Algol-like pair of stars.

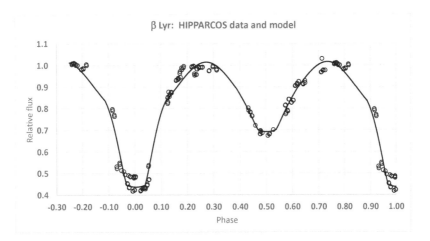

Figure 7.4 Light curve of β Lyr: an atypical member of the EB light-curve type. The photometry, carried out by the HIPPARCOS satellite, shows considerable additional variability about the basic eclipsing binary form.

The EB type light curve shows a continuous pattern of variability including the rounded inter-minimum regions going with more pronounced tidal interactions. It is often associated with earlier type binaries and relative separations closer than those of the EAD type. There are, however, some early type binaries having EB type light curves with absolute separations that are large compared to the radius of the Sun. β Lyrae itself, with its $\sim$12.9 d period and B8II spectral type, is an example (Fig 7.4). These latter systems, whose evolved character can be inferred from their relatively long, and often slowly varying, orbital periods, are physically different from the simpler situation of a pair of fairly close Main Sequence binaries. Such a grouping of massive stars associated with strong interactive effects has sometimes been associated with a separate 'Serpentid' class, after the prototype W Serpentis.

With the EW type light curves the proximity-induced inter-minimum rounding reaches such a scale as to merge into the eclipse minima that are no longer clearly distinguished. EW binaries show more or less equal depths of primary and secondary minima, implying similar mean surface temperatures. Hence, a light curve showing very pronounced out-of-eclipse rounding, but with minima depths differing by more than one or two tenths of a magnitude, would probably be assigned an EB classification.

Though there might be some overlap between EB and EW binaries at shorter periods, the EBs are preferentially found among systems of somewhat longer period and earlier type than the EWs. The relatively frequently observed EWs sharing similar properties with YY Eri (Fig 7.5), or W UMa, all appear to have periods appreciably less than one day in duration. They are predominantly made up of late type pairs, by far the largest proportion, by spectral class, being G type primaries and secondaries.

Thus, while EAD and normal (short-period) EB binaries can often be interpreted in terms of unevolved, Main-Sequence-like pairs, among the EAS, long-period EB and EW binaries we encounter a range of properties that have been considered puzzling, at least for a time. Perhaps the most well-known is the Algol paradox of the EAS binaries, that can be readily seen in the photometry. The deep primary minimum would imply a more evolutionarily advanced, and therefore, one could suppose, more massive companion, in contradiction to the low mass ratio indicated by the absence of tidal proximity effects. Another basic problem about at least some Algol systems concerns the difficulty which standard interactive binary evolution theory has in getting Algols into their observed 'gentle'

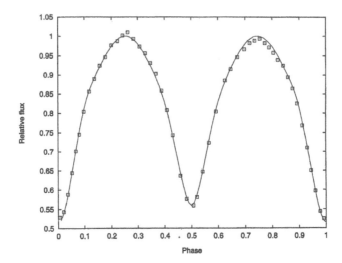

Figure 7.5 V light curve of YY Eri: of typical EW type, showing a smooth overall variation with closely similar eclipse depths. The difference in heights of the maxima is often referred to as the ' O'Connell effect'. The symmetry of the model's light-curve (continuous) shows that effects are present in the system that this modelling does not take into account.

configurations, without drastic run-away mass-loss leading to the spiralling in and merger of components. The critical role of angular momentum loss in the stellar evolution, hardly taken into account in early papers, features in more recent discussion.

A problem posed by the longer period EB stars or Serpentids relates to the lack of direct evidence for the large mass primaries that would go with the spectroscopic data if taken at face value. The interpretation is complicated by indications of a relatively large scale of mass transfer between the components. A favoured explanation is that the presently more massive primary stars are buried in a 'thick disk'.

EW binaries present their own set of problems. These are associated with the stars being so close as to come into contact with each other's surfaces, or indeed share a common photospheric envelope: an *overcontact* condition. A star in such a situation would have its outer boundary conditions fixed by constraints imposed by the contact configuration. The Roche lobe properties, studied in Chapter 3, have been used to determine such constraints. Although the Roche formulae are not strictly applicable to real stars with massive envelopes in a close binary configuration, something topologically similar would probably occur. But stars of given mass and composition fix their own outer boundaries according to the classical Vogt–Russell theorem of stellar structure. The ratio of masses to radii fixed by the contact configuration is, in general, incompatible with that of pairs of single stars of the same composition. Hence, such binaries are expected to have an inherent instability. Yet the W UMa type stars are observed in such high relative numbers as to indicate that any such instability does not have the disruptive effects one could expect.

In order to gain physical understanding of close binary systems, classification schemes that aim beyond a simple description of light curve morphology would be expected. A well-known scheme is that of the *detached, semi-detached* and *contact* systems, based on the relationship of the components to the Roche critical surfaces (Chapter 4). Stars that do not correspond exactly to the circumstances imposed by the mathematical formulation (e.g. they may not be rotating perfectly uniformly or in exactly circular orbits) still face an essentially similar physical situation if they continue to expand beyond a certain point. The Roche formulation, that depends only on the mass ratio for a given binary separation, is then applied, the relevant mass ratio being supplied from observations with the proviso that some degree of approximation is implied.

If a binary component, having expanded up to such a surface, were to grow still further it would start to lose mass. Detached pairs, not yet having evolved to this stage, can be largely identified in this way with the EAD type light curves, as well as some shorter period EB binaries. EAS type light curves are similarly closely associated with semi-detached systems, where one component pushes up against it limiting Roche surface. In almost all known cases this is the component of lower mass. The semi-detached condition need not always produce a light curve free from significant inter-minimum rounding, particularly if the evolved secondary happens to be of earlier than typical type — 68 Her is a frequently cited example of this: μ_1 Sco is another (Fig 7.3).

As suggested above, EW light curves and contact (or overcontact), binaries have been frequently identified with each other, though it is possible to produce an EW type light curve from a very close, but not actually contacting, pair. But, due to the relatively comparable luminosities, many EW systems are double-lined spectroscopic binaries, i.e. they have two radial velocity curves from which a mass ratio is independently derived. When these mass ratios are used to infer the locations of the Roche lobes, and the results compared with photometric analysis, there can be little doubt that the components must be very close to the surfaces of limiting stability. The earlier, contact description has frequently given way to that of a *common envelope* binary. In this picture, a common photospheric surface is posited to lie between the inner and outer critical Roche surfaces. This model, though not without theoretical difficulties, offers an empirical accounting for the EW type light variation. These issues are pursued in following chapters.

The fraction of all stars originating in binary systems is of first order. Among these, an appreciable fraction, perhaps ~5% are formed as pairs having orbital periods less than 10d. A fifth of these could be expected to show noticeable eclipses, suggesting that of the order of 1% of all stars are in eclipsing binary systems. This coarse estimate appears borne out by the data on the few thousand brightest and most well-studied stars. Of about 5000 stars brighter than magnitude 6.0, for example, around 50 eclipsing systems are known. However, among older populations of stars, such as are found in globular clusters, evolution effects alter the number of detected close binaries above a given magnitude, compared with the generally younger population of the nearby galactic field. If searches are continued to the faint, low mass, unevolved stars of globular clusters, close pairs show up again.

Evolution effects here refer not just to the physical conditions and interactions of the binary components themselves, but to dynamical effects peculiar to the dense stellar environment of globular clusters. The incidence of close binary systems in these environments is connected to a range of interesting astrophysical topics. Capture processes can result in the formation of otherwise unexpected high-energy sources, whilst binary incidence in the cluster core is strongly related to the dynamical evolution of the cluster as a whole.

Eclipsing binaries have been detected in the galactic fields of other galaxies. Eclipsing binaries thus offer a direct means for retrieving precise empirical information on stars over very large distances, so that there has been a developing interest in the increasingly faint star data-retrieval capabilities of more recent years. This allows further checking of the cosmological distance 'ladder' (at least over its lower rungs) by analysis of data from eclipsing binary systems.

7.3 Light Curve Analysis

Computer-based curve-fitting, discussed in general terms above, has successfully applied optimization procedures to close binary light curves. The fitting function to be considered in what follows builds on the theoretical basis presented in Chapter 2. It includes the proximity effects of tides and rotation ('ellipticity'), together with an appropriate description of radiative interactions ('reflection'). A solution set of parameters is again given formally by Eqn 5.22, though the simple appearance of that equation belies the potential complexity of the fitting function.

Photometric consequences of the Doppler effect, though relatively slight, can also be included in dealing with modern high-precision data. They include a term dependent on the orbital motion and its coefficient that relates to the temperature and observation wavelength. During eclipses, depending on the locations of the two stars, the remaining light may be intensified or diminished compared to an otherwise equivalent non-rotating eclipsed star. This *photometric Rossiter effect*, with allusion to the parallel, and historically well-known, spectroscopic equivalent, together with the orbital Doppler effect, result in a slight asymmetry of the light minima.

Proximity effects are often small in comparison to the main eclipse minima in the light curves of close binary stars. This consideration has led to a stepwise, or perturbation, approach to analysis. It should be noted, though, that Russell's perturbation-free spherical model is not the only starting point for such a programme. A popular alternative, that gains relevance for very close binaries, is that of the equipotential or Roche model discussed in Chapter 4. A number of authors have developed programs that follow this approach, but perhaps the most well-known is that of R. E. Wilson & E. J. Devinney (1971) (WD). This and other light curve modelling programs are referenced in Section 7.6.

In the early years of electronic computers, the speed of calculating the relevant integrals that account for the perturbed light curve explicitly may have been advantageous in fully exploring parameter space in reasonable amounts of time. But that is no longer so relevant. We follow the perturbations approach in subsequent discussion, though, not only for its formal interest and its increasing accuracy at low deviations from sphericity, but also because of self-consistency regarding structural effects. Thus, the structural coefficients that appear, for example, in accounting for apsidal motion are, in principle, the same parameters that factor the tidal and rotational distortions of the component stars.

Let us write, for the light flux l at a given phase ϕ,

$$l(\phi) = \int_A J_1 dA + \int_a J_2 da - \int_{a'} J_1 da' \ , \tag{7.8}$$

where the suffix 1 refers to the higher surface temperature primary object, having projected surface area A, while suffix 2 denotes the secondary, projecting area a perpendicular to the line of sight. The received flux scales linearly with the local projected flux J from either source. The area a' relates to the eclipses: it is zero when there is no eclipse and becomes a in annular phases of the transit (presuming here that $a < A$).

Eclipses of the secondary are also covered by Eqn (7.8), but with the eclipsing and eclipsed roles reversed. With the Js suitably re-arranged, the second and third terms would cancel each other out during totality phases of the secondary's occultation.

7.3.1   Zero Order Model

In general, the areas A, a and a' can all be regarded as functions of the orbital phase ϕ. Similarly, exposed or eclipsed regions of local flux J will generally vary with ϕ, though not in the simple model of circular projected disks with constant (undarkened) surface brightness. In such a zero order model, unperturbed, circular disks correspond to spherical stars (Fig 7.6). Such a model remains fairly direct for analysis even when components are darkened towards their limb according to normal expectation. Significantly more complication

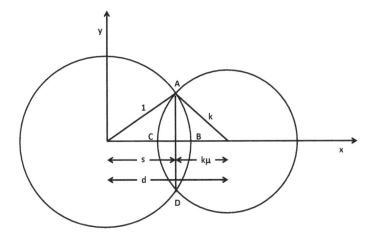

Figure 7.6 Zero-order representation of eclipse, marking quantities used in the text. The eclipsed star, on the left, is taken to have unit radius in the eclipse analysis.

arises, however, when we introduce the mass-related and radiative interactions, as well as the photometric consequences of the Doppler and light-time effects.

In Eqn 7.8, areas A, a and a' are directly related to a two dimensional (x, y) co-ordinate system in the tangent plane of the sky, where for convenience the origin is located at the centre of the star to be eclipsed and the x-axis is directed along the projection of the line of centres (Fig 7.6). Normal limb-darkening, to a good approximation, depends only on the z co-ordinate in this system, while more complicated formulations of the effect can be expressed as series in powers of z. The integrals of Eqn 7.8 then take the form $\int_{x_l}^{x_u} \int_{y(x)} J(z)\, dy\, dx$, where x_l and x_u are specifiable lower and upper integral limits.

In the unperturbed case, the limits of the inner integral in y are given from the circular boundaries of the components. If there is no eclipse these areal integrals simply amount to the projected luminosities of both components, equivalent to the first two integrals in Eqn 7.8. These are taken to be constant in the unperturbed model. Light variation arises during eclipses, when the third integral becomes non-zero and the relevant boundaries for the transit concern the overlapping 'shadow cylinder' from a, or that of the eclipsed region of A. We may write $y = \pm\sqrt{r_1^2 - x^2}$ along the boundary of A (arc ADB in Fig 7.6), or $y = \pm\sqrt{r_2^2 - (\delta - x)^2}$ along the boundary of the shadow cylinder (arc ACB).

The key variable δ represents the distance between the two centres projected along the x-axis as a fraction of the mean separation. δ is dependent on the orbital phase ϕ through

$$\delta^2 = \sin^2\phi\sin^2 i + \cos^2 i \ , \tag{7.9}$$

where i is the orbital inclination. The area a' is then formed, in the partial phases, by the outer segment of A from $x_u = r_1$ to the common chord at $x = (r_1^2 + \delta^2 - r_2^2)/2\delta$, and by the inner segment of a from the common chord down to $x_l = \delta - r_2$. For annular eclipses the integration goes from $x_u = \delta + r_2$ down to $x_l = \delta - r_2$.

Integrals involved in this approach are reduced to the form $\alpha_n^m = (1/\pi)\int_{x_l}^{x_u}\int_{y(x)} x^m z^n\, dy\, dx$, with x, y and z now scaled to the relative radius of the eclipsed star r_1, rather than that of the orbit. Integrands in y^i do not appear, since for odd values of i the integrals sum to zero, while for even values they can be rearranged in powers of x and z, using $y^2 = 1 - x^2 - z^2$. These α_n^m integrals descend from the original light loss formulations of H. N. Russell (1912). The simple spherical model, generalized to include a standard limb-darkening effect that is linear in z, involves only the integrals $\alpha_0^0(k, d)$ and $\alpha_1^0(k, d)$, where

$k = r_2/r_1$ and $d = \delta/r_1$. Eqn 7.8 is usually normalized so that the sum of the first two integrals, in the unperturbed version, is unity. The first term then becomes the primary's fractional luminosity L_1, the second L_2, and the zero order light curve equation is

$$l_0(\phi) = 1 - \alpha\{u_1, k, d(r_1, i, \phi)\}L_1 \ , \tag{7.10}$$

where α is an appropriate combination of integrals of the type α_m^0. For a linear law of limb darkening the photospheric intensity I at a direction θ to the outward normal is of the form

$$I(\theta) = I_0(1 - u + u\cos\theta) \ . \tag{7.11}$$

We then find

$$\alpha = \frac{3(1-u)}{3-u}\alpha_0^0 + \frac{3u}{3-u}\alpha_1^0 \ . \tag{7.12}$$

This α function for a spherical star, with a more general limb-darkening expressed as a series in powers of $\cos\theta$ ($\equiv$ z-coordinate of the corresponding surface point), eclipsed by a spherical companion, is given by a corresponding series in terms of α_n^0 integrals.

The more general integrals α_n^m appear when departures from sphericity are included. The relevant formulations appear complicated at first sight, but recursion formulae are available that enable the whole range of α_n^m integrals to be evaluated quickly, provided only two basic α and four auxiliary integrals are given. This range of integrals can be computer generated and rapidly applied to light curves that show proximity effects.

7.3.2   Limb-darkening

The effect of stellar limb-darkening can be described in a qualitative way on the basis of semi-transparency of the outer layers, or atmosphere, of a star, where there is a local temperature that increases inwardly. If we could look directly towards the centre of the star the line of sight through the outer layers to full opacity would take us further in, to hotter and brighter regions, than if the line of sight was traversing obliquely through layers towards the visible edge, or limb. This model was presented by P. Bouguer for the noticeable limb-darkening of the Sun in the 18th century To deal quantitatively with this effect we need to consider the transport of radiation through the star.

The transport of radiation through the atmospheric regions of a star has given rise to a highly developed field separate from our present range. We confine attention to the main points bearing on the present context. A useful formulation was made by Eddington, which, although using a preliminary analysis, provides insight into the essential issues. Techniques of the present day apply much more detailed modelling, though often requiring computer-based numerical methods rather than formal development.

Transfer of radiation through the outer layers of a star is summarized by the equation

$$-\frac{\mathrm{d}I}{\mathrm{d}s} = \kappa\rho(I - S) \ . \tag{7.13}$$

This equation refers to a small material cylinder of length ds, with axis in the direction $\hat{s}$, in an atmosphere where the density is ρ and opacity κ. In a steady state, the incremental difference between the incoming and outgoing intensity $-dI$, as a result of local absorption from a beam of intensity I radiating through the element, is offset by the local contribution of radiation sources S within the cylinder.

With regard to the radiation beam, real applications would refer to its wavelength range, or *effective wavelength* λ. We will assume that an equivalent form of transfer equation as (7.13) would apply at λ, with an absorption coefficient κ_λ averaged over a restricted wavelength range within the integrated beam. The useful concept of a 'grey atmosphere' bypasses the dependence of κ on λ, adopting that there is some constant κ that can apply to the whole atmosphere. Equilibrium of the radiation field means that the net source function S equates to the directionally averaged intensity J within the considered material element.

In interpreting this equation, an additional condition (local thermodynamic equilibrium) is often applied, allowing a local temperature T to characterize the source function within the element through the form $B(T)$, where B stands for black body radiation $(B = \sigma T^4/\pi)$. Also, use is made of an 'optical depth' variable τ such that $d\tau = -\kappa\rho\,dr$, where r corresponds to the star's outward radial direction.

A further simplifying assumption is that the scale-height of the semi-transparent atmosphere, in which the optical depth decreases from unity to $1/e$, is small compared to the radius of the star. This allows us to view the atmosphere as stratified locally into plane-parallel layers, i.e. neglect the star's sphericity. We then rewrite (7.13) as

$$\cos\theta \frac{dI}{d\tau} = \kappa\rho[I(\theta) - B(T)] , \tag{7.14}$$

where θ is the angle between the arbitrarily chosen direction of s and that of r. Finding the solution for this differential equation, with the attached conditions referred to above, constitutes the *grey atmosphere problem*.

The mean intensity J, integrating over the solid angle increments $d\omega$, satisfies

$$J = \frac{1}{4\pi} \int_{4\pi} I\,d\omega = S . \tag{7.15}$$

J is the first of a triad of useful similar forms, the second being the normalised flux H, given by

$$H = \frac{1}{4\pi} \int_{4\pi} I\cos\theta\,d\omega = \mathcal{F}/4\pi , \tag{7.16}$$

where $\mathcal{F}$ is the regular radiation flux of emitted power from unit area. A corresponding term for the normalized radiation pressure K is given by,

$$K = \frac{1}{4\pi} \int_{4\pi} I\cos^2\theta\,d\omega . \tag{7.17}$$

Integrating (7.14) over all solid angles, it then follows that

$$\frac{dH}{d\tau} = J - S = 0 , \tag{7.18}$$

demonstrating constancy of the outward flow of radiation.

Multiplying (7.14) by $\cos\theta$ and integrating again, we find

$$\frac{dK}{d\tau} = H \tag{7.19}$$

so that

$$K = H\tau + \text{const} . \tag{7.20}$$

The variation of I with θ becomes small for $\tau > 1$, and even a linear dependence on $\cos\theta$ in the semi-transparent layers becomes cancelled out in the integration for J over the whole sphere. A similar cancellation occurs with the weighting by $\cos^2\theta$ in the definition of K. We then find the approximation adopted by Eddington, using (7.15) and (7.17), that $K = J/3$ through the atmosphere, to be reasonable. In the outermost layers, however, where the photon mean free path starts to become comparable to the stellar radius, the plane parallel analysis we have followed loses validity. In those layers, the flux becomes

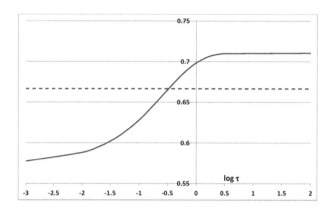

Figure 7.7   Plot of the theoretical Hopf function $q(\tau)$ for a grey atmosphere as appearing in Eqn (7.23). The difference from the Eddington approximation (dashed line), being greatest near the top of the atmosphere, anticipates departures from the linear form (7.11) to be greatest at the extreme limb.

essentially all outward and pre-determined: the intensity of the inward field tending to zero. The mean intensity over the whole sphere J, at the conceptual surface, drops to half that of the outward hemisphere alone. In accordance with this, (7.20) becomes

$$J = 3H(\tau + 2/3) = 2H(1 + \frac{3}{2}\tau) \ . \tag{7.21}$$

We had already, for thermal equilibrium $J = B = \sigma T^4/\pi$, so (7.21) furnishes an approximate connection between the temperature and optical depth as

$$T^4 = T_e^4(1 + \frac{3}{2}\tau) \ . \tag{7.22}$$

This means that when $\tau = 2/3$, $\sigma T_e^4 \equiv 4\pi H = \mathcal{F}$, i.e. this formulation satisfies the usual definition of the effective temperature T_e in terms of the astrophysical flux $\mathcal{F}$. Notice that the temperature T_0 at the idealized surface is given as $T_0 = T_e/2^{1/4}$ in this approximation.

The symbol F is often used for the surface mean intensity $(= 2 \int_0^1 I\mu \, d\mu \equiv \mathcal{F}/\pi)$, while a more detailed discussion of the transfer equation for the grey case results in

$$J = \frac{3}{4}F(\tau + q(\tau)) \tag{7.23}$$

where $q(\tau)$, known as the Hopf function (Fig 7.7), becomes the constant 2/3 in the Eddington approximation. The generally fair validity of that simple approximation can be understood by noting that the monotonic function q varies from $q(0) = 1/\sqrt{3} \approx 0.5774$ while $q(\infty) \approx 1/\sqrt{2}$, or 0.7104 (Fig 7.7). The fall in q towards the surface suggests that the corresponding temperature in more accurate representations would be lower than $T_0(\text{Edd}) = T_e/2^{\frac{1}{4}}$.

Returning to (7.13), this can also be written, setting $\cos(\theta) = \mu$, as

$$\frac{d\{I \exp(-\tau/\mu)\}}{d\tau} = \frac{-S \exp(-\tau/\mu)}{\mu} \ . \tag{7.24}$$

This can be formally integrated between arbitrary optical depths τ_1 and τ_2 to give

$$[I \exp(-\tau/\mu)]_{\tau_1}^{\tau_2} = \frac{1}{\mu} \int_{\tau_1}^{\tau_2} -S \exp(-\tau/\mu) \, d\tau \ . \tag{7.25}$$

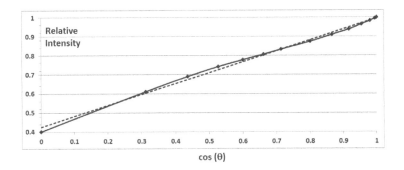

Figure 7.8 Normalized solar limb darkening (continuous) as determined from the near-total eclipse Apr 17, 1912 (Julius, 1913). The bolometric measures are compared with the linear form $I(\theta) = 0.43 + 0.57 \cos \theta$ (dashed).

By setting $\tau_1 = 0$ and $\tau_2 = \infty$, and with the previous assumptions, we can derive the intensity of the beam emerging from the surface as

$$I_0(\mu) = \frac{1}{\mu} \int_0^\infty S \exp(-\tau/\mu) \, \mathrm{d}\tau \ , \qquad (7.26)$$

i.e. the intensity dependence on the cosine of the projection angle θ turns out to be the Laplace transform of the source function's dependence on the optical depth τ. It follows that if $S = J$ is a linear function of τ, as in (7.21), the limb-darkening will be a linear function of μ, and indeed the ratio of the coefficients of the constant and first-power terms in both forms remains the same. With the limb-darkening given in (7.11) above, we find, in the Eddington approximation, that

$$\frac{1-u}{u} = \frac{2}{3} \ , \qquad (7.27)$$

i.e. $u = 0.6$. Measured values for the limb-darkening coefficient in integrated light for the Sun are, perhaps surprisingly, close to this value (Fig 7.8).

The limb-darkening coefficient derived from (7.27) depends on the ratio of the linear and constant terms, J_1/J_0 say, in τ in (7.21), so that from (7.27) we have $u = (J_1/J_0)/(1+J_1/J_0)$, leading to $0 < u < 1$, while u would become insensitive to J_1/J_0 if this ratio becomes large. This ratio derives, in the Eddington approximation, from the adopted conditions that $J = 2H$ at the surface and $J \to 3H\tau$ at depth, J being linear in τ. From the Laplace transform comparison, we can expect that a linear cosine law for the limb-darkening depends on a linear dependence of the mean intensity on optical depth.

But even if J, or B, is linear in τ in integrated light, this is not necessarily the case at wavelength λ. Indeed, while B has a simple dependence on T^4, in the Rayleigh Jeans region of the Planck (black-body) function the contribution is less sensitive to temperature than the full emission form. In the Wien region it has the more sensitive exponential, rather than power-law, behaviour. Since these radiation formulae directly involve temperature, the temperature having a monotonic, single-valued dependence on optical depth in the adopted conditions of local thermodynamic equilibrium, we may write:

$$\frac{\mathrm{d} \log B}{\mathrm{d} \log \tau} = \left(\frac{\mathrm{d} \log B}{\mathrm{d} \log T} \right) \times \left(\frac{\mathrm{d} \log T}{\mathrm{d} \log \tau} \right) = 1 \ . \qquad (7.28)$$

If we compare this with the variation at a particular wavelength λ, and use the grey atmosphere's uniform opacity model (i.e. $\tau_\lambda = \tau = $ const) to eliminate the factor $\mathrm{d} \log T / \mathrm{d} \log \tau$,

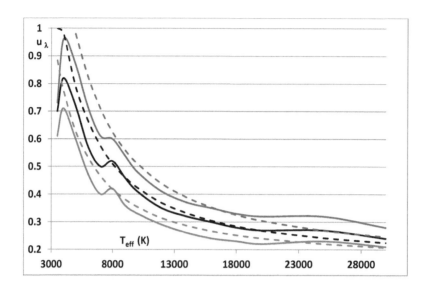

Figure 7.9 A comparison of the simple approximation for u_λ (7.31) (dashed), in the commonly used photometric bands B (dark grey – upper), V (black – central) and R (light grey – lower), with the results of more detailed numerical modelling (full), taken from the tables of Van Hamme (1993). The diagram shows the approximate validity of the linearized grey atmosphere model through most of the optical range. More detailed modelling points up the effects of discrete or sudden variations of transparency at particular wavelengths.

we find corresponding rates of change in the source function at wavelength λ and in integrated light as

$$\frac{\mathrm{d}\log B_\lambda}{\mathrm{d}\log T} \Big/ \frac{\mathrm{d}\log B}{\mathrm{d}\log T} = \frac{a}{4(1 - e^{-a})} = \frac{b}{4}, \text{ say,} \tag{7.29}$$

where $a = c_2/\lambda T$, c_2 being the second radiation constant $= 1.44$ cm K and b is always greater than a but usually of the same order in optical contexts.

If the integrand for emergent intensity in (7.26) at a particular wavelength λ behaves in an essentially proportional way to the integrated light, as may be expected for a wavelength range in the vicinity of the Planck-function maximum, then the ratio S_1/S_0 is unaffected, and so $u_\lambda \approx u$ in this region. The limb-darkening coefficient, written as the centre to limb deficit $\Delta I/I_0$, takes on the form of a logarithmic derivative in the linear approximation, so that, regarding the source function as B in (7.26), and for relatively small deviations of u_λ from the integrated light value u, we can write

$$\begin{aligned} \frac{u_\lambda}{u} &= \frac{\Delta B_\lambda}{B_\lambda} \Big/ \frac{\Delta B}{B} \\ &\approx \frac{\mathrm{d}\log B_\lambda}{\mathrm{d}\log T} \Big/ \frac{\mathrm{d}\log B}{\mathrm{d}\log T} \end{aligned} \tag{7.30}$$

This means an approximate value for the limb-darkening coefficient u_λ is given by

$$u_\lambda \approx ub/4 \ . \tag{7.31}$$

In Fig 7.9, we compare this approximation as applied to the standard B, V and R photometric ranges with the results of detailed numerical integrations using modern stellar atmosphere models. The centre to limb deficit, using (7.22), implies a relative temperature

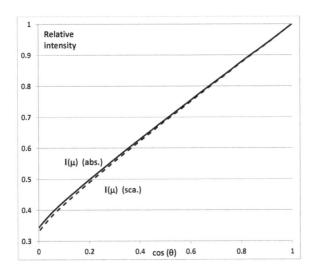

Figure 7.10 Comparison of the surface intensity distribution for pure absorption (continuous) and scattering (dashed) models in dependence on $\cos(\theta)$ $(= \mu)$, as tabulated by Kopal (1959).

variation $\Delta T/T$ of about 0.2. The ratio of the second to first order logarithmic derivatives, by carrying out the relevant differentiation, turns out to be $2b - 2 - a$, which is of order unity for optical range data on stars. Non-linearity of (7.30) should then start to become significant for $b/4 \gtrsim 3/2$, towards the upper left of Fig 7.9 where $\lambda T \lesssim 0.24$. Eqn (7.31) thus provides a quick guide to intermediate and lower values of observable stellar limb-darkening coefficients, pointing to the generally fair validity of the grey atmosphere approximation for stellar atmospheres in the optical range with effective temperatures $\gtrsim 5000$ K.

But it can be noticed in Fig 7.9 that while the results of the simple linear approximation and the detailed numerical calculations trend similarly through the broad, smoothly varying regions, they differ in regions where the effects of continuum discontinuities become strong. In other words, the differences between linearizations of the grey model and more accurate representations are generally smaller than, or comparable to, differences arising from discrete opacity effects on derived limb-darkening coefficients.

Although the association of a local temperature with a given location in a star seems natural, particularly when the optical depth increases to a point where the local radiation field becomes spherically symmetric to a high degree, the substitution of $B(T)$ for S in (7.24) is not necessary for the formal solution (7.26) to hold. Indeed, the argument that led to (7.27) would hold for any source function that was a linear function of τ. Similarly, the approximations that led to the linear form (7.21) did not depend directly on a local temperature, though (7.23) indicates the breakdown of that approximate model towards the edge of the radiation-transporting medium.

These points bear on what happens if we depart from the underlying supposition of thermalization of the radiation field and disengage the distribution of photons from the statistical dynamics of the particles they are interacting with in correspondence with the *scattering* of photons. In line with the foregoing remarks, the centre-to-limb variation of the integrated intensity is little affected (Fig 7.10), in spite of the basic physical difference of scattering from the absorption-emission at T scenario.

The known form of the source function in (7.23) has enabled a corresponding exact centre to limb variation to be specified (Fig 7.10). If we write a representative (linear) limb-darkening coefficient u_g as

$$u_g = \left(1 - \frac{I(\mu)}{I_0}\right) / (1 - \mu) \ , \tag{7.32}$$

the data used in Fig 7.10 determines that this u_g would have a $\sim 7\%$ variation across the disk from its mean value of ~ 0.63, but most of the variation arises in the last $\sim 10\%$ of the range. Any given theoretical dependence of the centre to limb intensity variation on μ can thus be developed as a power series in μ, using semi-analytical, numerical or curve-fitting techniques. Kopal (1959) produced expressions of this form, in which the successive terms of the series are strongly decreasing, although as more terms are included in the representation individual coefficients do not necessarily decrease. This raises the recurrent issue of parameter interdependence, touched on in the opening section of this chapter.

Various alternative forms for modelling the centre-to limb variation, influenced by theoretical studies, have been proposed, such as:

$$\frac{I_\lambda(\mu)}{I_\lambda(1)} = 1 - u_{1,\lambda}(1 - \mu) - u_{2,\lambda}(1 - \mu)^n$$

$$\frac{I_\lambda(\mu)}{I_\lambda(1)} = 1 - u_{1,\lambda}(1 - \mu) - u_{2,\lambda}\mu \ln \mu$$

$$\frac{I_\lambda(\mu)}{I_\lambda(1)} = 1 - u_{1,\lambda}(1 - \mu) - u_{2,\lambda}(1 - \sqrt{\mu})$$

$$\frac{I_\lambda(\mu)}{I_\lambda(1)} = 1 - \sum u_{k,\lambda}(1 - (\sqrt{\mu})^k) \tag{7.33}$$

Here, the additional coefficient $u_{2,\lambda}$ addresses non-linear behaviour of the centre to limb variation (or $u_{k,\lambda}$, with $k \neq 2$, in the last form due to A. Claret; 2000). In principle, Kopal's (1959) formalism

$$\frac{I_\lambda(\mu)}{I_\lambda(1)} = 1 - \sum u_{k,\lambda} + \sum u_{k,\lambda}\mu^j \tag{7.34}$$

can be adapted to such generalizations, with the advantage that the corresponding effects in the eclipse fitting function are known and directly available. Thus the decline of the Hopf function towards the surface, visible in Fig 7.7, keeping in mind the Laplace transform properties, would point to a small negative value for the higher order coefficient u_2 in (7.34). It seems clear that as stellar photometry becomes increasingly more precise with the evolution of technology, it will call for greater precision of the fitting function. However, increased parametrization at a given data accuracy risks the determinacy deterioration discussed in Section (7.1).

Regarding the practical consequences of limb-darkening on light curve analysis, it is not clear, in general, how much non-linearity in μ would affect parametrization in relation to the probable error of a data-point. Well-defined annular phases are certainly informative on the local emergent flux distribution across the bright central regions of an eclipsed star. Classical photospheric models predict a small drop below the linear-in-μ form in the vicinity of the limb. But as the opacity approaches zero the relationship of the photon field to the local kinetic energy of particles, that participate in other processes such as global circulation or magnetic field interactions, becomes complex. The outward flux directed into a ray grazing the photospheric limb could then differ from classical treatments, to an extent requiring extra parametrization beyond the information content of the data. In short, the linear approximation formula for the limb-darkening (7.11) allows a practicable basis for initial analysis. A more developed fitting function can be applied if the determinacy permits extra parametrization, or if an enhanced parametrization is on the basis of an adopted theoretical position for some subset of the parameters. Recent, satellite-derived, high accuracy data, such as from the Trasnsiting Exoplanet Survey Satellite (TESS), encourage such developments.

7.3.3   Gravity-darkening

The full nature of gravity-darkening appears still not conclusively established, both from theoretical and observational points of view, yet it is significant in the analysis of light curves of distorted stars, the more so as stellar components are closer together. But even the naming of the effect – sometimes darkening, sometimes brightening – reflects the lack of clarity. For continuity, we adhere to older usage of the term gravity-darkening, although it is clear that for a radiative atmosphere classical theory predicts that the surface flux increases in direct proportion to the local gravity on a distorted stellar surface.

Classical arguments were originally spelled out in the oft-cited work of von Ziepel (1924). From the theory of stellar structure, the energy flux though the stellar envelope should be proportional the temperature gradient – the 'diffusion law'. This can be shown to hold for (7.19), certainly as the optical depth becomes comparable to unity and greater. If now the temperature, as a state variable, is constant along equipotential surfaces (the Clairaut stability criterion, or barotropic condition), then the local flux will be inversely proportional to the local relative separation of those surfaces. But the local gravity itself has such an inverse proportionality, when the surface is in equilibrium. Hence, we deduce directly that the flux is proportional to gravity ('von Zeipel's law').

In practice, this result seems to give a reasonable accord with observations, at least for stars whose envelopes are dominated by a flux essentially propagated by radiative transport. From this context it has been common to write $H \propto g^\tau$, where H is the normalized bolometric surface flux $(= \mathcal{F}/4\pi)$, g is the gravity and τ is known as the gravity-darkening index. If energy transfer in the sub-surface layers is purely radiative, the bolometric value of τ is set as unity ($\tau = 1$) following von Zeipel's law.

The finding is substantiated by considering a pair of neighbouring equipotentials separated by the small distance $\Delta r(\theta, \phi)$. Now allow that the local flux H and gravity g, along with Δr, vary with the angular coordinates θ and ϕ. The diffusion condition follows from (7.19), recast as

$$H = -\frac{1}{3}\left(\frac{\Delta B}{\kappa\rho\Delta r}\right) \quad , \tag{7.35}$$

so that as Δr varies by $\delta\Delta r$, say, in moving a small distance along the lower equipotential surface from some arbitrarily selected initial position satisfying (7.35), ΔB, κ and ρ remaining constant over the two equipotentials, we have

$$\frac{\delta H}{H} = -\frac{\delta\Delta r}{\Delta r} \quad . \tag{7.36}$$

In a similar way, the magnitude of the local gravity, with movement from the initial position, satisfies

$$g + \delta g = \frac{\Delta V}{\Delta r + \delta\Delta r} \quad . \tag{7.37}$$

with the potential scalar V $(\equiv -\Psi)$, characterizing each equipotential surface, decreasing outward (cf. 2.1).

Combining (7.36) and (7.37), we have

$$\frac{\delta H}{H} = \frac{\delta g}{g} = -\frac{\delta \Delta r}{\Delta r} \ , \tag{7.38}$$

so that $\tau = \mathrm{d} \log H / \mathrm{d} \log g = 1$ for a diffusive energy flow.

For a convective atmosphere, the nature of gravity-darkening is not self-evident. In an oft-cited paper, L. B. Lucy (1967) proposed a lower value for τ for an envelope in which heat was transferred mainly by convection in the outer layers. Lucy utilized convective envelope models' data at a depth where the local radiative temperature gradient first becomes super-adiabatic. We may note, at this point, the use of the effective temperature T_e to characterize the heat flux. Since, by the definition of T_e, $\mathcal{F} = \sigma T_e^4$, we can write $T_e \propto g^\beta$, as did Lucy, where $\beta = \tau/4$. It is generally agreed that β should be 0.25 when the energy transfer is purely radiative. According to Lucy, however, $\beta \approx 0.08$ for stars with convective envelopes. The procedure discussed by Lucy has been followed by other investigators using other programs, who found similar results.

A. Claret (1998) made use of evolving stellar models to give the gravity-darkening exponent as a function of mass and age. He argued that the often adopted values $\beta = 0.25$ and $\beta = 0.08$ for radiative and convective envelopes should have a smooth transition between the two energy transport regimes. His findings paralleled observationally based studies that have found no sudden transition of light curve effects at spectral type $\sim$F5V, as having two distinct indices predicts.

Observational determinations of gravity darkening have often reported a difference between higher τ values for radiative envelopes and lower ones for cooler stars. The point is particularly significant for W UMa type binaries, where various authors gave values for the (bolometric) gravity-darkening index in close agreement with the prediction of Lucy. The papers at issue had, however, already presumed that the stars in question were overcontact systems, though the geometric parameters for a presupposed common envelope model are correlated with those of the photospheric light distribution. Different authors who have analysed the gravity-darkening exponent from observations of W UMa type binaries, without prejudice as to modelling requirements, have found a relatively wide spread in their β-values.

Regarding observational evidence in general, it can be reasonably asserted that light curves of eclipsing binaries, particularly older ones, may not have sufficient information content to discriminate between alternative effects of convection on the surface heat distribution or internal structure, keeping in mind the sinusoid-like light-curves. Parameter ambiguity is bound to arise when essentially similar photometric effects can be produced by different combinations of their values. Even so, the overall effect of gravity-darkening is not small: at optical wavelengths, it effectively doubles the scale of inter-minimum ellipticity effect that would be due to tidal distortion alone. It has thus a critical role to play, if independently determinable, in assessing the degree of contact, or overcontact of the photospheres in very close pairs of stars.

Calculating β

To follow the method discussed by Lucy, it is required to locate the lowest layer of the convective zone for a given mass, composition and age model. Models used for the calculation tabulate the (one-dimensional) envelope integration, layer by layer. There is general agreement that below this critical layer the temperature gradient reverts to a von Zeipel form. Lucy considered the adiabatic constant, also called the specific entropy, defined by

$$\log K = \log T - (\gamma - 1) \log \rho \ , \tag{7.39}$$

in the layer where the convection starts, and examined its dependence on both T_e and g, which are set as adjustable star-modelling parameters. If K is constant over a surface that determines the heat flow from the stellar surface, a numerical value for β follows as:

$$\beta = -\left\{\frac{\Delta \log K}{\Delta \log g}\right\}_{T_e} \div \left\{\frac{\Delta \log K}{\Delta \log T_e}\right\}_g \qquad (7.40)$$

Evaluation of the two fractions in (7.40) should pose no difficulty for suitably controllable stellar envelope programs, though care may be required to ensure that the two fractions correspond to numerical forms of truly independent partial derivatives. There is usually an empirically selectable 'mixing-length' parameter (α) for convective envelopes. Its effect on (7.40) can be checked, though that is generally found to be small for reasonably expectable ranges of α. Derived values of β in the range 0.05-0.1 are typical for a wide range of low mass stellar models.

Looking further around this, the flux should be proportional to the gradient of the gravitational potential in the deeper layers of the envelope, otherwise hydrostatic balance of the star would not be possible, and there would result visible fluctuations on a dynamical timescale. Eqn (7.40) for β can thus be written for any barotropic variable u as

$$\beta = -\left\{\frac{\Delta \log u}{\Delta \log g}\right\}_{T_e} \div \left\{\frac{\Delta \log u}{\Delta \log T_e}\right\}_g \quad , \qquad (7.41)$$

and from Eqn (7.39) and Section (2.1) we can see that K is such a variable. In a purely radiative regime, with constant luminosity L, locally increasing $\log T_e$ by an amount s_1, say, would have the same effect on u as decreasing $\log g$ by $4s_1$, since $\log L = 4 \log T_e + 2 \log R + const.$ Hence, $\beta = 0.25$ is confirmed for the ratio of partial derivatives in such a regime.

The outermost layers of the star present subtleties, however, that have featured in numerous papers. One argument is that the local temperature would grow in regions of greater local heating flux, which scales the local temperature in thermal equilibrium as in (7.23). This would change the local temperature gradient and thus have a feedback effect on the flux, as well as other local variables. The implied failure of equilibrium causes a global circulation of gas from the hotter poles to the cooler equator for a rotating star. This would warm up the outer parts of the atmosphere near the equator and tend to make less negative the temperature gradient there. There would correspondingly be a return flow from the equator to the poles lower down in the envelope, as in Eddington's meridional model, or perhaps in a more complex flow regime. Local variations of mean molecular weight are also a consequence.

It is feasible that a steady state may settle at a surface flux distribution having a lesser variation over the photosphere than the von Zeipel law. The $\tau = 1$ form would thus be softened to a weaker average dependence on position. Such a case was discussed by R. Connon Smith and R. Worley (1974) on the basis of a perturbation analysis for the effects of rotation in the outer layers of a star. The result $\tau \approx 1/2$ was derived for a classical Eddington-type atmosphere. However, circulating material high in the atmosphere moves at high speed, introducing dissipative turbulent local interactions that redistribute the emergent flux back to the von Zeipel formula.

Theoretical arguments may be raised in support of convection softening the gravity darkening index below unity. There are, however, still questions that can be raised about the method used to evaluate this index. It is accepted that, for stellar envelopes in general, heat fluxes settle to the radiative temperature gradient in the zone below the convective layers. The mean value of this flux provides the radiative luminosity from the photosphere. And, while the specific entropy remains constant over the equipotentials of a distorted static star, whether it has the same value all over the layer in which convection starts can be challenged; the onset of convection depending on the temperature gradient, rather than temperature itself.

Envelope modelling shows that K for a test-cell in the sub-convective boundary layer can be increased by decreasing the gravity parameter. This corresponds to moving the test-cell towards lower latitudes in a rotating star. The local effective temperature would have to decrease to restore K to constancy, but not by much, due to the high sensitivity of K to T_e around this layer. It is this high sensitivity that gives β its low value in (7.40). The high sensitivity also means that the layer in which convection starts cannot coincide with an equipotential. If we suppose it started at a mid-latitude on a particular equipotential, say, the increase of T_e toward the pole would move the convection's initial layer further out from the centre, while towards the equator it would be pushed lower down. The lower empirical values of β from the model data selected for the calculation thus cannot satisfy the $\log K = \text{const.}$ requirement for (7.40) to be valid. Use of the corresponding index implies an inappropriate comparison of alternative conditions of subphotospheric energy transfer.

In any case, the particular layer that was selected for this calculation is not the most typical of the convective flux environment. Excess heat that goes into the kinetic energy of convection at the base of the zone is later put back into the radiative luminosity of the photosphere. This point can be checked further by evaluating the value of β in the layer where the convection is most established, i.e. where the mean turbulent velocity is maximal. The value of $\log K$ in this layer is less sensitive to the layer's position and, on that basis, may be more appropriate for the evaluation. It is found that the numerical value of β for such layers is close to the von Zeipel value.

Observational evaluations of gravity darkening for close binary systems have not provided wholly independent checks of this discussion, due to the strong correlation between β and other parameters characterizing the light curve shape particularly regarding photometric 'ellipticity' effects. However, in this incipient age of very precise light curves from space-based observatories it is likely this situation will be remedied, or at least considerably improved, from high-accuracy exoplanet transit data. In the meantime, modelling parameters derived from older observations remain dependent on adopted priors.

7.3.4 α-integrals: Basic Formulae

The modelling of an eclipsing binary light curve so as to take into account proximity effects using generalized α-integrals was referred to in connection with Eqn (7.10) and subsequently. For quick reference, we now list usable formulae for the integrals α_0^0 and α_1^0. A guide as to how these forms are derived or checked is given in the subsection that follows. Readers interested in applications may skip directly to Section 7.3.6.

(i) Annular case $d \leq 1 - k$, $k < 1$ (trivial for α_0^0):

$$\alpha_0^0 = k^2, \tag{7.42}$$

$$
\begin{aligned}
\alpha_1^0 = \ & \frac{2}{3}\left(\Lambda(\kappa_a, \beta) \ - \right. \\
& - \frac{1}{3\pi\sqrt{1-(d-k)^2}}\left\{[d^2(d^2 - 2dk - 5 + 8k^2) + \right. \\
& + \ dk(8 - 14k^2) + (1 - k^2)(4 - 7k^2)]E(\kappa_a) - \\
& \left.\left. - [d^4 - (5 + 2k^2)d^2 + 6d - 2 + k^2(1 + k^2)]F(\kappa_a)\right\}\right) \ .
\end{aligned}
\tag{7.43}
$$

Here, we require the complete elliptic integrals E and F, with modulus κ_a given by $\kappa_a = \sqrt{4dk/[1-(d-k)^2]}$. $\Lambda(\kappa, \beta)$ is Heumann's Λ-function, given as

$$\Lambda(\kappa, \beta) = \frac{2}{\pi} \left(E(\kappa)F(\kappa', \beta) + F(\kappa)E(\kappa', \beta) - F(\kappa)F(\kappa', \beta) \right),$$

where the incomplete elliptic integrals E and F, with modulus $\kappa' = \sqrt{1-\kappa^2}$, have argument β given by

$$\beta = \arcsin \left(\sqrt{\frac{1+k-d}{1+k+d}} \right).$$

Special case expressions for α_1^0 are:
(1) $d = 1 - k$,

$$\alpha_1^0 = \frac{4}{3\pi} \left[\arcsin \sqrt{k} - \frac{1}{3}(1+2k)(3-4k)\sqrt{k(1-k)} \right].$$

(2) $d = k$, $\kappa_a = 2k$, and then,

$$\alpha_1^0 = \frac{1}{3} + \frac{2}{9\pi} \left[(8k^2 - 4)E(\kappa_a) - (1 - 4k^2)F(\kappa_a) \right].$$

(3) if $d = 0$ we can write, simply,

$$\alpha_1^0 = \frac{2}{3} \left[1 - (1 - k^2)^{3/2} \right].$$

(ii) Partial case $1 + k > d > 1 - k$. Auxiliary quantities are used here, thus (cf. Fig 7.6) $s = (1 + d^2 - k^2)/2d$, and $\mu = (d - s)/k$. The expression for α_0^0 is:

$$\alpha_0^0 = \frac{1}{\pi} \left[\arccos(s) + k^2 \arccos(\mu) - d\sqrt{(1 - s^2)} \right]. \tag{7.44}$$

The formula for α_1^0 is:

$$\begin{aligned}
\alpha_1^0 &= \frac{2}{3} (1 - \Lambda(\kappa_p, \xi) + \\
&+ \frac{1}{3\pi\sqrt{dk}} \{ 2(d^2 - 4 + 7k^2)dk E(\kappa_p) - \\
&- [d^3 k + 5d^2 k^2 - d(3 + 4k - 7k^3) + 3(1 - k^2)^2]F(\kappa_p) \}). \tag{7.45}
\end{aligned}$$

Again we encounter elliptic integrals, with modulus $\kappa_p = 1/\kappa_a = \sqrt{[1-(d-k)^2]/4dk}$, and, for Heumann's Λ-function, argument $\xi = \arcsin \left[\sqrt{2d/(1+k+d)} \right]$.
Another special case occurs if $d = k$, i.e. $\kappa_p = 1/2k$,

$$\alpha_1^0 = \frac{1}{3} + \frac{2}{9\pi k} \left\{ (16k^2 - 8)k^2 E(\kappa_p) - [k^2(16k^2 - 10) + \frac{3}{2}]F(\kappa_p) \right\}.$$

(iii) Total case $d \le k - 1$, $k > 1$ (trivial):

$$\alpha_0^0 = 1, \tag{7.46}$$

$$\alpha_1^0 = \frac{2}{3}. \tag{7.47}$$

7.3.5 α-integrals: Notes on Derivations

For completeness we present background on the α-integral formulae. Firstly, the partial form for α_0^0:— referring to Fig 7.7 we write

$$\pi \alpha_0^0 = 2 \int_s^1 \sqrt{1 - x^2} ds + 2 \int_{d-k}^1 \sqrt{k^2 - (d - x)^2} dx \ , \qquad (7.48)$$

i.e. two integrals that we can call A and B, say.

The complications to the explicit forms of the α integrals are found to arise essentially from component B. In the present example, component A is directly evaluated as a trigonometric integral, using the substitution $x = \cos\theta$. Component B is similarly evaluated by the substitution $v = (d - x)/k$. The additional relation, easily checked from Fig 7.7, that $\sqrt{1 - s^2} = k\sqrt{1 - \mu^2}$, allows the two radicals in A and B to be combined to produce the result given in (7.15).

For α_1^0, writing $\sqrt{1 - x^2 - y^2}$ for z, and then substituting $\sqrt{1 - x^2} \sin\theta$ for y yields

$$A = 2 \int_s^1 (1 - x^2) \int_0^{\pi/2} \cos^2\theta \, d\theta \, dx \ , \qquad (7.49)$$

which is readily evaluated as

$$A = \frac{\pi}{3} - \frac{\pi}{2}\left(s - \frac{s^3}{3}\right) \ . \qquad (7.50)$$

It will be useful to notice here that as $k \to \infty$ the area corresponding to the crescent ACB in Fig 7.7 tends to zero, i.e. $B \to 0$. This condition occurs in the context of atmospheric eclipses of a relatively small bright star by a companion with an extended atmosphere.

The substitution $y = \sqrt{1 - x^2} \sin\theta$ allows us to write

$$B = 2 \int_{d-k}^s (1 - x^2)\left(\theta_0 + \frac{\sin 2\theta_0}{2}\right) dx \ , \qquad (7.51)$$

where

$$\theta_0 = \arcsin(\sqrt{(k^2 - (d - x)^2)/(1 - x^2)}) \ . \qquad (7.52)$$

If we now split the integrand in B into its two components in θ_0 as B_1 and B_2, say, we find

$$B_2 = \int_{d-k}^s \sqrt{k^2 - (d - x)^2}\sqrt{(1 - x^2) - k^2 + (d - x)^2} \, dx \ , \qquad (7.53)$$

The substitution $v = (d - x)/k$ leads to

$$B_2 = k^2 \int_\mu^1 \sqrt{(1 - v^2)(a + cv)} \, dv \ , \qquad (7.54)$$

where $a = 1 - k^2 - d^2$, $c = 2kd$, while $a/c = -\mu$, so that

$$B_2 = k^2\sqrt{2kd} \int_\mu^1 \sqrt{(1 - v^2)(v - \mu)} \, dv \ . \qquad (7.55)$$

This type of integral can be worked out using the Jacobian elliptic functions sn u, cn u and dn u, with modulus $\kappa = \sqrt{(1 - \mu)/2}$, and whose amplitude u is the incomplete elliptic integral $F(\kappa, \psi)$ given by

$$u = \int_0^\psi \frac{d\psi}{\sqrt{(1 - \kappa^2 \sin^2\psi)}} \ . \qquad (7.56)$$

B_2 becomes

$$B_2 = 16k^2\sqrt{2kd}\,\kappa^4 \int_0^{F(\kappa)} \text{sn}^2 u\,\text{cn}^2 u\,\text{dn}^2 u\,du \ . \tag{7.57}$$

This is now in the form of a standard elliptic integral that can be evaluated by regular procedures or looked up in appropriate tables. The result is

$$B_2 = \frac{16}{15}k^2\sqrt{2kd}\left(2(1-\kappa^2+\kappa^4)E(\kappa) - (2-3\kappa^2+\kappa^4)F(\kappa)\right) \ , \tag{7.58}$$

where $F(\kappa)$ and $E(\kappa)$ are the first and second kinds of complete elliptic integral.*
  The component B_1 is integrated by parts to produce the rather complicated form

$$\begin{aligned} B_1 &= \left[\left(x - \frac{x^3}{3}\right)\arcsin\left(\sqrt{\frac{k^2-(d-x)^2}{1-x^2}}\right)\right]_{d-k}^s \\ &\quad - \int_{d-k}^s \frac{(x-x^3/3)\,\text{d}\{D\}}{\sqrt{(1-x^2-(k^2-(d-x)^2)/(1-x^2)}} \ , \end{aligned} \tag{7.59}$$

where*

$$\text{d}\{D\} = \frac{(d-x+k^2 x-d^2 x+dx^2)\,dx}{(1-x^2)\sqrt{(1-x^2)(k^2-(d-x)^2)}} \tag{7.60}$$

  The term in square braces reduces to $\pi(s - s^3/3)//2$, which removes part of component A in (7.20), so that we are left with

$$\pi\alpha_1^0 = \frac{\pi}{3} + B_1' + B_2 \ , \tag{7.61}$$

where B_1' is the rather cumbersome expression

$$B_1' = \frac{-d}{3}\int_{d-k}^s \frac{(3x-x^3)(1-2sx+x^2)dx}{(1-x^2)\sqrt{k^2-(d-x)^2(1-x^2-(k^2-(d-x)^2)}} \ . \tag{7.62}$$

The same substitutions used before, i.e. $(d-x)/k = v = 1 - 2\kappa^2\text{sn}^2\,u$ allows B_1', after some manipulation, to be converted into standard elliptic integral forms. The whole may then be put together with the results appearing in (7.43) and (7.45).

α_n^m and related integrals

  Writing $x = r\cos\theta$, $y = r\sin\theta$ and $z = \sqrt{1-r^2}$, we have

$$\pi\alpha_n^m = 2\int_{d-k}^c (1-r^2)^{n/2}r^{m+1}\int_0^{\theta_0(r)}\cos^m\theta\,d\theta\,dr \ , \tag{7.63}$$

where $c = 1$ for partial eclipses and $c = d + k$ for annular eclipses, while $\theta_0(r)$ is given by

$$\theta_0 = \arccos\left\{\frac{d^2-k^2+r^2}{2dr}\right\} \ , \tag{7.64}$$

---

*Standard algorithms for the elliptic integrals F and E, in both complete and incomplete forms are readily available; cf. e.g. Hofsommer, D.J. and Van der Riet R.P. *Numer. Math.*, Vol. 5, 291, 1963.
*Note the difference in these expressions between the differential sign d and the projected distance between the star centres d.

with $0 \leq \theta_0 \leq \pi$.

Now define an integral $I_{\beta,\gamma}^m$ as

$$\pi k^{\beta+\gamma+m+1} I_{\beta,\gamma}^m = \int_{d-k}^{c} (k^2 - (d-x)^2)^{\beta/2} (2d(s-x))^{\gamma/2} (d-x)^m \, dx \ . \tag{7.65}$$

Note here that $2d(s-x) = z$ along the (circular) boundary of the eclipsing star's shadow cylinder on the eclipsed star's (spherical) surface.

It is easily seen that for $\gamma = 0$, writing $\cos\theta = (d-x)/k$, we have

$$\pi I_{-1,0}^m = \int_0^{\theta_0} \cos^m \theta \, d\theta \ . \tag{7.66}$$

For this kind of integral the recursion formula

$$\pi(m I_{-1,0}^m - (m-1) I_{-1,0}^{m-2}) = \sin\theta_0 \cos^{m-1}\theta_0 \ . \tag{7.67}$$

is available. Putting (7.67) into (7.63), we find

$$m\alpha_n^m + (m-1)\{\alpha_{n+2}^{m-2} - \alpha_n^{m-2}\} =$$
$$2\int_{d-k}^{c} (1-r^2)^{n/2} (r\cos\theta_0)^{m-1} r I_{-1,0}^1 (\cos\theta_0) r \, dr \tag{7.68}$$

where we have used

$$\pi I_{-1,0}^1 = \sin\theta_0 \tag{7.69}$$

Consider now another form, similar to (7.65), written as

$$Y_{\beta,\gamma}^m = \int_{d-k}^{s} (k^2 - (d-x)^2)^{\beta/2} (2d(s-x))^{\gamma/2} x^m \, dx \ . \tag{7.70}$$

Let us write $x = r\cos\theta_0$, and, using (7.64), change the variable from x to r, noting that $dx/dr = r/d$. We then find

$$\int_{d-k}^{c} (1-r^2)^{n/2} (rx)^{m-1} (r I_{-1,0}^1 (x)) r \, dr = d Y_{1,m}^{m-1} \ . \tag{7.71}$$

Combining (7.71) and (7.68) we then obtain

$$m\alpha_n^m + (m-1)\{\alpha_{n+2}^{m-2} - \alpha_n^{m-2}\} = 2d Y_{1,n}^{m-1} \ . \tag{7.72}$$

This result is the first of a small set of recursion relations that can be used to determine the general α_n^m integral in terms of simpler integrals, that will reduce, eventually, to a small number of basic forms.

Now substitute (7.66) into (7.63) and integrate the latter by parts to find

$$\pi \alpha_{n-2}^m = \frac{2}{n+2} \left[-r^m I_{-1,0}^m (\cos\theta_0)(1-r^2)^{(n+2)/2} \right]_{d-k}^{c} \tag{7.73}$$
$$+ \frac{2}{n+2} \int_{d-k}^{c} (1-r^2)^{(n+2)/2} \frac{d}{dr} (r^m I_{-1,0}^m (\cos\theta_0)) \, dr \ ,$$

or, replacing n by $n-2$,

$$\alpha_n^m = \frac{2}{n} \int_{d-k}^{c} (1-r^2)^{n/2} \frac{d}{dr} (r^m I_{-1,0}^m (\cos\theta_0)) \, dr \ . \tag{7.74}$$

Combining the last two equations, the following can be derived:

$$(n+2)\alpha_n^m = n\alpha_{n-2}^m - 2\int_{d-k}^c (1-r^2)^{n/2}r^2\,\mathrm{d}(r^m I_{-1,0}^m)\ , \tag{7.75}$$

so that, differentiating the integration variable,

$$n\alpha_{n-2}^m - (n+m+2)\alpha_n^m = 2\int_{d-k}^c (1-r^2)^{n/2}r^{m+2}\frac{\mathrm{d}}{\mathrm{d}r}I_{1,0}^m\,\mathrm{d}r\ . \tag{7.76}$$

We introduce another similar kind of integral, denoted $J_{\beta,\gamma}^m$, satisfying

$$\begin{aligned}\pi J_{\beta,\gamma}^m &= (k^2-d^2)Y_{\beta,\gamma}^m + dY_{\beta,\gamma}^{m+1}\\ &= \int_{d-k}^c (k^2-(d-x)^2)^{\beta/2}(2d(s-x))^{\gamma/2}(k^2-d^2+dx)x^m\,\mathrm{d}x \end{aligned} \tag{7.77}$$

If we observe that

$$\frac{r\,\mathrm{d}I_{-1,0}^m}{\mathrm{d}r} = \frac{-1}{\pi}\frac{\cos^m\theta_0}{\sqrt{(k^2-(d-x)^2)}}\frac{(r^2+k^2-d^2)}{2d}\ , \tag{7.78}$$

the same substitution we used for (7.70) enables us to find a second recursion relation for the α_n^m integrals as

$$n\alpha_{n-2}^m - (n+m+2)\alpha_n^m = -2J_{-1,n}^m\ . \tag{7.79}$$

We can also observe, by combining equations (7.63), (7.66) and (7.71), that

$$\alpha_0^1 = 2dY_{1,0}^0\ . \tag{7.80}$$

Equations (7.72), (7.79) and (7.80) enable us to obtain any α_n^m integral, starting from only α_0^0 and α_1^0, if we can evaluate the integrals of the foregoing types Y and J. It will be shown that similar recursion relations to the above for the α integrals exist also for the Y and J integrals, and the whole framework of these auxiliary integrals reduce down to the explicit formulation of just four integrals of the type $I_{-1,n}^0$ ($n=0,3$).

Auxiliary integrals

If in Eqn (7.70) we write in place of x^m, $(d-(d-x))^m$, it is not difficult, using (7.65), to obtain

$$Y_{\beta,\gamma}^m = \pi k^{1+\beta+\gamma}\sum_{j=0}^m k^j d^{m-j}\binom{m}{j}(-1)^j I_{\beta,\gamma}^j\ , \tag{7.81}$$

From (7.70) we can also write

$$\begin{aligned}J_{\beta,\gamma}^m &= (k^2-d^2)k^{1+\beta+\gamma+m}\sum_{j=0}^m k^j d^{m-j}\binom{m}{j}(-1)^j I_{\beta,\gamma}^j\\ &+ dk^{1+\beta+\gamma+m}\sum_{j=0}^{m+1} k^j d^{m+1-j}\binom{m+1}{j}(-1)^j I_{\beta,\gamma}^j\ . \end{aligned} \tag{7.82}$$

This means that the network of α_n^m integrals can be reduced to α_0^0, α_1^0, and integrals of the type $I_{\beta,\gamma}^m$. Again recursion relations are available. The following are easily obtained

$$I_{\beta+2,\gamma}^m = I_{\beta,\gamma}^m - I_{\beta,\gamma}^{m+2}\ , \tag{7.83}$$

and

$$I_{\beta,\gamma}^{m+1} = \frac{k}{2d} I_{\beta,\gamma+2}^m + \mu I_{\beta,\gamma}^m \ , \tag{7.84}$$

while we already had from (7.67) that

$$m\pi I_{-1,0}^m = (m-1)\pi I_{-1,0}^{m-2} + \sqrt{1-\mu^2}\,\mu^{m-1} \ , \tag{7.85}$$

that exists for the simple class of I integrals with $\beta = -1$ and $\gamma = 0$).

Turning to the more general forms defined by (7.65), if we again use the substitution $(d-s)/k = 1 - 2\kappa^2 \mathrm{sn}^2 u$, we have

$$I_{-1,\gamma}^m = \frac{2^{\gamma+1}}{\pi} \left(\frac{d}{k}\right)^{\gamma/2} \int_0^{F(\kappa)} (1 - 2\kappa^2\mathrm{sn}^2 u)^m (\kappa\,\mathrm{cn}\,u)^{\gamma+1}\,du \ , \tag{7.86}$$

where the modulus κ is given (for partial eclipses) by

$$\kappa^2 = (1 - (d-k)^2)/4dk \ . \tag{7.87}$$

From this we see that the integrals of type $I_{-1,\gamma}^0$ take the form

$$I_{-1,\gamma}^0 = \frac{2^{\gamma+1}}{\pi} \left(\frac{d}{k}\right)^{\gamma/2} \left(\frac{1-\mu}{2}\right)^{(\gamma+1)/2} \int_0^{F(\kappa)} (1 - 2\kappa^2\mathrm{cn}\,u)^{\gamma+1}\,du \ , \tag{7.88}$$

where $(1-\mu)/2$ has been written for κ^2. Integrals of this type can be expressed in terms of the regular elliptic integrals for odd γ and 'pseudo elliptic integrals' (i.e. inverse trigonometric functions) for even γ. A recursion relation is known, which we can write as

$$(\gamma+2)\frac{k}{4\pi} I_{-1,\gamma+2}^0 = \gamma\frac{d(1-\mu^2)}{k} I_{-1,\gamma-2}^0 - (\gamma+1)\mu I_{-1,\gamma}^0 \ , \tag{7.89}$$

This same recursion relation (7.89) is found to hold for the $I_{-1,\gamma}^0$ integrals for annular eclipses, but here the modulus κ_a, say, becomes the inverse of that for the partial eclipses, i.e.

$$\kappa_a^2 = \frac{4dk}{(1 - (d-k)^2)} \ . \tag{7.90}$$

In evaluating (7.66) (with $\beta = -1$) for the annular eclipse we make the substitution $(d-x)/k = 1 - 2\mathrm{sn}^2 u$, so that

$$I_{-1,\gamma}^m = \frac{2^{\gamma+1}}{\pi} \left(\frac{2d}{k(1-\mu)}\right)^{\gamma/2} \int_0^{F(\kappa_a)} (1 - 2\,\mathrm{sn}^2 u)^m \,\mathrm{dn}^{\gamma+1} u\,du \ , \tag{7.91}$$

with $\mu = 1 - 2\kappa_a^2$. For the case $m = 0$ a recursion relation for integrals of the type $\int_0^{F(\kappa)} \mathrm{dn}^n u\,du$ then shows (7.89) to be valid also in the annular case.

We thus find that we need only the explicit form of the first four $I_{-1,\gamma}^0$ integrals to be able to construct a line of of them, with successive (even and odd) γ values (Fig 7.11). Mostly, it is the $I_{-1,\gamma}^m$ integrals that are required in relevant formulations, but (7.72) and (7.80) show that the $I_{1,\gamma}^0$ forms are also sometimes required. These can be derived from the basic set by (7.83).

The four basic $I_{-1,\gamma}^0$ integrals are then given by (partial phases)

$$I_{-1,0}^0 = \arccos(\mu)/\pi \ , \tag{7.92}$$

$$I_{-1,1}^0 = \sqrt{d/k}\,(2E(\kappa) - (1+\mu)F(\kappa))\,/\pi \ , \tag{7.93}$$

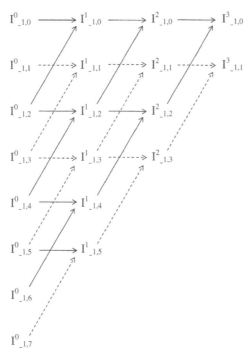

Figure 7.11 Recursive determination of the framework of auxiliary $I^m_{-1,\gamma}$ integrals from an initial set of $I^0_{-1,\gamma}$ ones.

$$I^0_{-1,2} = 2d/k \left(\sqrt{1-\mu^2} - \mu \arccos(\mu) \right) /\pi \ , \tag{7.94}$$

$$I^0_{-1,3} = 4(d/k)^{3/2} \left((1 + \mu(4 + 3\mu))F(\kappa) - 8\mu E(\kappa) \right) /3\pi \ . \tag{7.95}$$

For annular eclipses we have similarly

$$I^0_{-1,0} = 1 \ , \tag{7.96}$$

$$I^0_{-1,1} = 4\sqrt{d/k}\, E(\kappa_a)/\pi \ , \tag{7.97}$$

$$I^0_{-1,2} = -2(d/k)\mu \ , \tag{7.98}$$

$$I^0_{-1,3} = 8(d/k)^{3/2} \left((1 + \mu)F(\kappa_a) - 4\mu E(\kappa_a) \right) /(3\pi\kappa_a) \ . \tag{7.99}$$

7.3.6   Model Parametrization

The theoretical light curve corresponding to (7.10) is essentially similar to that in Fig 7.5, except for the slight rounding of the light levels between the two eclipses and higher level of the secondary eclipse compared to the primary. For the zero order model the light level outside of eclipses remains constant. The unknowns on the right of its defining equation can be seen to be the 5 parameters L_1, k, r_1, i and u_1. This extends to another 3, when we permit uncertainty in the fiducial quantities locating the x and y axes of the curve and also allow that the sum of the relative luminosities of the two stars in the model may not actually cover all the light in the system, i.e. there can be some extraneous 'third light'. The additional parameters U, $\Delta\phi_0$ and L_2 appear by replacing unity on the right in (7.4) by U, the curve's reference light level, initially taken as $U = 1$, and the independent phase variable ϕ by $\phi + \Delta\phi_0$, the shift in the phase scale $\Delta\phi_0$ being initially taken as zero.

Since there are, in general, two minima in a typical light curve we could initially expect that L_2 is constrained by $U - L_1$ in fitting the second minimum. However, allowing the possibility of additional light from some other source or sources to contribute to the data, by writing $U = L_1 + L_2 + L_3$, means that L_2 becomes unconstrained, i.e. it becomes another independent parameter. The second minimum also involves the coefficient of limb-darkening for the other star u_2. The zero order model is thus specified by 9 separate parameters.

This model omits any account of the rounding effect that becomes so noticeable in the light curves of observed systems with increasing proximity of the components. The logical next step is to consider appropriate development of this simple model to allow improved physical modelling and anticipate better agreement with observations. A perturbation approach is followed, that leads to the addition of at least another five parameters into the fitting function. We will follow details of the mass-related distortions, since the other departures from the unperturbed model are dealt with along essentially similar lines. For practical reasons, however, the fitting function is not spelled out in full in this book, but it is discussed and options provided in the freely available WINFITTER software package referenced in Section 7.6.

7.3.7   First Order Model

It was seen from (2.11) that the perturbed radius r' of a binary component, given as a fraction of the separation of the two centres of mass, may be expressed through a series of surface harmonics $r' = r\{1 + \Sigma_{j=2}^{\infty} Y_j(r, \theta)\}$, where r represents the mean radius of a sphere whose local perturbation r' is given through the zonal harmonics Y_j in the usual spherical polar co-ordinate θ (the physical conditions considered are axi-symmetric and the sectorial co-ordinate redundant). The radius r would be normally given in units of the mean separation of the two mass centres. This specification allows the appropriately developed form of (7.8) to be expressible as a series of integrals in powers of r, where mixed products of different order harmonics disappear.

The tidal and rotational disturbances were given, in Chapter 2, in the form $c_j r^j P_j$, where c_j is a known coefficient and P_j is the j-th Legendre polynomial in $\cos\theta$. The relationship between such perturbing terms and the responsive harmonics in powers of r showed dependence on the structure of the disturbed body, of mass m_1, as $Y_j = \Delta_j c_j r^{j+1} P_j / G m_1$. These responsive coefficients Δ_j are obtained from $\Delta_j = 2(j+1)/(j + \eta_j(r))$, where $\eta_j(r)$ is the surface value of the perturbation's logarithmic radial derivative, satisfying the structure-dependent differential equation, i.e. Radau's equation — Eqn (2.20). In practice, this equation is solved numerically for particular stellar models, allowing suitable values of Δ_j to be made available. In most cases of relevance, Δ_j is greater than, but close to, unity. In fact, $\Delta_j = 1$ for the well-known *Roche model* used in many binary light-curve modelling programs.

The order of magnitude of the proximity effect terms should be kept in mind for practical applications. Photometric Doppler effects may thus be expected to be an order of magnitude or so less than the classical proximity effects.

In evaluating the effects of the surface perturbations at a particular phase, or projected separation of the stars, the components of a given radius in the key directions of the other star and axis of rotation, written as $r_1\lambda$, $r_1\nu$ (for the star eclipsed at primary minimum) are decomposed into their equivalents in the observer x, y, z frame of reference (Fig 7.12). The corresponding surface distortions then appear in the relevant integrand.

It follows from (2.16) that the perturbations in r' start with terms of order r^3, and by the time we reach terms of order r^6 the interaction of tides on tides should be considered. The *first order* analysis, in which the mutual effect of perturbations on perturbations is neglected, therefore proceeds up to and including terms of order r^5. This implies three

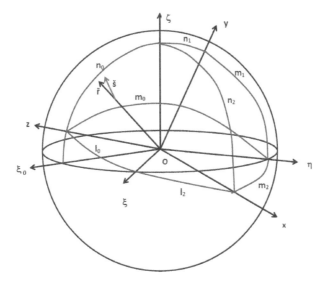

Figure 7.12 Arrangement of axes, reference directions and direction cosines (dcs) as in Fig 1.11. But now we add the arbitrarily selected radius vector $\hat{r}$, with dcs λ, μ, ν with respect to the ξ, η, ζ system. The perpendicular $\hat{s}$ to the local areal element δA, where $\hat{r}$ emerges through the stellar surface, is set with dcs l, m, n relative to the orbital axes; so that, for small surface perturbations, the angle (β) between $\hat{r}$ and $\hat{s}$ is small, and its cosine λl, μm, νn, would be close to 1. Over the visible surface, the cosine of the angle between $\hat{s}$ and the line of sight $\hat{z}$, i.e. ll_0, mm_0, nn_0, lies in the range (0-1).

tidal terms in the fitting function. Rotation with a constant angular velocity, as is usually assumed, requires only one term.

The first order form of (2.11) being specifiable, determination of the photometric effects of the mass-related distortions concerns calculation of the effects of the perturbations $\Delta r = r' - r$ on the integrals in (7.8). Specifically, (7.10) is modified to

$$l(\phi) = l_0(\phi) + \Delta l_A(\phi) + \Delta l_a(\phi) - \Delta l_{a'}(\phi) \ , \tag{7.100}$$

where the additional new terms Δl follow the same pattern as in (7.8).

Here we are essentially applying Taylor's theorem to the change in the integration of (7.8) that reduced to (7.10) in the unperturbed case. Leibniz' rule for the incremental change of a definite integral $\Delta \int_a$ gives (for the inner integral)

$$\Delta \left(\int_{y(x)} f(x,y)\,dy \right) = \int_{y_0(x)} \Delta f(x,y)\,dy + f(x,y_0)\Delta y(x)$$
$$+ \ h_1(x,y)\Delta y(x) \ , \tag{7.101}$$

where the terms on the right (at given x) are: the 'circular correction' from the effects of the perturbation of the integrand over the unperturbed ordinate y_0, the 'boundary correction' arising from the unperturbed integrand's contribution out to the perimeter perturbed by $\Delta y(x)$, and the 'second order' remainder h_1 that tends to zero faster than the other terms as $\Delta y \to 0$. The areas a, A and a', being double integrals in the adopted co-ordinates, imply that (7.101) is integrated across the range of x; whereupon we have

$$
\begin{aligned}
\Delta l &= \int_x \Delta \left(\int_{y(x)} f(x,y) \, dy \right) dx \\
&= \int_{x_0} \int_{y_0(x)} \left(\Delta f(x,y) \, dy + f(x,y_0) . \Delta y(x) \right) dx \\
&\quad + \int_{\Delta x} \left(\int_{y_0(x)} \Delta f(x,y) \, dy + f(x,y_0) . \Delta y(x) \right) dx \\
&\quad + \int_x h_1(x,y) \Delta y(x) \, dx \ ,
\end{aligned} \tag{7.102}
$$

where $\int_x$ refers to the perturbed limits of the range $(x_u - x_l)$, and x_0 in the lower limit refers to $(x_{u0} - x_{l0})$, the boundaries being formed by the circular arcs of the unperturbed model. The remainder integral of h_1 remains of second order compared with the first term when integrated over the same range. The term in $\int_{\Delta x}$ refers to the difference in the integration of Δf when the perturbed limits are used in place of $(x_{u0} - x_{l0})$. Since these further corrections $\Delta x_u, \Delta x_l$ introduce only higher order terms, the first-order version of Δl becomes the integral of (7.101) over the unperturbed range of x: i.e. the first of the three members in the lower right of (7.102).

Note that the integrand $f(x,y)$ in (7.102) is not simply the flux J of (7.8), altered by a change in local gravity associated with departure from sphericity (the gravity-darkening effect), but contains also the local areal distortion from (2.11) that would, locally, tend to counter that effect. Over the complete phase range, however, the integrated effect of gravity darkening enhances the curvature of the out-of-eclipse light variation above that of the projected areal change alone. The applied formula, usually taken to be a proportionality to local gravity, at a particular effective wavelength, is modified by a temperature effect on the distribution with wavelength governed by the coefficient $\tau = c_2 / \{ 4 \lambda T (1 - \exp(-c_2 / \lambda T)) \}$, c_2 being the second radiation constant (0.014388 m K). The integrand also accounts for the change in projection towards the line of sight of a local surface element from sphericity arising from the surface distortion.

Concerning the boundary correction, it follows from the relevant formulae defining the boundaries that it is only the deviation along the shadow cylinder's inner boundary on the eclipsed star that introduces a first order effect on $\Delta \int_a$. Boundaries formed by the intersection of the components with the sky plane have a negligible effect. The second term on the lower right of (7.102) therefore arises only from the inner boundary, while the quantities $\Delta l_A(\phi)$, $\Delta l_a(\phi)$ are obtained by integrating the perturbation over a complete hemisphere.

Ellipticity

The choice of harmonic function to represent the surface distortion allows the same type of function in the integrated result. The mass related (ellipticity) effects are usually greater for the primary term $\Delta l_A(\phi)$ in (7.100). The resulting expression for this reduces to the form

$$
\Delta l_A(\phi) = L_1 \{ c'_{r,2} P_2(n_0) + \Sigma_{j=2}^4 c'_{t,j}(j) P_j(l_0) \} \ , \tag{7.103}
$$

where the known c' coefficients (here we use suffix r for rotation and t for tides) incorporate the effects of limb-darkening, areal distortion, projection, gravity darkening, structure, and adjustment for wavelength and temperature-dependence. The direction cosine n_0 relates the line of sight to the rotation axis, while l_0 similarly relates the line of sight to the phase-dependent line of centres (see Fig 1.12).

We can write the end result, that will be verified later, as

$$
\begin{aligned}
\Delta l_{A,ell,0} ={}& L_1[(2+\Omega_{0,2}/4)P_2(n_0)v_2/3 \ - \\
& - (2+\Omega_{0,2}/4)P_2(l_0)w_2 \ + \\
& + (1+\Omega_{0,4}/24)P_4(l_0)w_4/8 \ ...]
\end{aligned}
$$

and

$$
\begin{aligned}
\Delta l_{A,ell,1} ={}& L_1[(8+4\Omega_{1,2}/5)P_2(n_0)v_2/9 \ - \\
& - (8+4\Omega_{1,2}/5)P_2(l_0)w_2/3 \ - \\
& - (3+\Omega_{1,3}/6)P_3(l_0)w_3/2 \ ...]
\end{aligned} \tag{7.104}
$$

in the linear arrangement

$$
\Delta l_{A,ell} = 3[(1-u)\Delta l_{A,ell,0} + u\Delta l_{A,ell,1}]/(3-u) \ . \tag{7.105}
$$

The rotational and tidal coefficients v_j, w_j, that were first encountered in connection with Eqn (2.12), require an integration of Radau's structural equation for the logarithmic density quantity η_j, where j is the order of the relevant spherical surface harmonic $P_j(\lambda)$. Now, the addition theorem for Legendre polynomials gives

$$
P_h(\lambda) = P_h(\cos\gamma)P_h(\cos\epsilon) + 2\sum_{m=1}^{n}\frac{(n-m)!}{(n+m)!}T_n^m(\cos\gamma)T_n^m(\cos\epsilon)\cos m\psi \tag{7.106}
$$

where γ is the angle between the outward normal to the equipotential $\hat{s}$ at a given point on the surface and the line of sight $\hat{z}$, ϵ is the angle between and $\hat{\xi}$ and $\hat{z}$ in Fig 7.12, and the sectorial angle ψ runs from 0 to 2π.* This theorem allows that integration of the distortion over the visible hemisphere will result in corresponding Legendre polynomials in $l_0 = \cos\phi\sin i$ for the tidal terms and $n_0 = \cos i$ for the rotation, ϕ here referring to the orbital phase of the star. In the present case we have, for the tidal terms $w_j = (2j+1)qr_1^{j+1}/(j+\eta_j)$, where q is the secondary to primary mass ratio m_2/m_1. For the rotational distortion we can similarly write $v_j = (2j+1)(1+q)\omega^2 r_1^{j+1}/[(j+\eta_j)\omega_0^2]$, where ω is the star's angular rotation speed, which is compared with that synchronized with the Keplerian motion ω_0.

The coefficients $\Omega_{i,j}$ in Eqns (7.104) are used to gather together contributions in the same surface harmonic. They involve both limb and gravity darkening as well as the structure, but their evaluation is fairly direct when the η_js for the star are provided. Kopal's (1959) formula is

$$
\Omega_{i,j} = (1+\eta_j)\tau - 2 - j(i+1), \tag{7.107}
$$

where τ is the gravity darkening coefficient and i is the limb-darkening index: 0 for undarkened, 1 for the linear term in $\cos\gamma$, 2 for the squared term and so on. Here j is the order of the relevant spherical harmonic, the full distortion being expressed as a convergent series of such harmonics.

With τ the bolometric index of gravity darkening as discussed in Section 7.3.3, its value at the wavelength of observation λ is given as

$$
\tau_\lambda = \frac{c_2}{4\lambda T[1 - exp(-c_2/\lambda T)]} \ . \tag{7.108}
$$

---

*This sectorial angle will be found to be largely redundant in the standard model expounded in the presented treatment.

The argument used here is essentially similar to that applied in Eqn 7.30, with changes of total energy having different results in the distribution of energy levels with wavelength, the Planck function forming a first approximation to this distribution. In modelling high accuracy light curves, such as from the TESS satellite, this coefficient may be regarded as an adjustable parameter, the Planck approximation being used for initial trials.

Reflection

The secondary contribution $\Delta l_a(\phi)$ in (7.8) tends to show more the consequences of its irradiation, which requires us to consider the *reflection effect*. In the first order model, effects of the reflection of reflected light are neglected and we have the form

$$\Delta l_a(\phi) = E_2 L_1 \Sigma_{j=2}^4 c'_{\text{ref},j} \Phi_j(l) \ , \tag{7.109}$$

where l contains the same phase argument as before, Φ_j being mixed trigonometric functions of l and $\sqrt{1-l^2}$, factored by powers of r_2^j in the coefficients $c'_{\text{ref},j}$. The 'luminouos efficiency' coefficient E_2 can be estimated from alternative theoretical models, but these involve information on the secondary's absorption and scattering properties that may not be directly available. E_2 would therefore often be treated as an adjustable parameter in the operation of an optimizing curve-fitting program. Another name for a parameter that plays an essentially similar role in the fitting function is the *geometric albedo* (A_g). A guide to the value of this parameter is obtained from the Planck approximation:

$$E_{1,2} = \tau_{1,2} \left(\frac{T_{2,1}}{T_{1,2}}\right)^4 \frac{\exp(c_2/\lambda T_{2,1}) - 1}{\exp(c_2/\lambda T_{1,2}) - 1} \ , \tag{7.110}$$

which assumes that the incident flux from the companion leads to a local increase of temperature, such that the bolometric emission at the new temperature is equal to the sum of the original plus received fluxes. The approximation may be useful for relatively low differences in the unperturbed surface temperatures of the components, but it becomes inaccurate when there are large differences in these temperatures, for example with the optical light curves of X-ray binaries containing cool dwarf and hot subdwarf components.

Δl_A also contains a term of the form (7.109) with leading coefficient $E_1 L_2$, where L_2, the often quite small intrinsic fractional light of the secondary star. E_1 may become large enough to enhance this contribution to a measurable level, however, because if the primary is a relatively cool star, like the Sun, or of lower temperature, a relatively low level of warming can increase the visible flux substantially. In a similar way, Δl_a contains an ellipticity contribution of the form (7.105), but the factoring by L_2 would often imply a rather low scale of effect.

Eclipses

During eclipse phases the integrands that went into $\Delta l_A(\phi)$ or $\Delta l_a(\phi)$ are integrated over the relevant eclipsed areas, where they form expressions in mixed powers of x and z factored by appropriate direction cosines. The ellipticity component of $\Delta l_{a'}(\phi)$, then produces a rather cumbersome-seeming summation requiring some 25 separate α_n^m-integrals to be calculated to construct the main first member of Eqn 7.104 for the case of linear limb-darkening. The existence of the simple recursion relations, presented in Section 7.3.5, make this summation not as daunting as it may first appear, and, in fact, the calculations are performed very quickly on modern computers

It is instructive to check through the integration of the right hand members for the ellipticity effect in (7.104), as an example, where we can write the integrand as made up of terms of the form $r^2 H \cos^h \gamma$. As before, $h = 1$ comes from the integration over an undarkened disk; $h = 2$ over a disk darkened towards its limb linearly with $\cos\gamma$; a quadratic darkening corresponds to $h = 3$, and so on; with h equivalent to $i + 1$ in (7.107). The local surface flux is H and r the local stellar radius.

Each of the three components r, H and $\cos\gamma$ is subject to variation due to the surficial distortion, as formulated in Chapter 2. For r we have from 2.11 with 2.16

$$r^2 = r_1^2(1 + 2\sum k_j P_j) \ , \tag{7.111}$$

the small quantities $\Delta_j a_1^{j+1} c_j / Gm_1$ being here abbreviated to k_j.

For the flux H, taking into account the gravity darkening from (2.48) and (7.38), we have

$$
\begin{aligned}
H &= H_0 \left\{ 1 + \tau \left(\frac{g - g_0}{g_0} \right) \right\} \\
&= H_0 \left\{ 1 - \tau \sum_j \left(\frac{2j+1}{\Delta_j} + 1 - j \right) \right\} k_j P_j \ .
\end{aligned}
\tag{7.112}
$$

The processing of $\cos\gamma$ is not so direct. We have, keeping in mind the orientation sketched in Fig 7.12,

$$\cos\gamma = ll_0 + mm_0 + nn_0 \ , \tag{7.113}$$

where the direction cosines (dcs) l_0, m_0, n_0 are those of the line of sight in the orbital system and l, m and n are those of the normal to the local equipotential surface $\nabla\Psi$, so that $l = -\Psi_x/|g|$, $m = -\Psi_y/|g|$, $n = -\Psi_z/|g|$, while $|g|^2 = \Psi_x^2 + \Psi_y^2 + \Psi_z^2$. The subscripts on Ψ refer to the partial (spatial) derivatives with respect to x, y and z. Given the predominating influence of the radial component of gravity in this model, it is useful to convert to a spherical polar co-ordinate system where we can write (noting that ν does not involve ϕ),

$$
\begin{aligned}
\Psi_x &= \lambda\Psi_r + \frac{\partial\lambda}{\partial\theta}\Psi_\theta + \frac{\partial\lambda}{\partial\phi}\Psi_\phi \\
\Psi_y &= \mu\Psi_r + \frac{\partial\mu}{\partial\theta}\Psi_\theta + \frac{\partial\mu}{\partial\phi}\Psi_\phi \\
\Psi_z &= \nu\Psi_r + \frac{\partial\nu}{\partial\theta}\Psi_\theta.
\end{aligned}
\tag{7.114}
$$

Here, Ψ_θ stands for $\partial\Psi/r\partial\theta$ and Ψ_ϕ for $\partial\Psi/r\sin^2\theta\,\partial\phi$. Hence, for $\cos\gamma$ we write

$$
\begin{aligned}
ll_0 + mm_0 + nn_0 &= -\frac{\Psi_r}{g}l_0\{\lambda + \frac{\partial\lambda}{\partial\theta}\frac{\Psi_\theta}{\Psi_r} + \frac{\partial\lambda}{\partial\phi}\frac{\Psi_\phi}{\Psi_r}\} \\
&\quad -\frac{\Psi_r}{g}m_0\{\mu + \frac{\partial\mu}{\partial\theta}\frac{\Psi_\theta}{\Psi_r} + \frac{\partial\mu}{\partial\phi}\frac{\Psi_\phi}{\Psi_r}\} \\
&\quad -\frac{\Psi_r}{g}n_0\{\nu + \frac{\partial\nu}{\partial\theta}\frac{\Psi_\theta}{\Psi_r}\}
\end{aligned}
\tag{7.115}
$$

with corresponding expressions for the products with μ and ν. Ψ_r differs from $|g|$ by the quantity $\cos\beta$ (Fig 7.12), where $\sin\beta$ is is taken to be a quantity of second order. This allows to rearrange the previous equation to write

$$\cos\gamma = l_0\{\lambda - l_0 \sum k_j \left[\frac{\partial\lambda}{\partial\theta}\frac{\partial P_j}{\partial\theta} + \frac{1}{\sin^2\theta}\frac{\partial\lambda}{\partial\phi}\frac{\partial P_j}{\partial\phi}\right]\} \tag{7.116}$$

where the small deviation of $\cos\gamma$ from $l_0\lambda + m_0\mu + n_0\nu$, the projection of the unit radius vector in the line of sight, involves only the derivatives of the surface harmonics P_j. At this point we should notice that the two main disturbances under consideration, i.e. rotation and tides, are separate functions of only ν and λ in the usual (aligned rotation) arrangement.

After a little manipulation, we find

$$\cos\gamma = z\{1 - \left[\frac{l_0}{z} - \lambda\right]\sum k_j P'_j(\lambda) + \left[\frac{n_0}{z} - \nu\right]k_2 P'_2(\nu)\} \tag{7.117}$$

where we refer to the unit sphere concentric with that of Fig. 7.11, with co-ordinates

$$\begin{aligned}
x &= l_2\lambda + m_2\mu + n_2\nu \\
y &= l_1\lambda + m_1\mu + n_1\nu \\
z &= l_0\lambda + m_0\mu + n_0\nu
\end{aligned} \tag{7.118}$$

We can also make use of the relation

$$\lambda P'_j(\lambda) = j P_j(\lambda) + P'_{j-1}(\lambda) \tag{7.119}$$

which allows the term in $P_j(\lambda)$ in $\cos^h\gamma$ to be collected with those coming from (7.111) and (7.112) into a main component with coefficient $\Omega_{i,j}$. We will then have, writing out the first (no limb-darkening) rotational term in ν,

$$r^2 H \cos\gamma = r_1^2 H_0 z\{1 + \frac{1}{3}v_2\left(\Omega_{0,2}P_2(\nu) + (\frac{n_0}{N}P'_2(\nu) - 1)\right)$$
$$(+ \text{ tidal terms}) \} \tag{7.120}$$

where the coefficient $\Omega_{0,2}$ incorporates the structural and gravity effects, as in (7.104), through the Δ_j and τ values appearing in (7.112), with $\Omega_{0,2} = (3/\Delta_2-1)\tau-4$, in accordance with (7.107) To deal with a general law of limb-darkening, the left side of (7.120) becomes replaced by $r^2 H \cos^h\gamma$, where $h = 1$ corresponds to (7.120) and $h = 2$ would introduce coefficients $\Omega_{1,j}$ as in (7.107).

Now, inverting (7.118), we have $\nu = n_0 z + n_1 y + n_2 x$ and $\lambda = l_0 z + l_1 y + l_2 x$. Note also that, in the rotational terms, odd powers of y in the integrand will drop out, as mentioned above, while, since $y^2 = 1 - x^2 - z^2$, even powers in y can be replaced by terms that involve only x and z. For the tidal terms, it can be seen in Fig 7.12 that l_0 and l_2 are complementary direction cosines in the same plane: the choice of axes ensuring that $l_1 = 0$.

To follow through the details, let us fix attention on the rotational term spelled out in (7.120), for which we write, in the case of zero limb-darkening:

$$r^2 H \cos\gamma = r_1^2 H_0 z\{1 + v_2[\frac{1}{2}\Omega_{0,2}(n_0^2 z^2 + n_1^2(1 - x^2 - z^2) + n_2^2 x^2 + 2n_0 n_2 xz - 1)$$
$$+ 3\frac{n_0}{z}(n_0 z + n_2 x) - 1]\} \tag{7.121}$$

This means that the rotation induced distortion, integrated over a given area, is made up of the principal second harmonic α-integral-containing components $A_{1,p}$, say, (containing the Ω terms) and that includes the remaining terms $A_{1,r}$ on the right of (7.121). The first lower subscript in A corresponds to the integer $h = i + 1$ where i is the leading power in the adopted limb-darkening law. In rendering the integration of (7.121) over the surface elements $r^2 \cos\gamma \, d\gamma \, d\phi$, for the visible hemisphere of unit sphere, as a two dimensional integration in $dx \, dy$ over the projected disk, the added $\cos\gamma \equiv z$ on the right drops from the formulation, as the now redundant projection factor. So, for the first ($h = 1$) integration of the $P_2(\nu)$ contribution, integrated in x and y over the eclipsed region, we have

$$A_{1,p} = 3(n_0^2 - n_1^2)\alpha_2^0 + 3(n_2^2 - n_1^2)\alpha_0^2 + (3n_1^2 - 1)\alpha_0^0 + 6n_0 n_2 \alpha_1^1 . \tag{7.122}$$

For the remaining terms, we have

$$A_{1,r} = 3n_0^2 \alpha_0^0 + 3n_2 n_0 \alpha_{-1}^1 - \alpha_0^0 . \tag{7.123}$$

Integration of the tidal terms proceeds in a similar way. We write:

$$B_{1,p} = 3l_0^2 \alpha_2^0 + 3l_2^2 \alpha_0^2 - \alpha_0^0 + 6l_0 l_2 \alpha_1^1 \tag{7.124}$$

and

$$B_{1,r} = 3(l_0^2 - 1)\alpha_0^0 + 3l_0 l_2 \alpha_{-1}^1 , \tag{7.125}$$

with the third harmonic given by,

$$C_{1,p} = 5l_0^3 \alpha_3^0 + 15l_0^2 l_2 \alpha_2^1 + 15l_0 l_2^2 \alpha_1^2 + 5l_2^3 \alpha_0^3 - 3l_0 \alpha_0^1 - 3l_2 \alpha_0^1 \tag{7.126}$$

and

$$C_{1,r} = (l_0 \alpha_1^0 + 2l_2 \alpha_0^1)\left(\frac{15}{2}l_0^2 - \frac{3}{2}\right) - \frac{3}{2}l_0(\alpha_1^0 - 5l_2^2 \alpha_{-1}^2 + \alpha_{-1}^0) . \tag{7.127}$$

For the fourth harmonic ($h = 1$) tidal term the corresponding components are:

$$\begin{aligned} D_{1,p} = \ & 35l_0^4 \alpha_4^0 + 140l_0^3 l_2 \alpha_3^1 + 210l_0^2 l_2^2 \alpha_2^2 + 140l_0 l_2^3 \alpha_1^3 + 35l_2^4 \alpha_0^4 \\ & - 30l_0^2 \alpha_2^0 - 60l_0 l_2 \alpha_1^1 - 30l_2^2 \alpha_0^2 + 3\alpha_0^0 \end{aligned} \tag{7.128}$$

and

$$\begin{aligned} D_{1,r} = \ & \frac{1}{2}(l_0(35l_0^3 - 15l_0)\alpha_2^0 + 15l_2^2(l_0^2 - 1)\alpha_0^2 - (15l_0^2 - 3)\alpha_0^0 \\ & + 15l_0 l_2(7l_0^2 - 2)\alpha_1^1 + 35l_0 l_2^3 \alpha_{-1}^3 - 15l_0 l_2 \alpha_{-1}^1) \end{aligned} \tag{7.129}$$

The higher terms in μ in the limb-darkening series (7.34) are dealt with by appropriate changes to the Ω coefficient according to (7.107), by increasing of the lower (n) suffix of the α-integrals, and suitably changing the coefficient of the remainder terms.

Putting things together, we find

$$\begin{aligned} \int_{x_0} \int_{y_0(x)} \Delta f(x,y) \, dy \, dx \ = \ & v_2/3\left(\frac{1}{2}\Omega_{h-1,2} A_{h,p} + h A_{h,r}\right) \\ & - w_2\left(\frac{1}{2}\Omega_{h-1,2} B_{h,p} + h B_{h,r}\right) \\ & - w_3\left(\frac{1}{2}\Omega_{h-1,3} C_{h,p} + h C_{h,r}\right) \\ & - w_4\left(\frac{1}{8}\Omega_{h-1,4} D_{h,p} + h D_{h,r}\right) \end{aligned} \tag{7.130}$$

We can check that this rather protracted development is in agreement with (7.104) by inserting the *total* forms of the α-integrals. These are given by

$$\alpha_n^m = \frac{\Gamma((m+1)/2)\Gamma((n+2)/2)\Gamma(1/2)}{\pi\Gamma((m+n+4)/2)} \tag{7.131}$$

for even m and $\alpha_n^m = 0$ for m odd. Table 7.1 lists the results for the lower m and n values.

Table 7.1 Total values of the α-integrals

m	0	2	4	6	8
n					
−1	2	$\frac{2}{3}$	$\frac{2}{5}$	$\frac{2}{7}$	$\frac{2}{9}$
0	1	$\frac{1}{4}$	$\frac{1}{8}$	$\frac{5}{64}$	$\frac{7}{128}$
1	$\frac{2}{3}$	$\frac{2}{15}$	$\frac{2}{35}$	$\frac{2}{63}$	$\frac{2}{99}$
2	$\frac{1}{2}$	$\frac{1}{12}$	$\frac{1}{32}$	$\frac{1}{64}$	$\frac{7}{768}$
3	$\frac{2}{5}$	$\frac{2}{35}$	$\frac{2}{105}$	$\frac{2}{231}$	$\frac{2}{429}$
4	$\frac{1}{3}$	$\frac{1}{24}$	$\frac{1}{80}$	$\frac{1}{192}$	$\frac{1}{384}$
5	$\frac{2}{7}$	$\frac{2}{63}$	$\frac{2}{231}$	$\frac{10}{3003}$	$\frac{2}{1287}$
6	$\frac{1}{4}$	$\frac{1}{40}$	$\frac{1}{160}$	$\frac{1}{448}$	$\frac{1}{1024}$

In discussing the light loss in eclipses expressed through the first term on the lower right (7.102) we have, with (7.130), written out only the first integral in that term. There remains the boundary correction $\int_{x_0} f(x,y_0).\Delta y(x)\,\mathrm{d}x$. These corrections, one for either star, can be evaluated in terms of integrals related to the α-integrals that we met already. The distortion of the inner boundary from circularity $\Delta y(x)$ can be shown to be (to the required accuracy)

$$\Delta y(x) = \frac{k^2}{y}\Delta r_2 + \frac{k^2 - d(d-x)}{y}\Delta r_1 \tag{7.132}$$

with the distortions $\Delta r_{1,2}$ being given, as before, in the form of

$$\Delta r_i = \sum_{j=2}^{4} w_i^{(j)} P_j(\lambda) - \frac{1}{3} v_i^{(2)} P_2(\nu) \ . \tag{7.133}$$

Attention is required here to the geometry of the secondary's shadow cylinder. Thus, we have $z^2 = 2d(s-x)$ along the undistorted boundary of the eclipsing star's shadow cylinder on the eclipsed star, where $y = (k^2 - (d-x)^2)^{1/2}$. The main integral in question, from (7.65), is of the form $I_{-1,\gamma}^m$, where

$$\pi k^{m+n} I_{-1,n}^m = \int_{d-k}^{c} (d-x)^m y^{-1} z^n\,\mathrm{d}x \ , \tag{7.134}$$

y and z here being set by the intersection of the shadow cylinder on the eclipsed star.

It turns out that the integral $I_{-1,\gamma}^m$ plays a basic role in setting up the whole calculation of proximity effects in eclipse. This may be gathered from the foregoing subsection on auxiliary integrals, where these I-integrals were used to link the recursion relations satisfied by the α-integrals. From the first term on the right in (7.132), we can anticipate that the Δr_2 component of the boundary correction $\int_{x_0} f(x,y_0).\Delta y(x)\,\mathrm{d}x = f_2$, say, is expressed in terms of the I-integrals. The integration proceeds along the boundary arc symmetrically above and below the x-axis, x and y being always positive. For even powers in y, however, the upper and lower members cancel each other out in the integration, and so $I_{\beta,\gamma}^m$ do not appear with even β. For the correction associated with the distortion of the secondary the direction cosines λ and ν in (7.133) are constrained by the projection of the secondary's limb into the sky plane where $z = 0$. Thus, $k\lambda = l_2(d - x)$, and $k\nu = n_2(d - x) + n_1 y$ for the secondary's boundary correction.

Considering just the rotational term, for example, it will be found that

$$f_2 = -k^{h+1} \frac{v^{(2)}}{3} \left(3n_1^2 I^0_{1,h-1} + 3n_2^2 I^2_{-1,h-1} - I^0_{-1,h-1} \right)$$
$$+ \text{ tidal terms.} \tag{7.135}$$

The tidal terms, involving the foregoing form for λ, are evaluated in an essentially similar way. Note here the appearance of the $I^m_{1,\gamma}$ integral, related to the basic set via (7.83).

The boundary correction arising from the primary is somewhat more complicated than this. It involves the integrals $J^m_{\beta,\gamma}$ defined in (7.77), whose connection to the the the Δr_1 component of the boundary correction, f_1, say, will be clear from the first line of (7.77) and the definition of the $Y^m_{\beta,\gamma}$-integrals given in (7.70). That the $J^m_{\beta,\gamma}$ can be given in terms of the $I^m_{\beta,\gamma}$-integrals follows also from (7.82). The direction cosines for λ and μ in (7.133) now revert to the previously used primary surface forms obtained from inverting (7.118): $\lambda = l_0 z + l_2 x$ and $\nu = n_0 z + n_1 y + n_2 x$; but with x, y and z required to be on the intersection arc. It follows that f_1 takes the form (spelling out just the rotational part):

$$f_1 = \frac{v^{(2)}}{3} (3n_0^2 J^0_{-1,h-1} + 3n_1^2 J^0_{1,h-1} + 3n_2^2 J^2_{-1,h-1}$$
$$+ 6n_0 n_2 J^1_{-1,h} - J^0_{-1,h-1})$$
$$+ \text{ tidal terms.} \tag{7.136}$$

The construction of the tidal terms goes in an essentially parallel way to the rotational ones.

For the final expression for the effect of surface distortions in eclipse summarized by (7.102) we have

$$\Delta l_{\text{ell}} = L_1(f_\star + f_1 + f_2) \ , \tag{7.137}$$

where $f_\star$ is given by (7.130), and f_1 and f_2 from (7.136) and (7.135), respectively.

Evaluating the effect of eclipses on the reflection function (7.109) involves a similar procedure to that for the ellipticity, but it turns out to be much shorter, at least up to the terms in r^4 factoring the reflected light. This is because, within the general scheme of approximation, we neglect the combination of perturbations on perturbations and integrate the reflection over the mean spherical surface. The loss of reflected light in eclipses can then be expressed by the formula

$$\Delta l_{\text{ref}} = E_1 L_2 \{ r_1^2 (l_0 \alpha_1^0 + l_2 \alpha_0^1) +$$
$$+ \ 3r_1^3[(l_0^2 \alpha_2^0 + 2l_0 l_2 \alpha_1^1) - \alpha_0^0] +$$
$$+ \ \frac{3}{2} r_1^4 (5l_0^3 \alpha_3^0 - 3l_0 \alpha_1^0) \} \tag{7.138}$$

This relatively simple, but expedient, formula overlooks the potentially great physical complexity of the subject in detail.

7.3.8   Frequency domain analysis

The use of Fourier techniques in information analysis is well-known. The coefficients of a Fourier series representation of uniformly distributed well-sampled data efficiently characterize the data's information content in some sense, though the number of Fourier coefficients that a typical close system light curve would confidently yield would be limited in comparison to the full list of parameters in a comprehensive physical model.

The normal procedure involves multiplication of the data with sines and cosines of $n\theta$, where θ is the phase and n are the regular integers 0, 1, 2 etc. The products are summed over the phase range in order to determine the corresponding Fourier coefficients. Efforts to analyse light curves in this way go back at least to the 1960's. Finding results generally involved numerical tabular comparisons rather than a direct literal connection between the sought physical parameters and the Fourier coefficients. But it can be shown that in the highly idealized circumstances of no photometric proximity effects, and a total eclipse minimum, accurately located with respect to the axes of phase and light flux, the limb-darkening of the eclipsed star being assigned, four Fourier-like 'moment' integrals A_{2m}, given as

$$A_{2m} = \int_0^{\frac{\pi}{2}} (1-l)\, d\sin^{2m}\theta, (m = 0,...3),$$

are necessary and sufficient explicitly to determine the four parameters of the zero-order problem: r_1, r_2, i and L_1. Other thoughts motivating this approach are (i) the idea of *separating out* proximity effects from the properties of the individual components, and (ii) that the integration process reduces the effects of noise on derived parameters, though parameters derived in this way are not necessarily optimal, in the χ^2 minimizing sense.

In the first order problem, moments A_{2m} are found from appropriate numerical quadrature of the light curve as given in the preceding formula. Corresponding proximity effect integrals are constructed using a series expansion in integer powers of $\cos\theta$ to match the light curve regions outside the eclipse minima. The required proximity integrals are then relatively simple expressions in the coefficients c_j of this series. These integrals are subtracted from the moments, to leave residues that are related to the sought parameters, along the lines of the zero order problem. The method is clear and suggests the foregoing advantages concerning separability and noise removal. Ambiguities have been found in practice, however, and a detailed exposition is omitted here. Relevant references are given in Section 7.6.

One particular issue concerns the predominating effect of the proximity integrals, because the $\sin^{2m}\theta$ weighting couples with the proximity effects carrying through all phases, unlike the eclipses. Hence, the proximity terms have to be very precise, and though the $\cos^j\theta$ form permits algebraic tractability, the corresponding coefficients are not well suited to accurate numerical derivation. The values of the eclipse residues, and therefore the sought parameters, become sensitive to the proximity representation, especially with the higher $2m$ components. While integrals higher than A_4 are not necessary to evaluate the basic geometric parameters if the moments of both minima are combined, a direct connection between eclipse moments and parameters applies only for total eclipses. For the annular case similar approximate expressions can be given, but the procedure definitely loses directness when eclipses remain partial.

Frequency domain investigations were pursued in a series of papers produced by Z. Kopal and his associates in the 1970s, and among the findings was a uniform expression for the generalized light loss function when expressed as a Hankel transform of the product of two Bessel functions. The first of these latter functions (J_ν) arises from the two-dimensional Fourier transform of the light distribution over the (presumed circular) outline of the eclipsed star, the other (J_1) is the equivalent transform for the opaque eclipsing star. Thus, for the eclipse integral α_n^0 we have

$$\alpha_n^0 = 2^\nu\, \Gamma(\nu) \int_0^\infty (kx)^{-\nu} J_\nu(kx) J_1(x) J_0(hx)\, dx \ ,$$

where $\nu = (n+2)/2$, $k = r_1/r_2$ and $h = \delta/r_2$. This integral can be expressed in known convergent series of hypergeometric functions. The formula demonstrates an interesting systematization of notation for the α-integrals, although, unlike the formulae of Sections 7.3.4-5, these are not closed expressions. Kopal summarized the relevance of his cross-correlation of apertures approach to eclipse integral determination as: (i) greater mathematical symmetry, (ii) wider application and adaptability to light losses from eclipses in general, (iii) more directness for numerical evaluation, and (iv) enhanced physical insights arising from parallels with general problems of optics. These points provided recurrent themes within Kopal's *Language of the Stars* (1979). Their wider implementation may be realized with future developments in information processing.

7.4 Atmospheric Eclipses

It was mentioned in connection with Equation (7.49) that the A component of the eclipse integral predominates when one star is very much larger than the other. This simpler 'straight edge' formula could become useful in such contexts, and it has been applied to eclipses involving a normal star and a giant. The luminosities in the visual range of two quite different sized stars can, in certain circumstances, be comparable, so that if the small star passes behind the large one there would be a noticeable loss of light.[*]

Classic stellar examples are ζ Aur and VV Cep. The very size of the large components in these binaries implies long orbital periods: some 972 d in the case of ζ Aur,[*] with phases of total eclipse lasting for almost 40 d. For several days on either side of the totalities, close monitoring of the light variation indicates semi-transparency of the outer layers of the eclipsing star probed by the dwarf's light rays: an *atmospheric eclipse*. Early spectroscopic observations of marked changes in the appearance of the hot component's spectrum during such phases alerted astronomers to the possibility of measurable photometric effects (see Fig 7.13). Comparable situations may occur in other systems involving planets, satellites or collapsed stars, but, historically, cases like ζ Aur prompted the kind of analysis that we now consider.

Hitherto, it has been adopted that observed eclipse effects can be associated with surfaces having distinct edges. For the solar photospheric standard model, at the optical depth unity surface, the opacity scale height is only $\sim 0.01\%$ of the radius (Fig 7.14). The absorptive path length in the line of sight near this limb would increase by an order of magnitude in moving in by such a small fraction of the radius. This implies that, for such a photosphere, atmospheric effects in eclipses would begin to affect phase intervals around the eclipse tangency points of the order of 10^{-5} of the orbital period, with effects well below what could be normally discerned even in precise contexts. To discuss the photometry of atmospheric eclipse effects meaningfully, therefore, a key factor is the magnitude of the opacity scale height near the optical depth unity surface in relation to the size of the star.

In an influential paper on the prototype system ζ Aur, D. H. Menzel found that the light decline in the early stages of the eclipse of the B type dwarf satisfied a linear relationship between the logarithm of the drop in magnitude $\log \Delta m$, and the separation h from an adopted reference point (external tangency) at phase 20 days ($\sim 7.4°$) from the mid-minimum. Such a relationship can be argued plausible, given the magnitude (m) scale

---

[*]The corresponding transit, when only a small fraction of the large star is eclipsed, would, of course, involve a much lower scale of light loss.

[*]More than 20 y for VV Cep. V777Sgr, with very similar parameters to ζ Aur, has been recently monitored by W. S. G. Walker et al. (see 7.6).

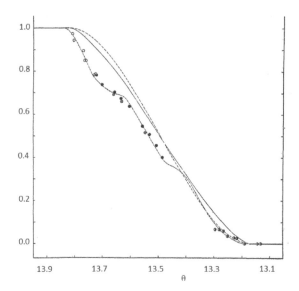

Figure 7.13   Eclipse of the system ζ Aur. The fractional loss of light of the early type component is plotted against phase angle θ (absolute value) measured from mid-eclipse. Four data-sources in the photovisual spectral range were compiled by Z. Kopal (from *Astrophys. J.*, Vol. 103, p310, 1946; ©AAS) and used to argue that the eclipse did not correspond to the standard model of two disks with well-defined boundaries between opaque and transparent regions.

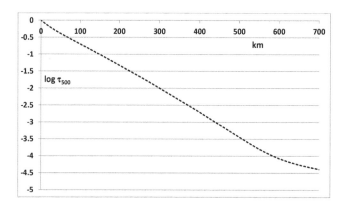

Figure 7.14   Logarithm of decline of optical depth in the visual range (τ_{500}) above the optical depth unity 'surface' in the outer 0.1% of the solar radius. The exponential fall of the optical depth is matched by a similar decline in the density, though the exponent for the latter is only about half that for the optical depth.

is already a logarithmic one. From Menzel's theoretical presentation, a double exponentiation can be expected to affect the light decline, since the emerging flux is reduced by the exponent of the optical depth, while the optical depth itself depends on an opacity that declines exponentially with height. The opacity's behaviour is here essentially similar to what is found in the solar photospheric model (Fig 7.14), but the length scale for the same proportional decline must be much greater in the case of giant atmospheres.

Menzel thus found an approximate relationship of the form

$$\log(m - m_0) = ah + b, \tag{7.139}$$

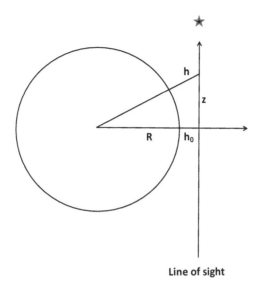

Line of sight

Figure 7.15   Schematic arrangement for the atmospheric eclipse model discussed in the text. Light from the eclipsed dwarf star passes above the limb of the eclipsing star at height h_0 above the reference radius R, where the atmosphere is taken to be fully opaque. Absorption is considered in dependence on the path length z.

with a and b constants, to accord tolerably well with the observations. From the slope a, Menzel found the photospheric scale height to be about $1/20$ of the adopted photospheric radius of the supergiant. We may then expect a difference between eclipses for ζ Aur type binaries and the classical ones with well-defined photospheric radii. The atmospheric region whose semi-transparency produces noticeable light loss could have a lateral extent several times that of the hot star undergoing eclipse.

In setting out the analysis, Menzel used an expression for the optical depth τ of the form

$$\tau = C_\lambda(T) \int N_i N_e \, dz, \qquad (7.140)$$

where C_λ, dependent on the temperature and wavelength, can be taken as averaged to a constant across the narrow photospheric layers. The main factor determining the optical depth is then the density arising from locally equal numbers of ions N_i and electrons N_e to be integrated along the line of sight z to the star undergoing eclipse, regarded as a point source in the first approximation.

Let us set z to be measured from the sky plane passing centrally through the eclipsing star and perpendicular to the line of sight as shown in Fig 7.15. We consider an absorbing column along z located at height h_0 above a nominal photospheric radius R. The relation

$$(R+h)^2 - (R+h_0)^2 = z^2 \qquad (7.141)$$

then holds. This allows us to deal with Eqn (7.140), adopting an exponential decline of density, so that we can write

$$\int_{-\infty}^{\infty} N_i N_e \, dz = 2N_0^2 \int_0^{\infty} \exp\{-2h/H\} \, dz \qquad (7.142)$$

where H is the density scale height. The positive root of the quadratic (7.141) is substituted for h, and the integral expressed in terms of the intermediate variable $x =$

$z/(R + h_0)$. The integrand can be developed as a power series in x^2 factoring the key term $\exp\{-[(R + h_0)/H]x^2\}$. This series, integrated term by term, is found to be dominated by its first member, due to the presence of rapidly decreasing coefficients in the small quantity $(H/(R + h_0))^i$ for the ith member of the series. The result is an expression for τ through the photospheric layers of the form

$$\tau = \sqrt{\frac{\delta}{R}}\,\exp\left(-\frac{2(\delta - R)}{H}\right) \tag{7.143}$$

where δ is as given previously by (7.9). The light curve through the occultation is then formulated as

$$l(\phi) = L_1 + L_2[1 - \exp\{-\tau(\delta)\}] \;, \tag{7.144}$$

which may be compared with (7.10).

The result of optimizing this fitting function to a data-set on ζ Aur extracted from the sources cited by P. D. Bennett et al. (1996) is shown in Fig 7.16. Similar results are obtained for Bennett et al.'s B and U data-sets. Although those eclipses are significantly deeper than the V, the scatter is also proportionately greater, so the net information content is comparable for the three filters. It appears that, though the difference between a standard two-disk model and the atmospheric attenuation one is not large relative to the scatter, the atmospheric model brings about a $\sim 25\%$ reduction in χ^2 relative to the disk model, which is significant at the 95% confidence level. We should expect that more detailed and impersonal space-based photometry will likely reveal more about the photospheric layers of such giant stars than is available from historic data.

Z. Kopal suggested two lines of development in this regard. Firstly, the darkening function $L_2 \exp\{-\tau(\delta)\}$ in (7.145) is only the point-source approximation to a light-loss function Λ that integrates over the finite surface area of the eclipsed star. Secondly, the form of the density function in the photosphere should be derivable from the data, rather than assumed in advance.

The point source approximation would tend, in the limiting case of the conventional two disk model of the eclipses, to a step function: the flux dropping suddenly from U to L_1. But in a more general case, the partial phases for the total eclipse of a small disk, having relative radius r_2, occur over a phase interval $\Delta\phi \approx 2r_2/\sqrt{1 - b^2}$, where the 'impact parameter' $b = \cos i/r_1$. The initial tangency is at $\phi = \arcsin(r_1\sqrt{(1 + k)^2 - b^2}/\sin i)$, where k is the small ratio r_2/r_1. The internal contact phase is at $\phi = \arcsin(r_1\sqrt{(1 - k)^2 - b^2}/\sin i)$.[*] For ζ Aur, $\Delta\phi$ corresponds to an interval of about $\sim 5\%$ of the eclipse's full phase range.

The discriminatory power of the data for the modelling then largely depends on a relatively narrow phase interval, comparable to $(H/Rk)\Delta\phi$, wherein measurable differences between the two models occur. The vertical difference between the models (Fig 7.16) maximises around the tangencies of the disk model. Optimal matching of the two models to the data in Bennett et al.'s eclipse photometry of ζ Aur incurs a quasi-sinusoidal difference of amplitude about 0.015 in relative flux units in Fig 7.16; barely above the level of general noise in the data.[*]

---

[*]If $b > 1 - k$ the eclipse remains partial and the foregoing approximation could be replaced by $\Delta\phi \approx 2r_1\sqrt{2k - \epsilon}$, where $b = 1 - k + \epsilon$, but this would be a rare occurrence in practice.

[*]The geometric parameters derived by Bennett et al. $r_2 = 0.00495$, $i = 87.3°$ are not very different from those of the 2-disk model $r_2 = 0.00478$, $i = 87.6°$. The ratios $r_1/r_2 = 32.9$ and 27.5 differ more appreciably, but Bennett et al.'s (larger) r_1 was defined from spectrsl line analysis rather than broadband eclipse photometry. The atmospheric model's results are in general agreement with other recent findings that $H/Rr_2 \approx 1.3$.

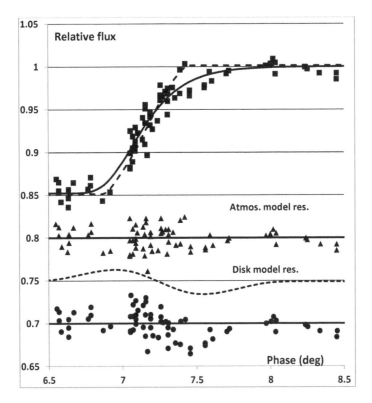

Figure 7.16 Atmospheric (full) and standard binary eclipse model (dashed) optimal fittings to the V light curve of ζ Aur, extracted from the data collected by Bennett et al. (1996) (upper part of diagram). Residuals to the former are shown as triangles centred about relative flux level 0.8. Those of the standard disk model are distributed about the level 0.7. The difference between the two models (atmospheric – disk) is shown as a dashed curve about the level 0.75. There are noticeable differences in the model predictions for small phase intervals at either end of the partial eclipse amounting to about 10% of the total light loss. These are sufficient to make a significant reduction of the fitting's χ^2-value in favour of the atmospheric model.

If we were to replace the point source model by, say, 3 equivalent sources at separations δ_1, δ_2 and δ_3, as the first development of the point source model to an areal quadrature for the loss of light function, we could write,

$$
\begin{aligned}
l(\phi) &= L_1 + \frac{L_2}{3} \left(e^{-\tau(\delta_1)} + e^{-\tau(\delta_2)} + e^{-\tau(\delta_3)} \right) \\
&= \frac{L_2}{3} e^{-\tau(\delta_2)} \left(e^{-(\tau_1 - \tau_2)} + 1 + e^{-(\tau_3 - \tau_2)} \right)
\end{aligned}
\tag{7.145}
$$

We can take advantage of the known form of the source, identifying δ_2 with the separation of the two star centres, i.e. simply δ in the point approximation, δ_1 and δ_3 corresponding to two laterally displaced centroid points in segments contributing equal thirds of the total amount of L_2. These would be at radial separations of $\pm c_l r_2$ from the centre, where the constant c_l is about 0.23 for a uniformly bright disk, and 0.16 for the fully limb-darkened one. As a possible improved approximation for the fitting function, we can then write

$$
l(\phi) = L_1 + \frac{L_2}{3} \exp\{-\tau(\delta)\}[1 + 2\cosh(\Delta\delta/H_o)] \ ,
\tag{7.146}
$$

where, in addition to the exponential term that coincides with the point source approximation in the limit $\Delta\delta \to 0$, we have new parametrization through the separation $\Delta\delta$ that depends on both the radius r_2 of the eclipsed star and its limb-darkening via the constant c_l. The contents of the square brackets in (7.146), since $1 + 2\cosh x \geq 3$, will slightly raise the model's relative light level in the initial tangency region. One might expect this intuitively, as the formulation moves towards a two disk model.

The problem is that the discriminatory power of the data for the modelling, already fairly small as seen from Fig 7.16, goes down as we seek greater parametrization. The point source approximation may be able to discern a value for the scale height of the photosphere from the information content of the available data, but the resolution of this will decrease as we apply a more detailed form for Λ, unless we have additional information that can help specify independently the values of r_2 and relevant limb-darkening coefficients.

The same issue affects the other suggestion of Kopal, regarding the atmospheric density function. Such issues are an aspect of the information mapping problem discussed in Section 7.1. In fact, apart from the empirical results of Menzel, Fig 7.14 shows the exponential form to be relevant for detailed studies of the solar photosphere. Indeed, the use of the scale height H in (7.142) presupposes such a form.

Even if such a form is relevant in a general way, the data discussed by Kopal (Fig 7.13), as well as more detailed inferences from more recent and comprehensive studies of ζ Aur and similar stars, point to significant local variabilities in the outer envelopes of giant stars. Such occurrences suggest that modelling the eclipses in ζ Aur and comparable binaries may be usefully approached on a simple *ad hoc* basis rather than applying a detailed generic model with a set of adjustable parameters as in conventional analyses of eclipsing binary light curves.

7.5   Polarimetry

The photometry considered so far has concerned itself only with the intensity I of the incident light. This neglects information in the flux that relates to the full description of its electromagnetic wave properties, that can be related to, in addition to I, the other Stokes parameters Q, U and V. Parameters Q and U determine the extent of any prevailing linear polarization in the incident beam and its orientation with respect to a set of reference axes, while the ratio V/I yields the nearness to circularity of the general elliptical form of the net incident wave-fronts. Measurement of the polarization state of an incident beam is a speciality that is largely outside the range of topics we address, but it involves an arrangement of optically active components that are able to determine the mean intrinsic linear polarization in at least two orientations (analyser), as well as account for residual effects not from the source itself (depolarizer). The sampling of the entrant beam is performed with the aid of a polarization modulator.

Procedures would normally involve calibrating the data with reference to observations of standard sources, typically using conventional broadband optical filters. A quarter-wave plate (retarder) placed in the wave train at different orientations allows the parameter V to be determined. In practice, it appears that optical polarimetry of stars has often concerned the relatively low, but informative, degree of linear polarization $p = \sqrt{Q^2 + U^2}/I$, though certain cataclysmic variables show rapidly variable polarization effects including circular. These sources are referred to again in Chapter 11. In general, measurements of p are less than 1%, but modern methods permit confident detections down to $\sim 0.01\%$, with the orientation angle $\theta = \frac{1}{2}\arctan(U/Q)$ measured to within 1 degree.

An early stimulus for stellar polarimetry came from a paper of S. Chandrasekhar in 1946, who predicted that the continuous radiation of early type stars should be locally polarized. Regarding the opacity of the atmospheres of O and Wolf-Rayet type stars to result from Thomson scattering by free electrons, he calculated that the radiation from the stellar limb

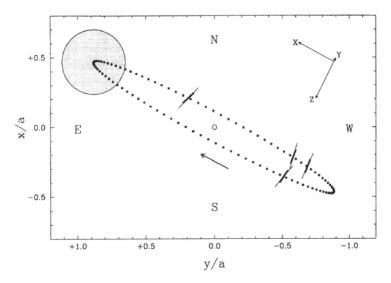

Figure 7.17 The 'polarimetric orbit' of the symbiotic binary SY Mus according to T. J. Harries & I. D. Howarth was published in *Astron. Astrophys.*, Vol. 310, p235, 1996; ©ESO. Data were obtained with the RGO spectrograph and polarimeter at the Anglo-Australian Telescope. The red giant component is on the left, shaded, and its relative orbit about the central O VI star (small circle) is indicated by the dotted ellipse. Most of the 7 parameters of the system are obtained from separate evidence. Best estimates for the nodal and inclination angles Ω, i are derived from fitting the 4 observed values of the direction of the linear polarization vector (long thin lines) with corresponding values from the orbit model (short thick lines). For further information, see Harries & Howarth (op. cit.).

could have a polarization of $\sim 11\%$. The direction of this polarization is tangential at the limb, so the net effect for the complete disc would register as zero. However, during eclipses, a significant level of linear polarization could become detectable due to the asymmetry in the illumination conditions. The idea was promptly responded to by Z. Kopal & M. B. Shapley, who urged attention to the early type binary V444 Cyg, noting (after a comment from H. N. Russell) that the nodal angle Ω could be revealed in such circumstances. In due course, anticipated effects for this star were confirmed.

It has been known since at least the work of D. N. Shakhovskoi and O. S. Shulov in the 1960s that there is a measurable level of linear polarization from certain binary stars. The sources should contain at least one high surface temperature component associated with the required ionization that yields the free electrons engaged in Thomson scattering. Binaries like the hot, wind-emitting O + Wolf-Rayet combination V 444 Cyg have received much attention. Systems engaged in Roche lobe overflow, notably classical Algols, have also been fruitful in the detection of polarization effects. The gaseous stream associated with such binaries is a plausible source of the light-scattering plasma that would introduce the observed dependence of Q and U on varying orbital phase.

A key result from these studies, noted in the PhD thesis of R. J. Rudy (1977), was the possibility to determine the inclination i for a binary orbit from observations of the polarization state. In particular, Q and U measurements with orbital phase ϕ exhibit trends of different amplitude in $\cos 2\phi$ and $\sin 2\phi$, respectively. The combined variation in the Q-U plane is a mapping of the projected orbit, tracing out the 'polarimetric orbit' twice as the astrometric radius revolves once. Physical modelling confirms this can be of elliptical form with eccentricity $(1 - \cos^2 i)/(1 + \cos^2 i)$. The orientation of the major axis to the Q-axis determines the nodal angle, although not uniquely, since Ω and $\Omega + 180$ produce the same observational effect. Conventional practice is to adopt $0 \leq \Omega < 180$ deg.

Concerning the modelling of polarimetric effects in binary systems, matters are greatly simplified by adopting that the scattering material is symmetrically distributed concentrated towards the orbital plane and axisymmetric about the line of centres. This allows the Stokes' parameters to approximate to the forms

$$
\begin{aligned}
Q &= \mathcal{K}[(1 + \cos^2 i) \cos 2\phi + \sin^2 i] \\
U &= \mathcal{K}[2 \sin 2\phi \cos i] \ ,
\end{aligned}
\tag{7.147}
$$

which shows the dependence of Q and U on $\cos 2\phi$ and $\sin 2\phi$, respectively. The factor $\mathcal{K}$ results from the integration of the scattered light, and is of the form $\mathcal{K} = I_0 \sigma_0 \int_V n/r^2 \, dV$ for an initial point source of intensity I_0 illuminating a volume V of plasma where the electron number density is n at distance r. The Thomson scattering coefficient, in the usual notation, is $\sigma_0 = \frac{1}{2}(e^2/m_e c)^2$. Typical optically thin conditions in scattering models entail $\mathcal{K}$ to be of order $0.01 \, I_0$. If we write the Q' and U' for the parameters scaled in terms of this quantity, then they will satisfy the elliptical form:

$$
\frac{(Q' - \sin^2 i)^2}{(1 + \cos^2 i)^2} + \frac{U'^2}{(2 \cos i)^2} = 1 \ .
\tag{7.148}
$$

The foregoing eccentricity value is confirmed from comparison of the denominators in (7.148) with those of the standard ellipse. It is easily checked, and intuitively clear, that this ellipse becomes a circle when the orbit is face-on, i.e. $i = 0°$, and reverts to a straight line along the Q'-axis when viewed edge-on with $i = 90°$, the polarization being perpendicular to the scattering axis at a given phase. In general, the ellipse would be observed at an arbitrary orientation. Conventionally, the reference direction for the observer's Q-axis is that of increasing declination. The orientation of the major axis with respect to the reference Q-axis thus determines directly the nodal angle Ω.

In recent work, the variation of the position angle, θ say, of the linear polarization with orbital phase has been compared with the comparable position angle of astrometric data. Eqn (1.10) expressed the inverse problem, which is dealt with by finding the optimal set of parameters in an appropriate fitting function that matches the variation of the astrometric θ. The polarimetric θ is set, for a bright primary with its companion observed in the projected relative orbit on the sky plane, by the condition that the linear polarization vector is perpendicular to the position angle of the companion star. The adopted astrometric position angle is given as $\arctan y/x$, where x and y are paired displacements parallel to the declination and RA directions (see Fig 7.17). The polarimetric value is then $90°$ added to this number.

The required measurements are obtained from polarimetry as

$$
\theta = \frac{1}{2} \arctan(U/Q) \ .
\tag{7.149}
$$

The inverse problem thus becomes: given a set of θ values, find the orbit that best matches their variation. Formally, the outcome is expressed by Eqn 5.22. The parameter a (orbital semi-major axis) cancels out in the ratio y/x, so is not available for determination in this exercise. But, as with the combination of spectroscopy and photometry to obtain a fuller information yield, polarimetric results can be added to other observational data. The example of SY Mus is presented in Fig 7.17, which is referred to again in Section 11.1.1.

7.6   Bibliography

7.1 The authors' previous book – *Introduction to Astronomical Photometry*, CUP, (2007) gives further background to the topics raised in Section (7.1). For the present context, these focus into the physics of the fitting function involved in modelling close binary light curves and the characterizing of the data through the model's parameters. J. B. Irwin's review in *Astronomical Techniques*, ed. W. A. Hiltner, Univ. Chicago Press, p584, 1962, illustrating the effects of varying the main parameters in eclipsing binary light curve models, still has a useful role for visualization. Specific technical issues in the optimal fitting of photometric data are discussed in that *Introduction* (op. cit.) pp219–20, and in the appendix to T. Banks & E. Budding's paper in *Astrophys. Space Sci.*, Vol. 167, p221, (1990). The relevant program WINFITTER is freely available from M. D. Rhodes' website at https://michaelrhodesbyu.weebly.com

P. R. Bevington's *Data Reduction and Error Analysis for the Physical Sciences*, McGraw-Hill, (1969), is still a useful introductory text to practical data-analysis, complete with numerous suitable program modules. For further information on Bayesian methods, a generally cited modern text is *Bayesian Data Analysis, (3rd Edition)* by A. Gelman, J. B. Carlin, H. S. Stern, D. B. Dunson, A. Vehtari & D. B. Rubin; Chapman & Hall, 2013. For further background to optimization techniques, a short general review, with some programming examples, is given in *Numerical Recipes* by W. H. Press, B. P. Flannery, S. A. Teukolsky and W. T. Vetterling, (1986). This book also includes a section on the Levenberg-Marquardt technique. The matrix formulation in 7.1.1 can be found in general texts, such as R. A. Horn & C. R. Johnson, *Matrix Analysis (2nd ed.)*, CUP, 2013; or G. Strang's *Introduction to Linear Algebra, Fifth Edition*, Wellesley-Cambridge Press, 2016.

Alternative views on formulating the fitting function were reviewed in the score of papers published in *Light Curve Modelling of Eclipsing Binary Stars*, ed. E. F. Milone, Springer, 1992. Use of the simplex technique for model optimization was discussed by J. Kallrath & A. P. Linnell, *Astrophys. J.*, Vol. 313, p346, 1987. The subject was developed further in J. Kallrath & E. F. Milone's *Eclipsing Binary Stars*, Springer, 1999. The triennial reports of Commission 42 (*Trans. Int. Astron. Union: Rep. Astron.*) has given background on the applications of light curve modelling procedures from time to time. Further references are given below.

7.2 The empirical classification of eclipsing binaries on the basis of light curve morphology is described as simple, traditional and suitable for observers in the 4th edition of the *General Catalogue of Variable Stars* (GCVS) of P. N. Kholopov et al., (11 authors), Nauka Publishing House, Moscow, 1985, and the system is followed for the several thousand variables of that type collected in the GCVS. It was somewhat elaborated in the *Catalogue of Approximate Photometric and Absolute Elements of Eclipsing Variable Stars* of M. A. Svechnikov & E. F. Kuznetsova, Univ. Urals, Sverdlovsk, 1990.

Early efforts to relate these groupings to physical properties were associated with such work as that of E. W. Pike, *Astrophys. J.*, Vol. 41, p73, 1931; V. A. Krat, *Astron. Zh.*, Vol. 21, p20, 1944; O. Struve, *The Observatory*, Vol. 71, p197, 1951; and others. However, the simple correlation between the classification of Algol (EA), β Lyr (EB) and W UMa (EW) light curve types and the physical basis of semi-detached, detached and contact groups introduced by Kopal (1955) has helped cement the retention of the empirical system, not that such general correlations are without a few issues of concern. A relatively simple example is the 'Algol', or EA type light curve, that can be produced by both detached (EAD) and semi-detached (EAS) configurations. There is further discussion of this subject in the review of K. Yakut & P. P. Eggleton, *Astrophys. J.*, Vol. 629, p1055, 2005.

Statistics of binary occurrence were discussed by E. Guenther, V. Joergens, G. Torres, R. Neuhäuser, M. Fernández, & R. Mundt, in *Proc. ESO Workshop*, Garching, ESO, ed. J. F. Alves & M. J. McCaughrean, Springer-Verlag, p43, 2002.

By the time of IAU Symposium 88 on *Close Binary Stars: Observation and Interpretation*, eds. M. J. Plavec, D. M. Popper, & R. K. Ulrich, Reidel, 1980, the classification of binary types had widened into about a half dozen groupings, though not without some misgivings (see e.g. J. Sahade's comments in *Comm. Astrophys.*, Vol. 18, p347, 1996). The same general arrangement is still recognizable in the ~ dozen more finely divided categories listed by R. W. Hilditch in his *Introduction to Close Binary Stars*, CUP, 2001. The main, light-curve-based, classes mentioned above, while important for introductory purposes especially where analysis of photometric data is concerned, are of limited relevance to the physical understanding of close binary stars.

The light curves shown in Figs 7.2-5 are from V. Bakış, E. Budding, A. Erdem, H. Bakış, O. Demircan, P. Hadrava, *Mon. Not. Royal Astron. Soc.*, Vol. 370, p1935, 2006, (δ Lib); C. van Antwerpen & T. Moon, *Mon. Not. Royal Astron. Soc.*, Vol. 401, p2059, 2010; see also E. Budding, R. J. Butland, M. G. Blackford, *Mon. Not. Royal Astron. Soc.*, Vol. 448, p3784, 2015, (μ^1 Sco); F. van Leeuwen, D. W. Evans, M. Grenon, V. Grossmann, F. Mignard, M. A. C. Perryman, *Astron. Astrophys.*, Vol. 323, L61, 1997, (β Lyr); E. Budding, C-H. Kim, O. Demircan, Z. Müyesseroğlu, K. Saijo , T. Banks, *Astrophys. Space Sci.*, Vol. 246, p229, 1997, (YY Eri).

The subject of binaries in globular clusters was reviewed by P. Hut et al., (9 authors) in *Publ. Astron. Soc. Pacific*, Vol. 104, p981, 1992. The likelihood that at least some of the 'blue stragglers' seen towards the upper left in colour-magnitude diagrams of globular clusters could be a manifestation of close binary evolution was discussed by M. Mateo (1996). This was in the context of *The origins, evolution, and destinies of binary stars in clusters* eds. E. F. Milone & J-C. Mermilliod, published in the *Astron. Soc. Pacific Conf. Ser.* Vol. 90, p21, 1996.

The occurrence of eclipsing binaries in other galaxies, particularly those of the Magellanic Clouds and Local Group, was reviewed by R. W. Hilditch in *Close Binaries in the 21st Century: New Opportunities and Challenges*, eds. A. Giménez, E. Guinan, P. Niarchos & S. Rucinski, Springer, pp201-204, 2006; and, in the same book, I. Todd, D. Pollacco, I. Skillen, D. M. Bramich, S. Bell, T. Augusteijn, pp221-223. See also J. Kaluzny, K. Z. Stanek, M. Krockenberger, D. D. Sasselov, J. L. Tonry & M. Mateo, *Astron. J.*, Vol. 115, p1016, 1998. A résumé of the properties of massive stars, deduced from binary star analysis, was presented by A. Moffat, O. Schnurr, A-N. Chené & N. St-Louis, in *J. Royal Astron., Soc. Canada*, Vol. 99, p141, 2005.

7.3 The analysis of light curves of eclipsing binary systems was a main theme in the work of Z. Kopal, over the years from the appearance of *An Introduction to the Study of Eclipsing Variables*, Harvard Univ. Press, 1946;* to the *Language of The Stars*, Dordrecht: Reidel, 1979. The discussion of Section 7.3 is largely built from this foundation, certainly as regards the development of the fitting function adopted for the problem presented in Section 7.1. The rather lengthy form of this function was checked by S. Söderhjelm, *Astron. Astrophys.*, Vol. 34, 59, 1974. The calculation procedure referred to, involving only 2 basic α-integrals and 4 I-integrals, was given by E. Budding, *Astrophys. Space Sci*, Vol. 29, 17, 1974; see also E. Budding & N.N. Najim, *Astrophys. Space Sci.*, Vol. 72, 369, 1980.

---

*Kopal's earlier detailed presentation in the *Proc. American Phil. Soc.*, Vol. 85, p399, 1942; referred to the similarly motivated studies of K. Walter *Königsberg Veröff.*, No 2, p28, 1931; S. Takeda *Kyoto Mem. A*, Vol. 20, p47, 1937; and H. N. Russell, *Astrophys. J.* Vol. 90, p641, 1939.

Regarding the two α-integrals: the first α_0^0 is elementary; α_1^0 was tabulated from quadratures in the work of H. N. Russell & H. Shapley in *Astrophys. J.*, Vol. 36, p239 and 385, 1912. Explicit forms were given by V. P. Tsesevich in *Bull. Astron. Inst. U. S. S. R. Acad. Sci.*, No. 45, 1939 and No. 50, 1940. *I*-integrals were evaluated by Z. Kopal in *Proc. Amer. Phil. Soc.*, Vol. 88, p145, 1944.

The Jacobian elliptic functions introduced after (7.55) are comprehensively described in P. F. Byrd and M. D. Friedman's *Handbook of Elliptic Integrals for Engineers and Scientists*, Springer-Verlag, 1954; where integrals of the type (7.57) are listed. Recursion formulae, such as (7.91), are derived in H. Hancock's *Elliptic Integrals*, John Wiley and Sons, p60, 1917. Efficient algorithms to calculate such integrals were given by D. J. Hofsommer & R. P. Van der Riet, *Numer. Math.*, Vol. 5, p291, 1963. Early historical background on the eclipse function was outlined by E. Budding and M. Rhodes in *The Observatory*, Vol. 136, p268, 2016.

The topic of limb-darkening follows the discussion of Z. Kopal in Chapter 4.1 *Close Binary Systems*, Chapman & Hall, (1959). The Eddington approximation, sometimes called the Milne-Eddington approximations, can be traced to A. S. Eddington's *Internal Constitution of the Stars*, CUP, (1926), p322; and E. A. Milne, *Handbuch der Astrophysik*, Springer, (1930), Chapter 2. Topics such as the grey atmosphere model, the Hopf function, non-grey and scattering atmospheres are discussed in such reference works as S. Chandrasekhar's *Radiative Transfer*, Oxford, 1950; V. A. Kourganoff's *Basic Methods in Transfer Problems*, Oxford, 1952; and D. Mihalas' *Stellar Atmospheres*, Freeman, 1978. W. H. Julius' examination of the Sun's limb darkening during the total eclipse of 1912 was published in the *Astrophys. J.*, Vol. 37, p225, 1913. The limb-darkening coefficient tables of W. van Hamme were presented in the *Astron. J.*, Vol. 106, p2096, 1993; and his maintained website http://faculty.fiu.edu/ vanhamme/limb-darkening/ makes them freely available. The alternative forms shown in (7.33) were collected together from literature cited by A. Claret, *Astron. Astrophys.*, Vol. 363, p1081, 2000; who had previously pointed out shortcomings in the use of the linear approximation for precise representation of the limb-darkening. The last formula in (7.33) is from Claret (op. cit.), whose paper is accompanied by some 46 tables of values for the 4 coefficients required for use with a dozen commonly used photometric filters, having given effective wavelengths, and a wide range of photospheric or *effective* temperatures, gravities, compositions and microturbulent velocities.

Reviews of the relationship of the surface flux to the local gravity were given in a series of papers by M. Kitamura & Y. Nakamura: *Tokyo Astron. Obs. Ann., 2nd Ser.*, Vol. 22, 31, 1988; *Lecture Notes Phys.*, Vol. 305, ed K. Nomoto, Springer-Verlag, 1988, p217; *Astrophys. Space Sci.*, Vol. 145, 117, 1988; *Space Sci. Rev.*, Vol. 50, 353, 1989; and also by M. J. Sarna, *Astron. Astrophys.*, Vol. 224, p98, 1989; and G. Djurašević et al. *Astron. Astrophys.*, Vol. 402, 667, 2003; and *Astron. Astrophys.*, Vol. 445, 291, 2006. The diffusion law for the radiative flux, related to Eqns (7.19) and (7.33), is clarified in *Stellar Atmospheres*, of D. Mihalas, 1978, p49, W. H. Freeman, San Francisco.

The circulation of stellar material was reviewed by L. Mestel in Volume VIII (Stellar Structure) of the series *Stars and Stellar Systems* eds. G. P. Kuiper & B. M. Middlehurst, Univ. Chicago Press, 1965. Further consideration of surface layer effects were given by Y. Osaki, *Publ. Astron. Soc. Japan*, Vol. 24, 509, 1972; R. Connon-Smith & R. Worley *Mon. Not. Roy. Astron. Soc.*, Vol. 167, 199, 1974; H. Kırbıyık & R. Connon-Smith *Mon. Not. Roy. Astron. Soc.*, Vol. 176, 103, 1976.

L.B. Lucy's paper in *Zeit. Astrofis.*, Vol. 65, 89, 1967; is often cited for its numerical derivation of relatively low 'convective' gravity effect coefficients ($\tau \sim 1/3$ in integrated light), but a still lower value (zero) was argued for in L. Anderson & F. H. Shu's study of contact binary light curves *Astrophys. J.*, Vol. 214, 798, 1977. This latter finding, however, is applied to a situation in which lack of hydrostatic equilibrium is intrinsic to the discussion.

Further calculations were given by A. Claret, *Astron. Astrophys.*, Vol. 359, 289, 2000; see also G. İnlek et al., *New Astronomy*, Vol. 12, p427, 2007.

In a similar way, S. M. Ruciński's lowered scale of reflection effect (effective bolometric albedo $\sim$ 0.5) in *Acta Astron.*, Vol. 19, 245, 1969, is also frequently used in light-curve fitting programs. This lowered reflection effect was previously noted by Y. Hosokawa *Sendai Rap.*, Nos. 56 & 70, 1957 & 1959. Apart from the sources provided in Kopal's cited text, numerous other studies of the reflection effect in close binary systems include I. Pustyl'nik, *Acta Astron.*, Vol. 27, p251, 1977; E. Budding & Y. R. Ardabili, 1978, *Astrophys. Space Sci.*, Vol. 59, p19, 1978; B. N. G. Guthrie & W. M. Napier, *Nature*, Vol. 284, p536, 1980; A. Periaiah, *Proc. Platinum Jubilee Symp. Nizamiah Obs.* Osmania Univ. 1984, p20; L. P. R. Vaz, *Astrophys. Space Sci.*, Vol. 113, p349, 1985; A. Yamasaki, *Publ. Astron. Soc. Japan*, Vol. 38, p449, 1986; M. Tassoul & J-L Tassoul, *Mon. Not. Roy. Astron. Soc.*, Vol. 232, p481, 1988; R. E. Wilson *Astrophys. J.* Vol. 356, p613, 1990; D. P. Kiurchieva & V. G. Shkodrov, *Dokl. B'lgarska Akad. Nauk.*, Vol. 45, p5, 1992; S. H. Choea & Y. W. Kang, *J. Astron. Space Sci.*, Vol. 9, p30, 1992; R. W. Hilditch, T. J. Harries & G. Hill, *Mon. Not. Roy. Astron. Soc.*, Vol. 279, p1380, 1996; N. T. Kochiashvili & I. B. Pustyl'nik *Astrophysics*, Vol. 43, p87, 2000; M. Srinivasa Rao, *ASP Conf. Proc.*, Vol. 288 , (eds, I. Hubeny, D. Mihalas & K. Werner, p645, 2003; and others.

Z. Kopal, having outlined much of the time-domain procedure considered in this chapter, retained a strong interest in the development of frequency-domain techniques, which were discussed in a series of papers published in the late 1970s, culminating in Kopal's *Language of the Stars*, (op. cit.). Earlier applications of Fourier analysis to light curves were those of H. Mauder, *Kleine Veröff., Bamberg, Remeis-Sternwarte*, Nr. 38., 1966; and M. Kitamura, whose *Tables of the Characteristic Functions of the Eclipse* was published by the University of Tokyo Press, Tokyo, 1967. See also articles in *Zdeněk Kopal's Binary Star Legacy: Astrophys. Space Sci.*, Vol. 296: P. Niarchos (p359), O. Demircan (p 209) and E. Budding (p371), 2005.

An alternative general approach to light curve analysis, with less presumption about the form of the fitting function but involving a comparable error minimization principle, was that of A. M. Cherepashchuk, A. V. Goncharskii & A.G. Yagola, *Soviet Astron.*, Vol. 12, p944, 1969. Other light-curve modelling programs are those of D. B. Wood, *Astron. J.*, Vol. 76, p701, 1971 (WINK); R. E. Wilson & E. J. Devinney, *Astrophys. J.*, Vol. 166, p605, 1971 (WD); S. W. Mochnacki, & N. A. Doughty, *Mon. Not. Roy. Astron. Soc.*, Vol. 156, p51, 1972; S. M. Rucinski, *Acta Astron.*,, Vol. 24, p119, 1974; G. Hill, *Publ. Dom. Astrophys. Obs.*, Vol. 15, p297, 1979 (LIGHT); W. M. Napier, *Mon. Not. Royal Astron. Soc.*, Vol. 194, p149, 1981; K. G. Strassmeier, *Astrophys. Space Sci.*, Vol. 140, p223, 1988; S. A. Bell, P. P. Rainger, R. W. Hilditch, *Mon. Not. Roy. Astron. Soc.*, Vol. 247, p632, 1990 (LIGHT2); P. Hadrava, *Publ. Astron. Inst. Acad. Sci. Czech. Republic*, No. 92, p1, 1994 (FOTEL); P. B. Etzel, in *Photometric and Spectroscopic Binary Systems*, eds. E. B. Carling & Z. Kopal, Dordrecht: Reidel, p111, 1981 (EBOP); J. Kallrath and E. F. Milone's *Eclipsing Binary Stars: Modelling and Analysis* (op. cit.) is a useful source of information on mathematical models for the light curves of eclipsing binary stars, particularly the popular WD code. This has been updated from time to time (e.g. E. F. Milone and C. R. Stagg, *Exp. Astron.*, Vol. 5, p163, 1994; J. Kallrath, E. F. Milone, D. Terrell & A. T. Young, *Astrophys. J.*, Vol. 508, p308, 1998), and adapted to certain 'user-friendly' sources such as STARLIGHT PRO, cf. http://www.midnightkite.com/index.aspx?URL=Binary; published by D. Bruton (2004). Other work on light curve modelling includes that of J. R. Barnes, T. A. Lister, R. W. Hilditch, A. Collier Cameron, A., *Mon. Not. Roy. Astron. Soc.*, Vol. 348, p1321, 2004; A. Prša & T. Zwitter, *Astrophys. J.*, Vol. 628, p426, 2005 (PHOEBE); J. Southworth, *Astron. Astrophys.*, Vol. 557A, p119, 2013 (JKTEBOP) and J.-V. Cardoso et al. (18 authors), *Astrophys. Source Code Library*, record ascl:1812.013, NASA, 2018, (Lightkurve).

The great usefulness of the stellar parametrization that comes from eclipsing binary light curves combined with reliable radial velocity data (Russell's 'Royal Road') was stressed in the contribution of J. Andersen, J. V. Clausen & B. Nordström, in their contribution to the informative *IAU Symp. 88: Close Binary Stars, Observations and Interpretation*, eds. M. J. Plavec, D. M. Popper & R. K. Ulrich, Reidel, 1980, p81; a theme later developed by J. Andersen *Astron. Astrophys. Rev.*, Vol. 3, p91, 1991. The Copenhagen school had by that time produced a large number of detailed studies of individual binaries, including a series of papers that started with K. Gyldenkerne, H. E. Jørgensen and E. Carstensen's paper in *Astron. Astrophys.*, Vol. 42, p303, 1975. These studies went on to check structural implications arising from using the absolute stellar parameters of binaries showing apsidal motion (see previous chapter), for example with papers like that of A. Giménez, J. V. Clausen, B. E. Helt & L. P. R. Vaz' study of the massive eccentric binary GL Car in *Astron. Astrophys. Suppl. Ser.*, Vol. 62, p179, 1985.

7.4 The discussion of atmospheric eclipses Section (7.4) starts from Chapter 4.7 of *Close Binary Systems* (op. cit.), but relates back to the pioneering paper of D. H. Menzel, *Harvard College Obs. Circ.*, No. 417, 1936. Fig 7.3 comes from Z. Kopal, *Astrophys. J.*, Vol. 103, p310, 1946. The work of P. D. Bennett et al. was published in *Astrophys. J.*, Vol. 471, p454, 1996. both of these papers dealing with the prototype ζ Aur. For a comprehensive review cf. *Giants of Eclipse: The ζ Aurigae Stars and Other Binary Systems*, T. B. Ake, E. Griffin (eds.), *Astrophys. Space Sci. Libr.*, No. 408, Springer, 2015; particularly Chapter 2, by R. E. Griffin et al. Other relevant papers are those of M. Kitamura, *Astrophys. Space Sci.*, Vol. 28, p17, 1974, discussing possible changes in the location of the supergiant's photosphere over the timescale of several years; R. D. Chapman, *Astrophys. J.*, Vol. 248, p1043, 1981; dealing with the wind from the supergiant and its interaction with the B-type dwarf; and K. G. Carpenter, in *Evolutionary Processes in Interacting Binary Stars*, IAU Symp. 151, eds. Y. Kondo et al., p51, 1992 (IAU); who introduced the topic of interferometry for measuring the angular size of the supergiant. A. M. Cherepashchuk discussed atmospheric eclipses involving Wolf-Rayet stars (including V444 Cyg, CQ and CX Cep, and CV Ser) in *Astrophys. Space Sci.*, Vol. 86, p299, 1982, in which the basic light curve equation appears in a form similar to (7.145). Cherepashchuk (op. cit.) derived the form of the absorption function by numerical processing of the optical depth integral. In applications of his analysis to V 444 Cyg he found the logarithm of the opacity to decline approximately linearly with log(radius), but with a significant increase from the optical to the infra-red ranges. A light curve of a less well-known system with similar parameters to ζ Aur, namely V777 Sgr, was presented by W. S. G. Walker in the *VSS Newsletter*, 2021-1, p15, 2021. Comprehensive background on Wolf-Rayet stars within the wider context of stellar evolution can be found in the proceedings of the 33rd Liège Intl. Astrophys. Colloq., on *Wolf-Rayet Stars in the Framework of Stellar Evolution*, Liège University, 1996, eds. J. M. Vreux et al. (summary article by P. *Conti,* p655).

7.4 D. Clarke, in *Stellar Polarimetry*, Wiley VCH, 2010, gave a very thorough reference book on the subject. A recent review is that of A. Berdyugin, V. Piirola & J. Poutanen in arXiv:1908.10431v1, 2019. The cited paper of S. Chandrasekhar was in *Astrophys. J.*, Vol. 103, p365, 1946. The response from Z. Kopal and M. B. Shapley appeared as *Astrophys. J.*, Vol. 104, p160, 1946. Chandrasekhar's prediction of polarization effects in the eclipses of hot stars was eventually confirmed for V444 Cyg by C. Robert, A. F. J. Moffat, P. Bastien, N. St.-Louis & L. Drissen in *Astrophys. J.* Vol. 359, p211, 1990. N. M. Shakhovskoi's early detection of stellar polarization from β Lyr was published in *Soviet Astron.*, Vol. 6, p587, 1963. The work of O. S. Shulov, appeared in *Trudy Astron. Obs. Leningrad*, Vol. 24, p38, 1967. For further discussion of V444 Cyg and its polarimetry see N. St. Louis et al. (7 authors), *Astrophys. J.*, Vol. 410, p342, 1993; and J. R. Lomax et al. (11 authors), *Astron. Astrophys.*, Vol. 573, p43, 2015; the latter combining spectropolarimetry with X-ray photometry to investigate the complex wind interactions in the system.

R. J. Rudy's PhD thesis was presented to the University of Oregon and can be accessed from *Dissertation Abstracts Internat.*, Vol. 38-10, Section: B, p4859, 1977. R. J. Rudy & J. C. Kemp presented observations of Q and U showing cyclic variations at twice the orbital period for 5 bright systems (including Algol) in *Astrophys. J.*, Vol. 221, p200, 1978. As a result they were able to provide inclination values using the formula given in Section 7.5. This formula, along with more general modelling parameters, is derived in the detailed analysis of J. C. Brown, I. S. McLean & A. G. Emslie, *Astron. Astrophys.*, Vol. 68, p415, 1978. A simple account was presented in Section (4.6) of E. Budding's *Introduction to Astronomical Photometry*, CUP, 1993, with an application to Rudy and Kemp's polarization data on Algol. It was implied that the scattering medium was the mass-transferring stream. Whilst axisymmetric, such a stream is diverted from the line of centres by close to $20°$, according to the seminal study on mass transfer in Algol binaries of S.H. Lubow & F.H. Shu, *Astrophys. J.*, Vol. 198, 383, 1975, due to the action of the Coriolis force. In principle, this could be checked if the nodal angle for Algol's orbit is separately determinable. The approach of T. J. Harries & I. D. Howarth, in which Fig 7.17 can be found, was published in *Astron. Astrophys.*, Vol. 310, p235, 1996.

8

Evolution Overview

8.1 Introduction

This chapter addresses the conditions wherein binary and multiple stars find themselves, against a backdrop of how these conditions vary over long periods of time. Some aspects of this backdrop are clearer than others. Just how the stellar forms and distributions we see originated from their diffuse interstellar matrix, for example, includes less clearly defined subject matter. This is touched on below, and will be discussed further in the next chapter. But the twentieth century introduction, of high-power electronic computers allowed certain topics to be addressed in great detail, particularly those relating to the structure and development of stellar models. This chapter gives an introductory outline to these topics.

Notions of very long periods of time, compared with human history, gathered force from studies within geophysics in the nineteenth century. That the Sun could have been more or less static for such long periods then became a subject of natural enquiry, and the term *evolution* was adopted also in the physics of stars.

Towards the end of that same century, particularly with the growth of interest in the thermodynamics of gases, the behaviour of matter in the extreme conditions in the Sun and other similar bodies aroused active discussion. This is associated with the rise of *astrophysics* as a subject in its own right, propounded by such pioneers as Lane, Emden, Schwarzschild and Eddington. By the present day, study of the internal constitution of a star and its slow changes with time has become a highly developed field. This chapter includes only a broad outline, mainly as a setting for the more selective discussions in the later chapters. C. de Loore and C. Doom's (1992) book on the evolution of single and binary stars comprehensively reviews the background while cautioning also about unsolved problems and controversies. There is a core of accepted principles, but subjects like the detailed nature of convection, the role of magnetic fields, internal mixing of the material, and the dynamics of stellar winds offer separate investigative challenges.

8.2 Single Star Evolution

In the early years of the twentieth century H. N. Russell, having appreciated the significance of the distributions of observed points in E. Hertzsprung's colour-magnitude diagram for stellar clusters and distance-calibrated field stars, accounted for the trends in the diagram in terms of stellar evolution over long periods of time. Long, here, originally referred

DOI: 10.1201/b22228-8

to the Kelvin-Helmholtz or *thermal* time-scale τ_{KH}, given by

$$\tau_{KH} = \frac{GM^2}{RL} \ . \tag{8.1}$$

Eqn (8.1) gives the time it would take for the reserve of gravitational potential energy, associated with a condensation of mass M from great distances to a presently quasi-steady radius R, i.e. GM^2/R, to be radiated away by the outward flux of radiation from the whole surface, i.e. by its luminosity L. For the Sun, this works out at about 30 My. Such a time-scale may have once seemed adequate to accommodate past events of relevance. However, the build-up of evidence, particularly from geophysics, showed that much greater time-spans were required to explain the Earth's prehistory within a fairly unchanging solar system environment.

Astrophysics, in due course, found additional sources of energy (thermonuclear) to bolster that of the gravitational reserve, and account for the rates of migration of stars in the colour-magnitude diagram. A corresponding thermonuclear time-scale τ_N appeared as

$$\tau_N = \frac{\kappa M}{L} \ , \tag{8.2}$$

where κ represents the mean available energy per unit mass from nuclear fusion processes within the star. In the case of the Sun, τ_N works out at about 10^{10} y.

The fact that $\tau_N \gg \tau_{KH}$ means that the above-mentioned idea of L being accounted for by a slowly condensing mass of gas can be by-passed: the star is *thermally stable* and the luminosity radiated away at the surface is balanced by an internal energy production that has very little immediate impact on the star's structure. It can be presumed that at some stage this thermonuclear energy generation took over from a previous condensation process, the structure at that time would have defined the initial conditions at which the release of the stabilizing nuclear energy switched on.

A key factor in evolutionary trends is the initial mass of the star. Early in the development of understanding of stellar properties it was shown that the luminosity L should have a rather sensitive dependence on M: typically $L \sim M^n$, where n is about 3-4. This means that the foregoing time-scales should decrease rather swiftly with increasing mass. The well-known star Rigel, for example, with a mass of almost twenty times that of the Sun outputs energy at more than sixty thousand times the solar rate. It could not stay in a quasi-steady state for more than several million years.

By the middle of the twentieth century, the early modelling of stellar structure by algebraic formulae was giving way to numerical integration methods using digital computers. Formally, stellar structure can be regarded as a two point boundary-value problem. The integration of four ordinary differential equations in four unknowns is required. Two of the four basic boundary conditions are specified for the centre of the star; i.e. for the runs of mass and luminosity internal to radius r, M_r and L_r, at $r = 0$ we have $M_r = 0, L_r = 0$. For the other pair – the pressure P_r and density ρ_r at the external boundary R – we can write, at least formally, $P_R = 0, \rho_R = 0$. In the numerical procedure, the star is partitioned into a given large number of thin layers, and the differential equations replaced by layer-to-layer finite differences for the four dependent variables. In principle, if approximate starting values for the unspecified parameters total mass and luminosity M_R, L_R are provided, together with the central pressure and density P_0 and ρ_0, numerical integration of the four underlying equations of condition, from the centre outward and the surface inwards, can be continued up to an internal matching point. The four dependent variables are compared at the matching point and average values taken. The direction of the numerical difference calculations is then reversed so as to obtain new and improved starting values at the original boundaries. The whole procedure may then be repeated until the differences at the matching point become negligibly small.

This conceptual formulation of the problem is, in practice, replaced by a more practically manageable one, in which the mass at radius r, M_r becomes the independent variable and logarithmic units are used. Other quantities appear in the structure problem apart from those already mentioned, such as temperature, composition, absorption coefficient, thermonuclear energy generation rates; but these either follow from ordinary non-differential equations or are specified as parameters of the model. With layer-to-layer numerical integration suitably set up, the convergent iterative procedure leads to a continuous and stable run for all the variables through the stellar interior. The four *eigenvalues* for the initially not precisely known boundary values that sustain the continuity and stability of the internal run of variables are then also determined. As indicated above, stability of structure implies that thermonuclear energy generation makes up for the arrest of gravitational energy release from continued contraction. However, in that process, the chemical composition of the stellar material must slowly change. With that change, in essence, lies the basis of stellar evolution.

Application of this modelling procedure to the steady changes due to stellar evolution was significantly advanced by a technique put forward around 1960 by L. G. Henyey. This oft-cited procedure involves matrix inversion to determine the small corrections to the time-dependent variables through the adopted number of stellar layers. Finding the required small corrections by matrix inversion is fast compared with the previous matching point procedure. This 'relaxation' method was incorporated in computer programs to follow successive time steps, so that an evolutionary sequence of models, starting from the convergent initial result, could be constructed relatively quickly.

The flexibility of control and versatility of adjustable conditions for such programs soon gave the advantage to computer-based stellar modelling over literal methods. What is briefly reported below largely depends on such numerical work and its success in accounting for the properties evidenced in colour-magnitude diagrams and comparable observational data. This developed historically from various groups pioneering the subject with earlier generations of computing machines. In the present day, internet sources allow user-specified modelling to be done with highly refined procedures and results communicated quickly and accurately online.

Fig 8.1 schematizes the main evolutionary stages of a sunlike star with an observational viewpoint in mind. Of course, there are differences in the behaviour depending on the mass and composition of the star, though a few general features persist. We met already the Main Sequence and Giant Branch in Chapter 1. The former is indicated by the heavy dashed line running from blue and bright to dim and red in the colour-magnitude plane. Long-enduring thermal stability on the Main Sequence entails that relatively many stars are located in this region of the diagram. Giant stars are associated with the region of the dotted curve D-E-F — particularly the longer-lasting upper (E-F) reaches of that curve. The initial movement from the Main Sequence (D-E) takes place in a relatively small amount of time; so that relatively few stars are found in the *Hertzsprung gap* at the bottom of the Giant Branch. In this stage, the original Main Sequence (MS) power supplies having failed, the luminosity has to be accounted for by a resumption of thermal time-scale gravitational contraction, unless further nuclear energy generation sources arise.

Before all this, a full account of stellar history should include the early phases of matter accretion, condensation and contraction of gaseous material from the interstellar medium. Some ambiguity is apparent in the use of the terms 'protostar' and 'pre-Main-Sequence' (PMS) for these stages. A protostar is more recently understood to start with a steady accumulation of material from its galactic environment. Of necessity, there must be many potential variables to specify for such conditions even in a single condensation. Analysis

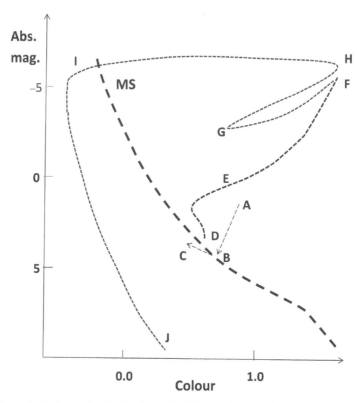

Figure 8.1   The main features of evolution for a solar-like star in the colour-magnitude plane are sketched. Numerical scale values are indicative only. Luminosity increases as absolute (bolometric) magnitude decreases along the vertical axis. Colour may be measured by the difference in magnitude of a star in photometric blue (B) and visual (V) spectral ranges. This is arranged so that stars that are relatively brighter in blue (lower colour value) are leftward on the horizontal axis. Such colour measures have an approximately linear relationship with the inverse of a star's effective surface temperature. This way of summarizing stellar long-term behaviour, associated with E. Hertzsprung and H. N. Russell, has played a fundamental role in stellar astrophysics.

of the situation with more than one simultaneous matter accumulation will clearly offer challenges, though, in fact, evidence has grown that stars are usually formed at least in pairs, and frequently with higher multiplicity. Gravitational energy generation emanating from the contraction process and related interactions should be generally more intense towards the centre of the accumulation, and at some point the protostar develops enough luminosity to halt and reverse further infall. Conventional stellar modelling might start from this point, though thermonuclear energy generation may not yet have reached a significant and steady level for such an object. PMS models thus cover the structure and development of a fixed accumulation of interstellar material of specified mass (A-B-C) up until the commencement of stable *hydrogen fusion* in the cores of MS stars of 'zero age', i.e. ZAMS stars.

Stars of the *T Tauri* type have been associated with PMS formations when the mass is less than about 2 $M_\odot$. PMS objects in the mass range 2-8 $M_\odot$ are identified with Herbig Ae/Be type stars. Accumulations with masses greater than about 8 $M_\odot$ are not observed in this PMS state, since their core temperatures are sufficiently high to start hydrogen fusion as they are already condensing to a visible stellar form. As with Hertzsprung gap stars, PMS objects are relatively infrequently observed. When detected, they tend to be picked up from their spectral appearance, since their radii, larger than ZAMS stars of the same mass, entail lower surface gravity. Hence, their spectral lines show low pressure broadening.

Of special interest is that young PMS stars may also show accretion structures that remain from the condensation process, and so are potentially informative about the *angular momentum problem* that arises in the context of star formation. This concerns the condensation of galactic material from its initial tenuous distribution within volumes of order $\sim 10^{20}$ times greater than the stellar end-product. Such a process would be expected to produce an enormous increase in rotation rate. This is so high that centrifugal forces, unless braked in some way, would prevent the condensation reducing to stellar dimensions. This angular momentum problem is thus concerned with the way in which the condensation to a regular star can occur. Disk-like structures around PMS objects that can lead to the formation of planetary systems form an important topic within this field of study.

Although the Main Sequence is depicted as a single trend in Fig 8.1, it should be recognized that a star continues to evolve in this region, i.e. the MS has a finite width. The point of arrival (C) concerns the ZAMS, mentioned above, while that of departure (D) is identified with the 'terminal age' or TAMS. From detailed modelling, D would be placed a little higher and to the right of C.

During the ascent of the Giant Branch* (E-F) hydrogen fusion moves outwards to a spherical shell surrounding the condensed accumulation of helium nuclei at the star's centre. The extreme conditions within the core give rise to an enhanced pressure, not following the standard gas law behaviour, but arising from a *degeneracy* in the available quantum energy states of electrons. At this stage, there appears a 'mirror effect': the envelope of the star swelling out, while central regions steadily fall in. The photospheric radius by now may have reached $\sim$100 times that during the Main Sequence stage. Subsequent events depend critically on the star's initial mass and composition.

For stars of comparable mass to the Sun, the central temperature may, in due course, become sufficiently high to trigger a sudden onset of *helium burning* in the core (F). This may have an explosive character at first, due to the degenerate state of the core material that would assist helium burning by the *triple alpha* process. The ensuing rise of temperature is expected to return the core to the condition of a normal gas, when the heating parallels the previous hydrogen-fuelled one; now with the fusion of helium nuclei into those of carbon.

The new store of energy propels the structure back towards configurations that are smaller but hotter at the surface (G). The location of G is sensitive to the composition. Stars with a relatively lower proportion of metals (*Population II*) can move well over towards the blue region, making up the *Horizontal Branch*, as observed in the colour-magnitude diagrams of globular clusters. Later, ongoing shell burning of hydrogen is joined by another shell in which helium fusion takes place. The star's condition, at this stage, is essentially similar to what it had been on the previous classical Red Giant Branch: the aggregation towards point H is thus called the *Asymptotic Giant Branch*.

The high sensitivity of nuclear energy generation processes to temperature is associated with the occurrence of structural instabilities on the thermal time scale of the envelope. A contraction of the star's photosphere subsequently occurs, as strong winds start to move matter out from the visible surface (H-I). By the end of this stage (I), there is an accumulation of matter surrounding the stellar photosphere that gradually expands to form a fluorescent mass observationally recognized as a *planetary nebula*. In a relatively short time, this diffuses out into the interstellar medium: the condensed remains of the core gradually cooling to form the prolonged and relatively inert white dwarf stage (J).

---

*Identified here as the Red Giant Branch (RGB) to distinguish it from the similar later evolutionary stage of the Asymptotic Giant Branch (AGB) in an adjacent part of the diagram.

Stars of lower mass than the Sun are expected to follow a generally similar pattern to that of Fig 8.1 in this kind of modelling, but at a correspondingly slower rate. More-massive stars, however, after their relatively brief sojourn on the Main Sequence, can simultaneously fuse heavier elements in concentric shells about the core up until its central accumulation of nuclei consists mostly of the element iron together with a group of comparable heavy elements, whereupon the packing of hadrons in nuclei reaches a limit and no further energy is available from nuclear fusion. Stars of mass an order of magnitude greater than that of the Sun then face a serious stability problem with the collapse of the core, due to the inability of electron degeneracy alone to sustain the high pressure required to support overlying layers. This condition is linked with the occurrence of *supernovae* (of type II) and the production of highly compact objects: neutron stars or stellar black holes.

While models of stellar evolution as briefly depicted in the foregoing have been extensively calculated for single stars, their predictions are illuminated in peculiar and interesting ways by observational data on binary and multiple stars.

8.3   Binary Evolution Essentials

In Kopal's (1959) book, an insightful attempt at a basic classification of binary stars was made by referring to three essential parameters, styled C_1, C_2 and q. These correspond to suitably scaled binding energies of the two stars (Section 2.1) and their mass ratio (lesser to greater) $q = M_2/M_1$. A particular point was made by introducing the quantity C_0, with which observed values of C_1 and C_2 might be compared. The significance of C_0 is that it corresponds to the limiting *Roche* surface that passes through the inner Lagrangian point L_1. The special properties of this surface were examined in Chapter 4. Its existence leads to strikingly different stellar forms and behaviour than are found for single stars.

Configurations having C values greater than C_0 can be found among bright young pairs of stars: U Oph was the representative (if rather massive) example cited by Kopal. This type of close binary is often found together with other stars in a young star association or cluster. Independent evidence confirms that these stars are essentially similar to single stars of similar mass, composition and age on the Main Sequence.[*] Kopal called such binaries *detached*.

Another type of close binary was described as *semi-detached*, in which one of the C values was greater than C_0, but the other measurably equal to it. From such observational evidence as was available, it appeared that there was an accumulation of systems that were either semi-detached or close to that condition, with the restriction that the star for which $C = C_0$ was always the less massive of the two. The bright star β Persei (Algol) is the probably best-known example. Close observation of Algol, the first eclipsing binary star to be so recognized (Section 1.1), has confirmed that the photosphere of the less bright (at present) secondary star is in more or less the same location as the surface of limiting stability corresponding to a Roche model. The name Algol has been assigned to numerous other close binary systems sharing this characteristic, though not unambiguously.

Apart from such genuinely semi-detached 'classical Algols', Kopal distinguished two other groups of close binaries, ostensibly similar to the classical Algols though with noticeable differences. These were the 'undersize subgiant' group: the star in front of the brighter one at primary eclipse being enlarged above what would be normally expected, though not yet to the C_0 surface; and the 'R CMa' type, wherein both stars were abnormal in the sense

---

[*]This point has interesting implications on the independence of the stellar form to the particular path the accumulated material followed to reach its stellar situation.

of being overluminous for their apparent masses. The first subgroup has become associated with a recognized class of detached close binary system, now usually called the 'RS CVn' type. The second group originally contained a number of systems whose parameter estimates were probably significantly affected by errors. It has become clear, however, that R CMa itself, whilst likely to be a classical Algol, has a puzzling combination of very low mass ratio and short period. A small number of comparable systems have been identified in more recent catalogues. These categories are discussed in more detail in Chapter 10.

A third main group of binary exists in relatively plentiful supply in which both components have C values that are close or equal to C_0. These systems have been described as *contact* binaries.* W UMa is an oft-cited representative of the class. This type of binary, along with the other two kinds of eclipsing system, were associated in the previous chapter with morphologically distinct types of light curve, though this association has ambiguity as physically different configurations sometimes show quite similar forms of light variation. That being said, the physical condition of these main groupings: detached, semi-detached, and contact systems, with their corresponding light curve types, presents the basic question as to why this arrangement prevails? The answer is generally taken to be found in applying stellar evolution theory to the close binary situation.

One of the early discussions of binary evolution was that of Morton (1960), who gave particular attention to the second, or Algol, group; seeing the configuration as resulting from the evolution of the originally more massive star in a detached binary away from the Main Sequence along the track DEF in Fig 8.1. At a relatively early stage in this development the expanding star would encounter the Roche limit. Appreciable amounts of matter would then spill over the saddle-point at L_1 and into the corresponding *Roche lobe* of the less massive star.

This idea, rather startling when first proposed, has proved to be a pivotal concept in binary star evolution. The mass transferring stage in classical Algols would proceed relatively quickly at first, the two stars approaching each other with net angular momentum conserved. In time, the originally more massive star would become less massive than its companion, whereupon things should slow down. The two stars would separate and the rate of mass transfer decline considerably. The star receiving the influx could then appear similar to a Main Sequence object, but with mass appreciably increased over its original value. But that situation is not assured, since overflowing material does not necessarily arrive at the receiver star: even if gravitationally bound, it may have sufficient angular momentum to form a separate *accretion structure* about that star, particularly at increased orbital period. The situation of the gainer star in such an arrangement, occupying a proportionately reduced fraction of the available lobe and surrounded by a rapidly rotating disk-like accumulation of matter, is clearly different from a typical single Main Sequence object with the same overall mass, initial composition and age.

The star shedding matter would remain close to its Roche limiting surface, its mass loss rate – in the frequently observed slower and later stages of mass transfer – appreciably curtailed on that of the early phase. This is not just related to angular momentum conservation allowing the loser's Roche lobe to expand, but a shell-burning stage, associated with the more evolved star's condition, would tend to place its ongoing evolution back towards a regime of thermonuclear stability. Morton likened the lack of observed Algol systems with the more massive component at its Roche limit to the Hertzsprung gap effect in the evolution of single stars, keeping in mind the relevant time-scales in relation to observed evolutionary effects.

---

*Closer inspection has given rise to the terms 'over-contact' or 'near-contact' in some cases.

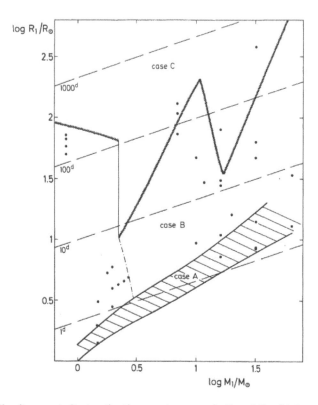

Figure 8.2   The diagram indicates the three main cases: A, B and C, of interactive binary evolution as depicted by H-C. Thomas, (1977), ©Annual Reviews Inc. Case A occurs when the primary arrives at the Roche limit (RL) while still on the Main Sequence (shaded region). Case B is when the primary reaches its RL between the Main Sequence and the onset of core helium burning. The zig-zag upper boundary for this region draws from models for the onset of helium burning for single stars. Case C allows for evolution beyond helium burning. There appears a discontinuity in locating conditions for the onset of helium burning at a mass of about 2.25 $M_\odot$ associated with the 'helium flash' in stars of lower mass. The points indicate just a few of the early models of different groups, and the dot-dash line indicates that the models to the left of it were treated as powered by a hydrogen-burning shell. Corresponding orbital periods (using Kepler's Law with $M_2/M_1 = 0.5$) are indicated by the sloping wide-dash lines.

The foregoing scenario, applicable to the classical Algols, involving a mass losing-star that already left the Main Sequence, is now commonly referred to as 'Case B' of interactive binary evolution, following the terminology of early modellers (see Fig 8.2). A situation in which the primary becomes close to its limiting Roche surface whilst slowly expanding, but still on the Main Sequence, is known as 'Case A'. This mass-losing star might, under certain circumstances, continue with further transfer after leaving the Main Sequence; the succession of events then being modelled under the heading 'Case AB'. A further alternative, applying to massive stars, initially well-separated from their binary companions and evolving on the Asymptotic Giant Branch when they reach their Roche lobes, has been considered as a 'Case C' mode of binary evolution. There may be still other possibilities.

In considering relevant principles, Morton referred to a *dynamical* (or *pulsational*) timescale

$$\tau_P = \frac{1}{2\pi}\sqrt{\frac{1}{G\bar{\rho}}} \; , \qquad\qquad (8.3)$$

which is of the order of 10 minutes for the Sun, and increases to a little over 2000 s for a 10 $M_\odot$ star. Morton found that when a small amount of mass is removed from stars around this mass ($\sim$10 $M_\odot$), the immediate response (on a dynamical time-scale) is a general reduction in size. In other words, a potentially mass-losing star on reaching its Roche limit would remain essentially close to hydrostatic equilibrium, from the point of view of the equations describing its internal structure as a whole. Its instability would be on the relatively slower time-scale referred to above: an evolutionary urging of its expansion into *Roche lobe overflow* (RLOF), that should account for the observed Algol type binaries[*] if this picture is valid.

We follow this line of thought in what follows. Consider the mass element $\delta M = 4\pi R_1^2 \rho_s \delta R_1$ from the *loser* star, of mass $M_1 = 4\pi R_1^3 \bar{\rho}/3$, whose radius is R_1 and whose density at the losing surface is ρ_s. We suppose this surface to be expanding at a mean rate s, of the order of 10^{-7} solar radii per year. Let us write $x = M_1/(M_1 + M_2)$ for the loser's mass as a fraction of the total mass of the system M, here assumed constant. So now, with $\delta x/x = \delta M_1/M_1$, and a little rearrangement, we can write a *mass transfer equation* as

$$\dot{x} = -3\eta \frac{x}{R_1}(s - \dot{R}_1) \; , \tag{8.4}$$

where we have written $\eta = \rho_s/\bar{\rho}$ for the ratio of the effective mean surface density to that of the whole star. Given hydrostatic equilibrium of the loser on a dynamic time-scale, we could expect

$$\eta = \frac{1}{3}\left(\frac{\partial \log M_R}{\partial \log R}\right)_{\text{surface}} , \tag{8.5}$$

where this logarithmic mass gradient might be available from the modelling of comparable stars. In practice, the surface layers of a mass-losing star are not at rest, whereas classical stellar modelling would normally refer to static conditions. The structure of a mass-losing star may well remain similar to that of a static model given the dynamical stability. However, the limiting form of $\partial \log M_R/\partial \log R$ for a static model at the nominal surface would not reflect the disequilibrium of the mass-loss there. As for x ($0 < x < 1$), it is initially greater than $\frac{1}{2}$, since the more massive star evolves faster to reach its Roche lobe first.

The formulation in Eqn (8.4) concerns long-term, or evolutionary, behaviour. More detailed consideration of the RLOF process itself was given by S. H. Lubow and F. H. Shu (1975), who produced a feasible hydrodynamic scenario in which gaseous matter emerges at sonic speed v_s through a narrow nozzle of area A_n in the vicinity of the inner Lagrangian point L_1. That discussion makes use of a small parameter ϵ, defined as the ratio of v_s in the outer regions of the Roche Lobe filling component to an orbital speed Ωd, where Ω is the angular rotation rate of the binary orbit (presumed circular) and d is the separation of the two stars. A_n is then found to be of order $(\epsilon d)^2$. Lubow and Shu relate a reference density for transferred matter $\dot{M}/\Omega d^3 = \rho_t$, say, to the actual mean density ρ_0 in the stream at L_1, $\rho_0 = \dot{M}/A_n v_s$ (from continuity), as $\rho_0 = \epsilon^{-3}\rho_t$. The stream density in the subsonic region of flow in the surface regions of the contact component becomes of order $\rho_1 = \epsilon^{-4}\rho_t$. The reference density would be quite low in practice: for a notional value of $\dot{M}$ of order 10^{-7} $M_\odot$ yr^{-1} or less, ρ_t would be of order 10^{-10} or less, for typical Algols, with ϵ typically of order $1/30$.

This implies that matter transferring through the outer regions of the semi-detached component is at densities comparable to that of the photosphere, or less. We can then suppose that sub-photospheric structure is not seriously affected by the rather fast horizontal

---

[*]Dynamical time-scale RLOF has been considered relevant for circumstances other than those presently discussed.

wind just above the visible surface. This wind may have a complicated structure when considered in more detail. Lubow and Shu make analogies with global wind patterns familiar from the terrestrial example, though this topic is rather outside the evolutionary trends presently addressed. In the evolutionary context, the material slowly pushing up from below the photosphere that sources the outflow drifts radially outward at about 9 or 10 orders of magnitude below photospheric sound speed. To conserve mass this material would have a density several orders greater than that in the wind. For discernible effects on period variation, this should match with η having values typically in the range $10^{-1} - 10^{-3}$.

Eqn (8.4) can be integrated to give

$$\frac{1}{3\eta} \int_{x_2}^{x_1} \frac{R_1}{x} \, dx - [R_1]_{x_2}^{x_1} = \int_{t_1}^{t_2} s \, dt \ . \tag{8.6}$$

Let us suppose that suitable average (constant) values for s and η can be found to represent a time interval $(t_2 - t_1)$ that might be relatively short, for example in the early stage of mass transfer, when we expect R_1 to be decreasing with time. Assume also that the angular momentum is conserved. If we consider only the orbital component of the angular momentum and neglect, for now, the relatively small rotational and other components, the separation A of the two stars, when moving as mass points in circular orbits, satisfies

$$A = \frac{A_0}{16x^2(1-x)^2} \ , \tag{8.7}$$

where A_0 is the minimum separation, occurring at equal masses.

The radius R_1 can be set as $Af(x)$, where $f(x)$ here expresses the mean radius of a Roche lobe in terms of the separation of the two mass centres in dependence on the fractional mass, so that

$$R_1 = \frac{A_0 f(x)}{16x^2(1-x)^2} \ . \tag{8.8}$$

The form of $f(x)$ was considered in Section 4.4, where the approximation $f(x) = 0.38 + 0.09 \ln q$, with $q = x/(1-x)$, was cited. This allows Eqn (8.6) to be directly evaluated, permitting an overview of the mass transfer process.

We then have, for the time-scale $t_2 - t_1 = \tau_{1\,2}$, say,

$$\tau_{1\,2} = \frac{A_0}{s}[\{R_s(x_2) - R_s(x_1)\} - \frac{1}{3\eta}\{I_s(x_2) - I_s(x_1)\}] \ , \tag{8.9}$$

where R_s is simply the loser's radius scaled in units of the minimum separation.

$$R_s = \frac{0.0238 + 0.0056 \ln q}{x^2(1-x)^2} \ . \tag{8.10}$$

The related integral $I_s(x)$ follows as

$$
\begin{aligned}
I_s(x) \ = \ & 0.0238 \left(\frac{6x^2 - 3x - 1}{2x^2(1-x)} + 3 \ln q \right) \\
+ \ & 0.0014 \ln q \left(6 \ln q + 6 + \frac{12x^2 - 6x - 2}{(1-x)x^2} \right) \\
+ \ & 0.0014 \left(\frac{6x^2 - 9x - 1}{(1-x)x^2} + \text{const.} \right)
\end{aligned} \tag{8.11}
$$

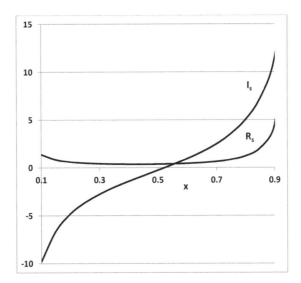

Figure 8.3   Underlying functions (8.10, 8.11) in interactive Algol-type evolution.

where we have written q for the argument of the logarithms $x/(1-x)$. The behaviour of $R_s(x)$ and $I_s(x)$ is shown in Fig 8.3. Since we are only interested in differences of $I_s(x)$ for particular values of x, the arbitrary constant of integration can be set to zero. It can be seen from Fig 8.3 that while I_s is a monotonic function of x, the radius R_s is bitonic, passing through a minimum at $x \approx 0.44$. This will lead to a basic asymmetry in the differences in (8.9): on the one side ($x > 0.44$) the trends of the two functions tend to reduce their differences over a given interval, on the other side ($x < 0.44$) the differences become enhanced.

A key role is played by the parameter η. If η becomes very small, the integral term in I_s becomes large and τ_{21} becomes long. This implies that if the mass-loss source layer is of low mean density the process becomes extended in time. The time intervals $\tau_{1,2}$ become relatively small in the central region of x when the masses of the components are comparable.

8.3.1   Classical Algols

The formalism of (8.4) can be developed to apply to more specific situations. It is useful to have practical examples in mind: Fig 8.4 shows the upwards parabolic form of the difference between observed and calculated times of minimum (O – C) for the 'classical' Algol type binary WX Sgr (present mass-ratio ≈ 0.3). This kind of O – C trend is suggestive of a steady increase in period, that could be associated with mass transfer from the Roche-lobe-filling former primary to its near-Main-Sequence primary.

Following regular stellar modelling, the original primary should have left the Main Sequence, its core becoming like that of a low-mass RGB star. A thin shell source above this degenerate core is able to maintain the tenuous envelope with its throughput of relatively excessive luminosity, compared with the normal mass-luminosity formula. The basic reason for the loser's luminosity excess is that core, or near-core, processes are uncoupled from the removal of the envelope material due to RLOF. There would be, though, an ultimate closedown of the shell source when the envelope has been sufficiently depleted. Thereafter, the evolved star should sink to a white dwarf condition with a separation between components of around an order of magnitude greater than that which existed originally. But before that, the semi-detached Algol condition can exist for a long time for evolved binaries of low

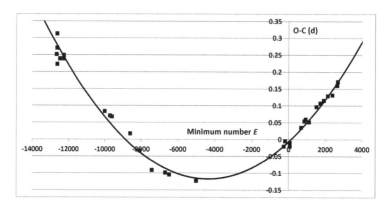

Figure 8.4 Upward parabolic form of O–C diagram in the Algol system WX Sgr, suggestive of a steady increase in period.

mass, permitting comparisons of theoretical and observed properties with regard to basic parameters. An uncertainty in this scheme is the nature and extent of angular momentum transfer and dissipation.

Using the foregoing formulation, we find, for the orbital angular momentum J,

$$J = \frac{2\pi x(1-x)MA^2}{P} \ , \tag{8.12}$$

where P is the orbital period. Kepler's 3rd law gives

$$4\pi^2 A^3 = GMP^2 \ , \tag{8.13}$$

from which we deduce

$$P = P_0/64x^3(1-x)^3 \ , \tag{8.14}$$

where the minimum period $P_0 = 128\pi J^3/G^2 M^5$, with $A_0 = 16J^2/GM^3$.

Differentiating (8.14) with respect to time, we find

$$\dot{P} = 3g(x)P\frac{\dot{x}}{x} \ , \tag{8.15}$$

where the function $g(x) = (2x-1)/(1-x)$ in 'conservative' conditions (J, M constant). In the normally encountered 'slow phase' of Algol evolution $x < 0.44$, so $g(x) < 0$, while $\dot{x} < 0$ is a necessary condition of mass loss. Under these circumstances, as x becomes small we could expect $\dot{P}$ to increase in a rather sensitive way to the mass transfer, depending on the behaviour of $\dot{x}$.

Since $\dot{x}/x = \dot{M}_1/M_1$ and $\dot{P}$ may be available from analysis of a curve like Fig 8.4, Eqn (8.15) has been used to gain insight on the scale of mass transfer, particularly using the simplified conditions mentioned, the variation of period being regarded as evidence of RLOF.

Consider a sequence of events (such as times of minimum light due to eclipses) E_1, E_2 ... etc., following a regular progression in time $T(E)$. If the time interval ΔT between such events is an integral number of eclipse minima, say $\Delta E = E_{n2} - E_{n1}$, then the ratio $\Delta T/\Delta E$ is the mean period P for the interval in question, i.e. at $E = (E_{n1} + E_{n2})/2$. We would not normally expect P to change rapidly for low values of ΔE.

The same idea can apply to non-integral separations of the cyclic process, so that a period can be associated with all points along the E-axis. Such a period corresponds to the limit of the ratio $\Delta T/\Delta E$ as the interval ΔE about the point $E, T \to 0$.

Let us now assume that such $T(E)$ data follows the parabolic form

$$T(E) = A + BE + CE^2 \tag{8.16}$$

with A, B and C positive constants. The period at E, usually given in days, is then given by the uniformly increasing slope

$$P(E) = B + 2CE , \tag{8.17}$$

where, in practice, C is very small compared to B. It follows that the mean period in the interval $T_2 - T_1$ is equal to the gradient at the mean value of E in the same interval, i.e. $(E_1 + E_2)/2$.

After selecting some reference epoch T_0 where $E = E_0 = 0$, the O – C diagram is formed from the difference between $T(E)$ and corresponding T-values calculated along the line $T_c = P_c E + \text{const.}$, where $P_c = B_c$, say, and the constant term in the expression for T_c is A_c, say. The constants A_c, B_c would be assigned to give the best position of the line to match the observed trend in the given E, T interval. We would expect them to be close to A and B in practice.

The O – C values are then

$$\text{O} - \text{C} = T(E) - T_c = (A - A_c) + (B - B_c)E + CE^2 , \tag{8.18}$$

this being usually given in days. This is again of parabolic form, and with the same coefficient of the E^2 term. Confidently determinable values of C, for an O – C of order 0.1 d or greater, involve a suitably covered interval with a half-range for E reaching to typically ~ 10000 periods, i.e. C would need to be greater than about 10^{-9} for a good determination.

The change of period ΔP with each event ($\Delta E = 1$) is then $\mathrm{d}P/\mathrm{d}E = 2C$. The rate of change of period per year at E becomes $730.5\,C/P(E)$. From (8.15), this is related to the corresponding mass change ΔM_1, per year, as

$$\Delta M_1 = 243.3 \frac{M_1 C}{g(x)P^2} \ \mathrm{y}^{-1}. \tag{8.19}$$

In other words, annual mass transfer rates should, in most cases, be greater than $\sim 10^{-7}$ solar masses per year to have discernible period change effects over a few decades.

In the case of of WX Sgr, shown in Fig 8.4, we have $x = 0.23$; M_1 (present) $= 0.68$ M$_\odot$; ($M_{10} \approx 2$M$_\odot$, say); $C = 5.79 \times 10^{-9}$, $g(x) = -0.70$, and $P = 2.129629$ d. The corresponding rate of period change per year is $2C/P = 1.99 \times 10^{-6}$, and the implied ΔM_1 is $\sim -3.0 \times 10^{-7}$ M$_\odot$y^{-1}.

The scope of Equation (8.15) can be broadened to include the case where angular momentum of the overflowing material is not transferred back into the orbit. At wider separations, the time-scale for such angular momentum transfer, even assuming no matter is actually lost from the system, may well become appreciable in comparison to Case B evolution time-scales. An empirical approach to such effects is to write

$$J/J_{\text{init}} = (x/x_{\text{init}})^k , \tag{8.20}$$

where $k = 0$ corresponds to the conservative case, and we expect $0 < k < 1$ in practice. Physically, this expresses the difficulty of putting the increasing proportion of specific angular momentum in the RLOF material at low x back into the orbit. $|g(x)|$ then becomes reduced in size to $(1 - 2x)/(1 - x) - k$. It might even be possible, in some range of x, for the coefficient of $\dot{x}$ in (8.15) to become positive, i.e. the period can decrease with decreasing mass of the loser. In any case, systemic angular momentum loss will decrease detectability

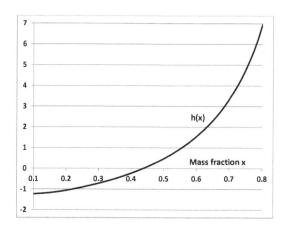

Figure 8.5   The function $h(x)$, given by Eqn 8.22), will influence the rate of orbital period change in classical Algols.

of the period increases expected for conservative mass transfer in classical Algols. It is clear from observations that not all Algols exhibit an O – C curve like that in Fig 8.4.

Eqn (8.4) can be rearranged as

$$\dot{x} = -3\eta s x / \left(R_1 \{ 1 - 3\eta \frac{\mathrm{d}\log R_1}{\mathrm{d}\log x} \} \right) \quad . \tag{8.21}$$

Although the logarithmic derivative on the right resembles the inverse of that in (8.5), it is important to notice that the variable R_1 is here not the static radius of (8.5), but is specified by its adherence to the Roche lobe, as in (8.8). This can be formally evaluated, so that $\mathrm{d}\log R_1 / \mathrm{d}\log x = h(x)$, say, where

$$h(x) = \frac{0.09}{f(x)(1-x)} + 2g(x) \tag{8.22}$$

The form of this function is shown in Fig 8.5, where it can be seen that for the range of loser mass fractions in Algol systems usually encountered $(0.1 < x < 0.4)$ we find $h(x)$ to be of order negative unity.

Hence, on substitution for $\dot{x}$ in (8.15), we have

$$\dot{P} \approx \frac{-9\eta s g(x) P}{R_1} \tag{8.23}$$

in view of the expected smallness of the second term in the denominator in (8.21). Using (8.8) and (8.14) we can rewrite (8.23) as

$$\dot{P} = -\frac{9\eta s g(x) P_0}{4x(1-x)f(x) A_0} \tag{8.24}$$

For small x, $f(x) \to 0.462 x^{1/3}$ (Eqn 4.19), and since $h(x)$ becomes nearly constant, we have,

$$\dot{P} \sim \mathrm{const.}\, \eta s / x^{4/3} \quad , \tag{8.25}$$

i.e., depending on η and s, a rather sensitive response of the period to the declining low fractional mass of the loser. We will consider appropriate values for η and s shortly.

Evolutionary time-scales are much longer than those of observational coverage hitherto, so we may be cautious regarding direct evidence on long-term trends for $\dot{P}$; however, the

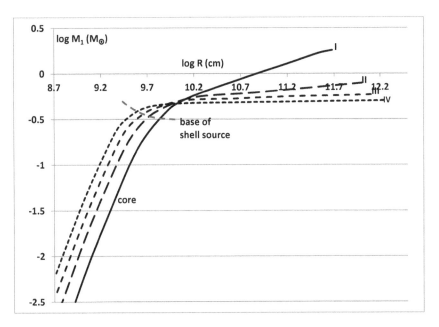

Figure 8.6 Schematic structure of semi-detached subgiants derived from the sequences of P. Harmanec (1970). The mass distribution becomes more centrally condensed in the sequence I-IV as time progresses. Use of the formula (8.26) yields for curve I, $\eta = 0.083$; II, $\eta = 0.055$; III, $\eta = 0.017$; IV, $\eta = 0.009$. The hydrogen mass fraction (X) in the shell source meanwhile diminishes as: curve I, X = 0.27; II, X = 0.14; III, X = 0.09; IV, X = 0.08.

integrated effects of short-term period variation may be determined from O – C diagrams. A positive value of $\dot{P}$, would be expected for an Algol in a near-conservative regime with $x < 0.44$, even if relatively small, and this can show up in an upturned parabolic variation of O – C values over a few decades, as seen for WX Sgr. This will allow numerical checks on whether the period variation is in keeping with the mass-transfer equation.

Regarding η, though numerous examples of mass losing stars in similar conditions (low mass, Case B) have been worked out, relatively few numerical details of the loser's ('subgiant') structure were published to allow appropriate values of η to be directly assessed. Among such calculations, those of P. Harmanec (1970) are useful for the present purpose, and plots of the subgiant structure in the $\log M_r, \log r$ plane, based on Harmanec's data, are reproduced in Fig 8.6.*

A fairly clear subdivision into core and envelope regions can be seen in Fig 8.6. The apparent constancy of η through the envelope region accords with general results for the structure of hydrogen shell-burning subgiants, believed to apply to the losers in many classical Algols evolving in the Case B mode. Stars in such conditions were studied by S. Refsdal and A. Weigert (1970), who found the mass of the degenerate core M_c to play a dominant role. They showed the basic structural variables: pressure, density, temperature and luminosity; to have a simple power-law type behaviour through the envelope, of mass M_{env}, say, where $M_c + M_{env} = M_1$. RGB stars of different mass would satisfy certain homology relations, involving M_c and R_c as parameters. A linear trend of $\ln M_r$ with $\ln R$, as in Fig 8.6, could then be expected through the range of RGB-like loser structures.

---

*Stellar composition, referred to in Fig 8.6, is usually summarized using the proportion by mass of hydrogen (X), helium (Y) and other elements (Z).

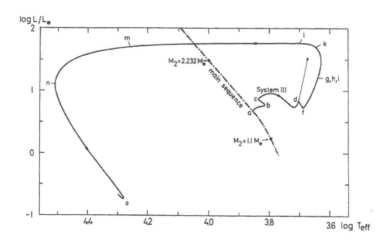

Figure 8.7   Evolution of a classical Algol model in the Hertzsprung-Russell plane; after S. Refsdal & A. Weigert *Astron. Astrophys.*, Vol. 1, p167, 1969; with permission ©ESO. Instead of continuing up the RGB (dashed line with arrow) at point d, RLOF intervenes and the subgiant is diverted towards point f. Classical (low mass) Algol former primaries can be expected to be found in the g, h, i-k region of the track. The gainer M_2 proceeds as indicated in the general direction of the Main Sequence.

On that basis, η can be estimated as:

$$\eta = \frac{1}{3} \left(\frac{\log M_1 - \log M_c}{\log R_1 - \log R_c} \right) . \tag{8.26}$$

Typical values of R_1 and R_c yield the difference in their logarithms to be ~ 2. While the envelope to core mass ratio M_{env}/M_c could start with a value of order unity it may sink to $1/10$ of this or less, while RLOF is still effective. We can thus expect classical Algols showing evidence of mass transfer to have η decreasing through the range $\sim 0.1 > \eta > \sim 0.01$, as a result of envelope expansion and mass loss. If we apply the results of Refsdal and Weigert and use the observed values of mass and radius for WX Sgr, we have $M_c = 0.178$, $M_{env} = 0.502$, $(M_\odot)$; $\log R_1 = 11.28$, $R_c = 0.032$ $(R_\odot)$, so that from (8.26) $\eta = 0.076$. This compares reasonably with inferences from Fig 8.6 at a corresponding stage of the evolution and the previous general estimates of densities from Lubow and Shu (1975). The decrease of η from ~ 0.1 is initially relatively slow, but in the later stages before eventual envelope collapse the decline to $\eta \sim 0.01$ speeds up. This results from the controlling influence of the degenerate core, that continues to accumulate mass as the envelope is depleted.

The envelope expansion rate s is related to the behaviour of the core, however it appears sensitive to internal structural changes, becoming markedly more rapid after convection starts up in the outer envelope. The expansion rates also show a strong dependence on the evolving star's original mass (M_{10}): thus in the evolution sequence of a 2.25 $M_\odot$ star, as published by I. Iben (1967), the surface expansion proceeds at a mean rate of about 3.6×10^{-7} $R_\odot$ y^{-1} during the first phase of shell burning while the envelope is still radiative, but increases to 1.4×10^{-6} $R_\odot$ y^{-1} during the convective outer envelope phase. This maximum rate of expansion is reached relatively early on, after which there is a long period during which the expansion rate declines as the convective region of the outer envelope narrows down, with a steep dependency on the remaining mass. This phase of slow decline can last several times as long as the lead-up to the convective maximum.

The fact that η is a small number in the final slow stages of mass transfer allows us to combine the foregoing equations in an alternative way to the way Equation (8.15) was

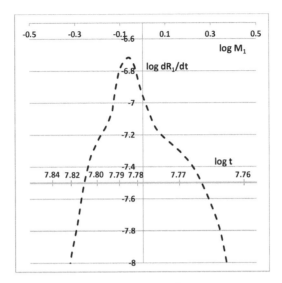

Figure 8.8 Variation of the evolutionary radial motion ($\dot{R}_1$, vertical axis) of the mass-losing component (M_1, upper horizontal axis) in a classical Algol binary. The time-scales (lower horizontal axis) would be appropriate for a star of initial mass $M_{10} \approx 3 M_\odot$. We would expect to observe most Algols in the lower left part of the dashed curve, that charts the slow expansion of the mass-losing star.

derived for the period, to show the relationship of s to $\dot{R}_1$, thus $s \approx \dot{R}_1/|3\eta h|$. The behaviour of $\dot{R}_1$ can be followed from calculated interacting binary model sequences. Fig 8.7 illustrates such Algol system modelling from Refsdal & Weigert (1969).

While there are differences of detail for individual cases, the overall pattern is as shown in Fig 8.8, which uses data from classical Algol modellings mentioned above. The lower axis records the logarithm of selected times after the primary's time of reference, during which it is engaged in mass transfer. The rate of expansion of the envelope material depends rather sensitively on the loser's initial mass but falls in later stages as the rate of core mass increase declines. Using previous models of low-mass Algols, a value of $s \sim 5 \times 10^{-6}$ $R_\odot$ y^{-1} can be reasonably estimated for the present condition of WX Sgr. From Eqn(8.23) we then find a rate of period variation as

$$
\begin{aligned}
\dot{P} &= 9 \times 0.076 \times 5.0 \times 10^{-6} \times 0.7 \times 2.129/2.7 \\
&= 1.89 \times 10^{-6} ,
\end{aligned} \tag{8.27}
$$

which compares favourably with the observed value of 1.99×10^{-6} obtained after (8.19). The condition-sensitive dependencies in the coefficients η and s make it difficult to determine a precise rate of period variation using the mass transfer equation, in order to compare theoretical binary evolution with an O – C curve results in general. However, the degree of agreement for systems with higher mass ratio, such as WX Sgr, gives confidence about the Case B model for this kind of Algol evolution.

WX Sgr is one of 25 Algol systems showing steady period increase in the compilation of Budding & Demircan (2007). On the other hand, the same review included almost double this number of Algols showing either period decrease or inconclusive, complex indications of period variations in the O – C diagram. There are various reasons why a steady period increase may not be observed from a system undergoing RLOF, including non-conservation of angular momentum. But it should also be noted from Fig 8.8 that it is only in the early stages of RLOF that the stellar expansion rate becomes relatively large and in the range where a trend of regular period increase would be shown in observations that extend over a few decades. A typical Algol system may well spend a larger proportion of its semi-detached life undergoing mass transfer below the detectable level of 10^{-7} M$_\odot$ y^{-1}.

This point relates to the more general issue of *observational selection*. Our understanding of the arrangements of stars is naturally based on how things are perceived, but this may be misleading. For example, there are more than 30 times as many M-type as A-type Main Sequence stars among the essentially completely sampled 100 nearest stars in Allen's *Astrophysical Quantities* (see Fig 5.2), whereas among the 100 brightest stars listed in the same source there are some 15 A-type dwarfs compared with the single K-type subgiant θ Cen, other late type cool dwarfs being fainter than the limiting magnitude for this sample of $V \approx 2.6$. If the primaries of detected Algol binaries were to distribute like the single Main Sequence stars they resemble, then the spatial occurrence of low mass Algols at short period would be much greater than that apparent from the observed incidence. This has a bearing on our next topic.

8.3.2   Contact Binaries

It was mentioned in this chapter's introduction and also Section 5.2, that ambiguity can arise in associating a physical configuration with a given light curve type. Kopal identified *contact binaries* with those binaries, essentially all eclipsing, for which the C parameter, as inferred from estimates of the relative radii matching the light curves, was indistinguishably close to the Roche limit C_0, but some such systems might just be detached pairs whose components happen to be very close to this limit. It was noted that there exists a particular well-populated group of eclipsing binaries, the EWs — of which W UMa is representative, characterized by short periods and relatively low masses. The fairly high incidence of these systems (of order 1 in $\sim$300 stars), together with their fast light cycle times has made them popular observational targets, and a large literature has grown since the mid-twentieth century. The quasi-sinusoidal W UMa type light curve, closely associated with this group, has been divided into A and W types, the former interpreted as having the larger, more massive star eclipsed at the slightly deeper light minimum.

Light curves of the A type can be produced, in principle, by a very close pair of Main Sequence stars: the young massive pair in V831 Cen, for example, would fit. An 'H' type has been proposed for such atypical high mass ratio systems. Unevolved pairs that are very close but not deeply eclipsing, due to a low orbital inclination, could also present a low-amplitude quasi-sinusoidal light variation (e.g. PU Pup).

The W type contact binaries, however, where the lower mass component has also the slightly higher mean surface temperature, pose a *prima facie* paradox, comparable to that of the Algols. Such systems look to have a high degree of mutual interaction. The secondaries of many A-type systems are also found to be too bright for their mass, suggesting, like the W-types, histories of interactive evolution. This gives rise to the point that, as well as very close yet under-contact binaries, 'over-contact' ones exist. The introduction of an over-contact close binary scenario, having two separate stellar cores embedded in a common envelope, is usually traced to G. P. Kuiper's (1941) discussion of the complex system β-Lyrae. Kuiper pointed out that the conditions imposed on a star filling its Roche Lobe differ from those that apply for a single star to remain in equilibrium for nuclear time scales (Fig 8.9). Contact conditions then point to thermal instability with structural changes taking place on the thermal time scale (8.1).

From another tack, certain wide binary stars, such as Z And, show evidence of being embedded in a gaseous plasma. In fact, the nature of star formation regions, exhibiting multiplicity and clustering in their condensation, points directly to common envelopes, at least in the initial circumstances. When the local gravitational binding energies of the condensing fragments are great enough they survive to become, in due course, Zero Age

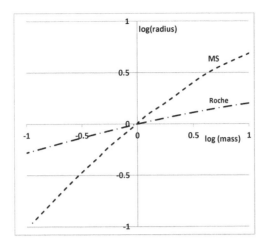

Figure 8.9 Mass radius relation trend for Main Sequence (dashed) and Roche-lobe-filling stars. For stars of mass lower than 10 M$_\odot$ it can be seen that $R \propto M^n$, where $1 > n > \sim 0.7$. The trend for Roche-lobe-filling (dot-dashed) pairs however has significantly lower slope: in that case $n \sim 0.46$.

Main Sequence objects. It has been surmised that proto-stars losing significant angular momentum in such condensation processes could produce a contact binary at zero age: TX Cnc within the Praesepe cluster being cited as a possible example.

In spite of the plentiful evidence of common envelope states in general, the physics of the condition remains indistinct. The envelope may expand and be shed from the system as it acquires angular momentum from the spiralling in of the stellar cores. The time scale for this process seems to be long from observational evidence on symbiotic binaries, though theoretical scenarios often involve dynamical time-scale mass loss. These mismatches may reflect differences in the initial conditions of formation of the common envelope: whether primordial, or as a result of a Roche lobe overflow, or some other process during the binary evolution.

Here, however, a distinction may be made. The relatively high incidence, together with photometric analyses of EW light curves, have been understood to indicate the particular condition of a narrow or *marginal* contact with a continuous common photosphere. This is distinct from more general states of local plasma interaction that all stars pass through. The possibility of a primordial origin for contact binaries, while present, represents but one issue in a still unsettled wider discussion, in which EWs can be produced from a range of possible evolution options including normal detached binaries.

Whatever the formation circumstances may have been, the components of short-period common envelope systems like W UMa are found to be not far from the Main Sequence, particularly the primary components. The secondaries are somewhat overluminous for their mass, while tending to be underluminous for their temperature. The suggestion of underlying thermal instability then raises the question about why should there be so many of them, relative to other close binaries, if their configuration is essentially unstable on a thermal time-scale?

Modelling background

Numerical models of W UMa type binary structures, incorporating a common photospheric surface lying between inner and outer critical Roche surfaces, started to appear within a decade of M. Schwarzschild's famous (1958) book on stellar structure and evolution. Such models were aimed at matching light curves to a degree comparable to the scatter of observations, although phenomenological agreement with W UMa type light curves does not give a complete explanation for their properties, particularly those of the lower mass EWs.

Practical procedures developed to cover the incompletely known details of convective heat transport in the outer layers of normal cool stars, as presented in Schwarzschild's book, allow the value of the adiabatic constant* (K in the relation $P = KT^{\gamma/(\gamma-1)}$, γ being the ratio of specific heats of the gas in question) to become a free modelling parameter. The surface boundary conditions for either star having convective sub-surface layers include the constraint $P = KT^{2.5}$; and as $T \to 0$ the radial variable $R \to R_\star$. L. B. Lucy applied this to a common photospheric boundary for a contact binary, setting one surface value of K to the internal integrations for both stars' separate energy regimes. He argued that in the convective common outer envelope the local photospheric flux would be independent of the core-generated heat flow, and the flux would be free of constraint from hydrostatic equilibrium at the surface.

For certain ranges of stellar mass, it was possible to find pairs of radii satisfying the mass-radius relation of contact binaries using the same value of K applied to the inward integrations of both stars, which were tailored into separate energy generation rates for either core. The range of possible agreement with observational data was very narrow, however. Other researchers extended the scope of the modelling by modifying the outer boundary conditions so that differences in the assigned values of K could be selected so as to allow convection to transfer luminosity from one star's envelope to the other. While such models were consistent with corresponding light curves, they still did not reproduce the full range of observed colour-period combinations, particularly at short periods, where models were generally too hot.

The lack of agreement with observations led to abandonment of imposed equilibrium conditions for W UMa type stars. The incompatibility of the mass-radius relation for normal Main Sequence stars and that required for hydrostatically stable stars in the contact arrangement (see Fig 8.9), apart from pairs of equal mass, implies that at least one of the components would not be stable over thermal time scales. The underlying problem gathers force from the relative plenitude of contact binaries. Evolution may have led to this, as with the Algols, and this is not incompatible with loss of stability on a thermal time-scale, but the apparent endurance of the configuration through time intervals comparable to the solar age poses a challenge to theoretical explanation.

In a development from the earlier, special-case, zero-age contact binary model, Lucy produced an evolutionary, interactive scenario in the following way. While relaxing the constraint of thermal equilibrium overall, allow thermal equilibrium in the interior of the more massive, and, by adoption, MS-like component, to be retained. With freedom to set initial conditions, imagine now that star's Roche-lobe-filling companion is compressed to the regular MS mass-radius relation shown in Fig 8.9. The compression would increase the secondary's relative rate of energy production, causing an expansion beyond the original Roche-lobe-filling radius on the secondary's thermal time scale. This expansion would bring about mass transfer to the primary: consequently, the mass ratio M_2/M_1 would decrease and, with conservation of total mass and angular momentum, the separation and period would increase (Eqns 8.7 and 8.14). The same tendency of decreasing mass ratio ought to follow if there were an initially normal MS secondary and an expanded primary, since the primary's relative lapse of energy production would cause it to contract within its lobe

---

*Also called the specific entropy.

and draw mass from the companion. The trend of decreasing $q = M_2/M_1$ would have the slowing-down feature characterizing the late stages of classical Algol evolution discussed in the previous section.

This mass transfer in contact conditions, however, fails to bring the binary into thermal equilibrium, and the decrease of M_2/M_1 is expected to continue until increasing separation breaks the binary contact. The gainer would still expand with time, it was argued, and, in due course, the cycle repeat in the manner of a thermal time scale *relaxation oscillator** (TRO). This interesting possibility was thought to be especially relevant to the W type subgroup, since many of the more well-studied examples do appear to show period irregularities suggestive of thermal time scale disequilibrium. However, clear observational proof of TRO behaviour is inconclusive, in particular regarding the paucity of confirmed close pairs in semi-detached or detached phases of the TRO cycle given the large amount of data.

Part of the problem is that it has been difficult to have sufficiently precise parametrization for EW binaries. There are two aspects to this. One is that the light curves often show additional effects that obstruct direct interpretation in terms of just the common outer equipotential surface model. Perhaps the most well-known is the *O'Connell effect* that refers to a distinct difference in the two out-of-eclipse levels of the light curve. Starspots, hot or cold, or gaseous stream effects, are invoked in this context. The other aspect is that the form of the data usually permits a significant degree of correlation between the parameters involved in the modelling. Thus the scale of the photometric ellipticity effect depends on the combination of the relative radii r_1, r_2, inclination i, mass-ratio q, and the light distribution over the surfaces $u_1, u_2; \tau_1, \tau_2$; or other parameters referred to in the previous chapter.

On the other hand, a system that cannot settle because of the differences between its hydrostatic and thermal requirements presumably either oscillates about its near equilibrium configurations or disintegrates, according as the prevailing forces acting during disturbance are restorative or disruptive. The plentiful supply of observed contact binaries thus lends support, in a general way, to the TRO proposition.

Comparison with Algols may be relevant, since the mass ratios of contact binaries, in general, are systematically higher than those of Algols. Feasibly, detached former-contact binaries could become low mass Algols. Observed classical Algols tend to have systematically higher overall mass than the W UMa type binaries, though, so that the former are mostly above 2 M$_\odot$ and the latter below.

The TRO scenario's lack of comprehensive observational support gave scope to alternative ideas on possible behaviour of contact binaries. An interesting development making use of the option of K as a free parameter to set different values of the specific entropy at the surfaces of two near contact stars, was the 'contact discontinuity' model of F. H. Shu, S. H. Lubow and L. Anderson. In this theory, the more rapidly evolving primary fills its lobe to overflowing, but the material of the primary can be set with a higher value of K, and therefore more buoyancy than the surface layers of the secondary. In a short time, matter from the primary surrounds the secondary, and fills, to some extent, the space between the inner and outer contact surfaces of the binary. The flux radiated from an optically thick photosphere, located in this common envelope, depends on only the potential gradient. When the common envelope's 'fill-out' increases to a sufficient extent, the flux determined by this gradient is not significantly perturbed by material flows. A more or less uniform flux is then emitted from the contact binary's lobe-engulfing outer surface. Given the range of available modelling parameters, the temperature and density can be set to the same values over the critical surface of both lobes, in accordance with the *barotropic* condition, i.e. the

---

*Relaxation oscillations are discussed in the context of dwarf nova type binaries in Section 11.2.3.

requirement for bulk hydrostatic equilibrium, in which the state variables are all single-valued functions of the potential. The temperature and density *gradients* across the contact surface differ for the two components, however, entailing discontinuous near-surface conditions (DSC). Such models have considerable versatility in producing light curves to match observed data; however, aspects of their underlying physics have met with controversy.

Statistical data on Algols show that systemic angular momentum loss is a feature of observed binary evolution. If angular momentum loss (AML) can thus be significant for the evolution of such semi-detached systems, we might expect it to have a role *a fortiori* for contact binaries.

The W UMa binaries are known to involve low mass stars that are normally associated with convective envelope regimes in a state of rapid rotation. These are the conditions in which magnetic fields are generated. Such fields can be effective in driving AML, since they can 'stiffen' stellar winds, so that even low-density material becomes effective in transporting angular momentum out from the wind sources. This line of thought forms a natural background to consideration of mass transport in the context of close binaries of low mass. Cool, detached, short-period binaries could feasibly become the progenitors of W UMa binaries. Comparability of the masses and period of a Main Sequence system such as UV Leo to many contact systems may be pointed out in this connection. There appears, however, to be a significant deficit in the supply of such 'parent' binaries to account for the high observed frequency of EWs.

We may note here the odd distribution of contact binaries maximizing around spectral type G (Fig 8.12). These pairs are relatively numerous, compared with the much more uniform trend of detached MS binaries against spectral type (Fig 8.10). If AML is a basic feature of the binary condition, either through interactive evolution or magnetic braking, we should be able to relate that to the peculiar distribution of contact binaries with spectral type.

This anomaly may point to a suitable supply of progenitors that are not easily distinguished observationally: an unperceived population of parent systems. This could be associated with the once difficult-to-detect, MS pairs of low mass ratio (say $q \sim 0.1$). Astronomical literature shows different assessments of the occurrence of such binaries. The comprehensive review of V. Trimble (1990) noted that the issue of which systems are selected for study, rather than which method of data-analysis is used, contributes to the difference of results from different investigators.

Many, if not all, of such investigations have shown that the mass ratio distribution is more uniform than the initial distribution of the masses of MS field stars, particularly close pairs with near equal masses. But there are also indications of a non-uniform distribution for unevolved binaries, with one peak near to $q = 1$ and a noticeable increase as q falls towards 0.2. Since the majority of field stars are in a binary state, binary primaries, at least, could be expected to distribute like field stars. We can thus reasonably expect more lower mass binaries in the spatial population of stars. Some component of these numerous lower mass binaries are found in the low mass semi-detached condition: they may also account for at least some contact binaries.

Feasible scenarios

Long after the early papers on contact binaries, a clear and full explanation of their condition still evades universal adoption. It has been said that although we may have acceptable parametrizations of individual systems, we are not certain just where the EW binaries come from, nor their ultimate fate. The kind of picture sought has been set out by K. Yakut and P. P. Eggleton, for example, as:

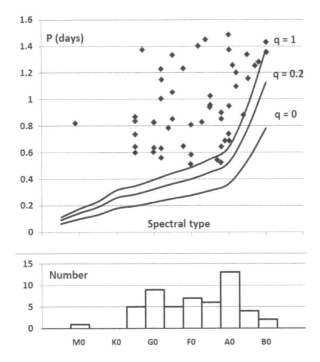

Figure 8.10 Distribution of 52 primaries of detached MS pairs in the spectral type : period plane, having periods less than 1.5 d. The incidence of primary spectral type is shown in the histogram at the bottom. The continuous curves trace the period at which an intermediate MS primary of the given spectral type accompanied by a companion of relative mass q would just fill its Roche lobe, with q taking the values 1 (top curve), 0.2 (middle), and 0 (lowest). They demarcate a possible near-contact condition. For the most part these EAD binaries lie outside this region. The two spectral type A primaries within it (EG Cep and RU UMi) could have rather low mass companions, or they may have been misclassified. YY Gem is the low mass star at the extreme left.

$$\text{DCB} \rightarrow \text{NCB} \rightarrow \text{LTCB(W)} \rightarrow \text{LTCB(A)} \rightarrow \text{coalescence}$$

where DCB stands for detached close binary (MS pair); NCB means near contact binary; LTCB is a low temperature $\sim$G type typical EW binary; and (W) and (A) are the W and A types of the latter. The complete scenario starts from a selection of DCBs out of the population of regular close binary stars, sufficiently close that AML may bring about the NCB condition of an incipient LTCB.

Regarding the NCB $\rightarrow$ LTCB step: — there are numerous close binaries with well-rounded out-of-eclipse regions, comparable eclipse depths and short periods but quite different masses. From the early days of modelling binary evolution, such systems have been identified with an Algol type process, but in Case A rather than Case B that the majority of classical Algols that are thought to follow. The relatively slow Case A process may be particularly relevant to the production of contact binaries. C. A. Nelson and P. P. Eggleton explored this issue in great detail, constructing a library of 5500 binary evolution tracks with a wide selection of initial conditions, while retaining conservation of systemic mass and angular momentum. They identified eight distinguishable sub-cases of case A, five of which led to contact. Two cases of special relevance to contact binaries were later identified as AR and AS, denoting relatively rapid and slow evolution into contact, respectively. These modelled patterns of behaviour depend on the imposed conditions of initial primary mass, mass ratio and period. Naturally, it would be desirable if the differing model features could

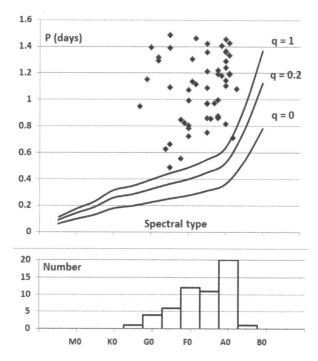

Figure 8.11   Distribution of primaries of 55 EAS systems, with the same layout as for Fig 8.10. One B8 type primary appears to come close to filling its Roche Lobe (RR TrA), though there may be some doubt about its assigned spectral type.

be unambiguously linked to observed characteristics, such as the W or A type classification of contact binaries. But since, observationally, a supposed LTCB could in fact be an NCB the appellation 'W type', for example, may be misapplied in practice.

The mass-transferring contact binary configuration can be considered along parallel lines to the previous discussion of Algols. Let us assume that a pair of closely orbiting stars could enter a common envelope condition while in the relatively fast phase of approach, the loser being still the more massive component. Matter from the original primary expands to fill the Roche lobe of its companion as well as its own. As the separation continues to decline, the common envelope becomes filled with matter shed from the primary. At some point angular momentum loss could occur by mass loss around the Lagrangian point L_2 that might produce an unstable runaway into coalescence. On the other hand, the volume of the common envelope is always appreciably greater than that within the Roche lobes, as may be seen from Fig 8.13, and we know from the Algols that at least some binaries in close interactive evolution must survive to the slower separation phase, where $M_{\mathrm{gainer}}/M_{\mathrm{loser}}$ is appreciably greater than 1.

In correspondence with the little-known fast phase of Algol evolution, EWs of the cool low-mass accumulation are not observed with mass ratios near unity: they are more likely to be found with $M_{\mathrm{gainer}}/M_{\mathrm{loser}} \sim 2$. These binaries, in this scheme, testify to a more active short-lived phase long passed. They still contain a slowly expanding, Roche-lobe-filling erstwhile primary, that, in the absence of AML, will slowly separate the system (Eqn 8.7), until a low mass, post-common-envelope Algol is formed. Fig 8.13 suggests that a conservatively evolving pair, entering into contact at mass ratio $M_{\mathrm{gainer}}/M_{\mathrm{loser}} \sim 0.5$ would, without compression of the transferred material, fill the outer limiting surface at equal masses. This would be in a relatively short time; thereafter, the increasing volume

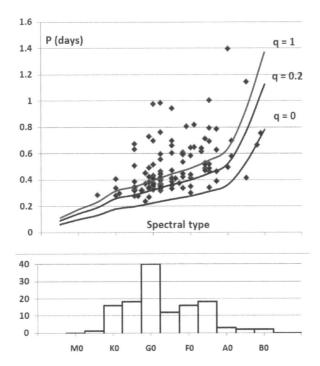

Figure 8.12 Distribution of primaries of 118 contact binaries, having the same layout as Fig 8.10. The A-type systems tend to lie to the upper right, while the accumulation of cool systems in the contact region are predominantly made up of W type EW binaries.

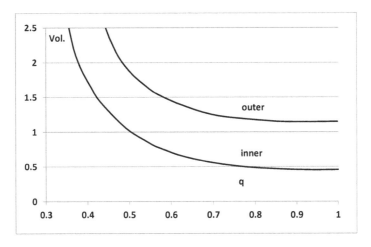

Figure 8.13 Volumes within inner and outer limiting Roche surfaces at different mass ratio, evolving conservatively.

of the gainer lobe would slowly deplete the common envelope. If AML is included, the conspicuous common-envelope phase would endure to the more slowly changing state beyond $M_{\text{gainer}}/M_{\text{loser}} \sim 2$.

The mass transfer formula (8.4) can be used to summarize the building up and retention of a common envelope, of mass M_3 say, which occurs if $\dot{M}_3 = -(\dot{M}_1 + \dot{M}_2) > 0$. So, $\dot{M}_1$ being negative, it is required that $|\dot{M}_1| > \dot{M}_2$. The behaviour of $\dot{M}_2$ depends on how the gainer lobe responds to the infalling matter. In the case where the orbital angular momentum is conserved, it can be shown that $|\dot{M}_1| > \dot{M}_2$ can happen while q is still moderate, but only if condensation of the transferred matter is implausibly low. In other words, if this slow interaction phase was taking place in a near-conservative angular momentum regime, the volume of the gainer Roche lobe would rise to comfortably accommodate the incoming material without forming a common envelope. It is necessary to introduce AML in order that the common envelope be retained for appreciable time-scales. Moreover, this AML has to be well-tuned to allow retention of the orbit while avoiding disappearance of the common envelope. Mechanisms commonly proposed for this include tidal dissipation and magnetic braking. The latter's range of available parametrization provides interesting scope for theoretical discussion. Growth of M_3 would, in any case, taper off as M_1 progresses to the slow phase of its expansion.

Apart from these effects that favour the detection of W UMa stars in a low over-contact regime, this Algol-like scenario naturally explains the over-luminosity of the secondaries compared with MS stars of the same mass at the level of a few mag. We should, though, keep in mind the significant uncertainties of the the parametrization of W UMa type light curves. Under-luminosity of the secondaries with respect to temperature, for a given single star comparison, may be related to their luminosity being spread out over the larger area of the common envelope. The primary would then not be conforming to the outer boundary conditions of a single MS star of the same mass. By implication, stars whose boundary conditions are fixed by alternative common envelope requirements cannot be compared directly with single stars, though some analysts refer to *corrections* that account for the differences in terms of transferred luminosity. The actual mechanism for transferring luminosity has stimulated much theoretical discussion. Yakut and Eggleton, in the work referred to above, raise the novel possibility that differential rotation may provide the required high efficiency heat engine.

The interplay between various contributory effects in the etiology of EWs: envelope expansion, mass transfer, AML and oscillatory effects in the mass ratio; complicates a clear comparison of modelling predictions and observational data. These complications offer both a warning and a challenge. The warning concerns the interpretation of data, when circumstances permit the "right combination of the wrong parameters" (to borrow a phrase from Kopal). The scope of modelling options has by now been made very extensive, as is the mass of accumulated data on EW systems. But the connections are still unsettled and ambiguous. It has been hard to make discriminatory testing of models thoroughly convincing. Even so, in recent years the quantity and quality of observational data has increased enormously. This fact renews the challenge that earlier studies left open.

This chapter has focused on the key role of Roche Lobe overflow in interactive stellar evolution. The mechanism finds its most directly traced and understood demonstration with the classical Algols, but we find that essentially the same process recurs, in one form or another, to link many, if not all, categories of close binary system. Subsequent evolutionary stages can lead to dramatic episodes of energy release, coupled with the formation of objects containing matter in unusual and extreme conditions. Naturally, interest surrounds the questions posed by such circumstances, and we conclude the chapter by briefly reviewing a selection of noteworthy topics. Further information will be found in the sources cited in Section 8.5, while various of these topics are pursued in following chapters.

8.4   Advanced Evolutionary States

The most commonly occurring binary stellar systems are of the kinds discussed in the preceding sections. Interactive evolution starts when the more massive star swells to reach its surface of limiting stability, thereafter losing mass, and later this leads to the drawn-out phase of slow mass loss. But the envelope of the loser does not expand indefinitely: at some point, the pressure gradient in the extenuating envelope becomes insufficient to maintain its structure. The donor collapses to become a helium rich white dwarf, sometimes referred to as hot subdwarf, or sdB star. At this point, conservation of angular momentum may have widened out the pair to the order of at least $\sim$0.1 AU, as with S Cnc. The mass of the resulting sdB star would be typically $\sim 0.5 M_\odot$.

This end-point of classical Algol evolution is not the only way in which sdB stars may be formed. The common envelope phase of close Algols, like R CMa, appears to have been very deep, by evolving the present configuration backwards in time while conserving angular momentum. The survival implies a considerable loss of systemic angular momentum in the system's previous history, so the possibility of merger of the stellar cores for similar but initially even closer pairs cannot be ignored. In a case where there is a third wide companion, that wide-orbit star becomes the partner in a new binary, where one component is a hot sdB. The evolution of contact binaries may also lead to mergers, with comparable resulting configurations if there is a third star.

Merging stellar cores would release a great deal of the potential energy of the original orbit in a relatively short time. Transient high energy sources have become more conspicuous in the era of space-based continuous photometry, providing evidence to support merger models. Such models have a large number of free parameters, so the rate of spiral-in can be tuned to observed luminosity changes. The merger of stars of higher mass has a relatively longer lead-in time due to lower envelope densities, but the final, dynamical time-scale collapse is catastrophic. A high mass merger like this could account for certain types of supernova, SN 1987A being a plausible candidate.

A luminous blue variable (LBV), comparable to the progenitor of SN 1987A, is one component of the well-known southern variable η Car, which is thought likely to produce a supernova explosion in the not-too-distant future. Depending on the main core-fusion element, this could be a matter of just years or up to the order of 10^5 y. The LBV star is in a wide and eccentric binary (period about 5.5 y) located near the centre of the large and complex Great Carina Nebula. η Car has a long history of meandering brightness, sometimes reaching to levels that imply a remarkably high intrinsic luminosity. Its quiescent brightness, at around 7th mag in V, would be consistent with that of a blue supergiant at the accepted distance of around 2300 pc, i.e. a luminosity of around 10^{39} erg s^{-1}. The normal internal energy reserves of the star could cover such outgoings for some hundreds of thousand of years, but not prolonged brightenings by a factor of $\sim 100 - 1000$, like those of the mid-nineteenth century. Such episodes, lasting for a decade or so, bespeak some additional energy source, and an internal process of merging cores in the LBV primary of the presently observed wide binary offers an explanation, not just of the brightenings, but also the energy of the ejected nebulosity and the high rotation velocity of the primary.

The configuration of a subdwarf star with one of more typical stellar proportions engaged in interactive binary evolution is met in a number of different contexts, from very wide binaries to the dwarf novae mentioned above. In fact, all the various kinds nova-like variables are now believed to involve interaction with a subdwarf. The very large potential well into which transferred matter falls when approaching a subdwarf provides a basic clue to the energy source of observed eruptions.

Besides dwarf novae, there are 'classical' and 'recurrent' novae; any of which may be referred to as a cataclysmic variable. These distinctions may be somewhat artificial, because the effects are understood all to result from essentially similar processes of heating of infalling hydrogen. Temperatures and densities then occur at which there is runaway thermonuclear fusion in the surface regions of the subdwarf. With recurrent novae appropriate conditions repeat themselves over human time-scales, that is to say, from about a decade or two to a century. A classical nova whose repetition period was a couple of centuries would probably not yet have been recognized as recurrent.

The infall is not directly onto the subdwarf surface, as there must be a diversion due to the action of the Coriolis force. The result is an accretion disk around the subdwarf. Frictional forces operate within the disk to remove angular momentum and allow the lower reaches of the disk to penetrate into the gainer photosphere. The impulsive energy release, when the temperature becomes high enough, expels a shell of gas, whose mass has been estimated at typically 10^{-4} that of the Sun. In the first few days after the explosion the spectrum has a continuum energy distribution like that of a regular photosphere, but after a week or so the spectrum would normally display a prominent set of emission lines – typically bright Balmer lines, but sometimes mainly helium – superposed on that of the original stellar source.

The rate of brightness increase and fall in a nova provides another distinguishing characteristic. 'Fast' novae such as CP Lac (1936) decline by 2 mag in less than a month. The corresponding decline is greater than 80 d for a 'slow' novae like DQ Her (1934). The overall change in brightness for well-documented novae is on the order of 10^5 from peak to quiescent levels. The peak emission may extend into the high energy range of the electromagnetic spectrum, even to the 100 MeV range, in keeping with the very high-temperature generation mechanism.

A relatively small group — $\lesssim 1\%$ of observed binary systems — consists of a hot subdwarf in an interactively evolved relationship with a distant, very swollen, giant donor; a combination known as a *symbiotic* star, typified by the previously-mentioned Z And. These systems share certain of the features alrady met, e.g. binary periods of the order of years with punctuated episodes of brightening taking place over decades. The scale of luminosity increases are significantly less than for the classical novae, however. The donor stars in these interacting binaries are so extended that the outer parts of their envelopes have become very tenuous. Wind effects are relatively strong, and the notion of 'Roche lobe overflow', in the sense discussed in the preceding section, loses relevance. The term 'case D' has been applied to this mode of interactive evolution..

In a comparable condition to the symbiotics are the barium, CH and S stars. The names of these groupings refer to their relatively unusual spectral features. There is a general expectation that these classes of star are binary systems, usually confirmed from periodic radial velocity variations. Their spectral oddities are interpreted as the results of an interactive evolution, often involving a previously massive star that has passed the Horizontal Branch stage (beyond point I in Fig 8.1) and is settling as a relatively inconspicuous white dwarf to the left of the Main Sequence. A so-called dredge-up episode may have occurred for the erstwhile donor, triggered by instability in its helium-burning shell. At that time, convection reaches down to regions containing the more massive elements from fusion processes and pulls them up to surface layers. The more massive elements in this surface enrichment result from the slow fusion of massive nuclei at depth: the 's-process'.

The discovery that barium stars are members of binary systems solved the difficulty of explaining how such heavy elements could appear in the spectra of stars whose parameters would not match those required for the dredging models. They present a kind of fossil evidence of what has occurred with their partner stars, passed by mass transfer to the photospheres of those now seen more directly. The detailed characteristics of these observed groupings are dependent on the initial parameters of the source binary systems.

8.5 Bibliography

8.1 Historical background to this chapter mentioned in its introduction is well set out in A. S. Eddington's *Internal Constitution of the Stars*, CUP, 1926, and the complementary *An Introduction to the Study of Stellar Structure*, Univ. Chicago Press, 1939, by S. Chandrasekhar. The many and varied developments since the mid-twentieth century have given rise to a large and continuing literature, well beyond our present scope. However, M. Schwarzschild's classic *Structure and Evolution of the Stars*, Princeton Univ. Press, 1958; set the stage for many of the developments of the computer age.

E. Hertzsprung's early appreciation of the role of colour as a basic stellar parameter is recognized in his paper in *Publ. Astrophys. Obs. Potsdam*, 1, Vol. 22(63), 1911. The relationship between that parameter and other characteristics of stars was considered by H. N. Russell in *Pop. Astron.*, Vol. 22, p275, 1914.

8.2 The general overview in Section 8.2 recalls much from the outline of M. A. Zeilik, & S. A. Gregory, in their *Introductory Astronomy & Astrophysics*, Saunders College Publishing, 1998. In turn, that draws from such well-cited sources as I. Iben Jr's reviews in *Ann. Rev. Astron. Astrophys.*, Vol. 5, p 571, 1967, and Vol. 12, p215, 1974; or K. -H. A. Winkler & M. J. Norman *Atrophys. J.*, Vol. 236, p201, 1980; D. A. Vandenberg *Astrophys. J. Supp. Ser.*, Vol. 58, p711, 1985; see also *Principles of Stellar Structure*, J. P. Cox, & R. T. Giuli, Gordon and Breach, 1968; or *An Introduction to the Theory of Stellar Structure and Evolution*, D. Prialnik, CUP, 2000.

The 'Henyey method' is usually associated with L. G. Henyey et al.'s papers in the *Astrophys. J.*, Vol. 129, p628, 1959; and Vol. 139, p306, 1964. The early history of computerized modelling is neatly summarized in the paper of S. W. Stahler, *Publ. Astron. Soc. Pacific*, Vol. 100, p1474, 1988. An apposite review for our present context is that of R. Kippenhahn in *Binary and Multiple Stars as Tracers of Stellar Evolution: IAU Coll. 69*, eds. Z. Kopal & J. Rahe, p3, 1982.

8.3 This Section 8.3 picks up the role of *interaction* between the binary components once the surface of limiting stability has been attained, as studied by D. C. Morton, *Astrophys. J.*, Vol. 132, p146, 1960. The introduction of *interactive evolution* modelling in binary systems is associated with G. P. Kuiper, *Astrophys. J.*, Vol. 93, p133, 1941; with important early contributions from O. Struve, *Comm. 5 Coll. International d'Astrophys.*, Liége, p236, 1954; and J. A. Crawford, *Astrophys. J.*, Vol. 121, p71, 1955. Z. Kopal's basic classification scheme was introduced in the *Ann. d'Astrophys.*, Vol. 18, p379, 1955; and fleshed out in the catalogue he produced with M. B. Shapley in *Jodrell Bank Ann.*, Vol. 1, p141 , 1956. In connection with this background, Z. Kopal's note in *Close Binary Systems*, page 545, referring to K. Walter's early recognition (cf. *Königsberg Veröff*, No. 2, 1931) of the semi-detached status of Algol systems, and apparent prescience of what that might entail, is of interest. The well-cited work of S. H. Lubow & F. H. Shu on the RLOF mechanism appeared in *Astrophys. J.*, Vol. 198, p383, 1975.

A good review of the development of binary evolution studies was given by B. Paczyński in *Ann. Rev. Astron. Astrophys.*, Vol. 9, p183, 1971; while a thorough bibliography was provided in *Interacting Binary Stars*, eds. J. E. Pringle & R. A. Wade, CUP, p201, 1985. Among the early contributions to the subject is that of R. Kippenhahn & A. Weigert *Z. Astrofiz.*, Vol. 65, p251, 1967; who later authored the comprehensive *Stellar Structure and Evolution*, Springer, 1990. Fig 8.2 comes from the review of H-C Thomas in *Ann. Rev. Astron. Astrophys*, Vol. 15, p127, 1977. An interesting summary of some key problems in stellar evolution was that of O. Vilhu in *Astrophys. Space Sci.*, Vol. 78, p401; where scenarios for the possible evolution of a binary system, from an initial fragmenting cloud to a neutron star containing X-ray burster, were neatly summarized by a general sketch.

With similar themes is *Structure and Evolution of Single and Binary stars*, of C. W. H. De Loore & C. Doom, *Astrophys. Space Sci. Lib.*, Vol. 179, (Kluwer), 1992. Observational aspects of the subject featured in *Interacting Binaries*, eds. P. P. Eggleton and J. E. Pringle, Springer, 1985. P. P. Eggleton's monograph *Evolutionary Processes in Binary and Multiple Stars*, CUP, (Cambridge Astrophysics), 2006, discusses interactive processes between two or more stars relating to conservative and non-conservative mechanisms of various kinds. The subject is updated in P. Podsiadlowski's Chapter 2 in *Accretion Processes in Astrophysics*, eds. I. G. Martínez-País, T. Shahbaz & J. C. Velázquez, CUP, Cambridge, 2014.

K.Pavlovski & H. Hensberge, in *Binaries – Key to Comprehension of the Universe*, eds. A. Prša and M. Zejda, Publ. *Astron. Soc. Pacific*, p207, 2010, discussed reconstructed spectra of binary star components. They analysed such data in order to probe elemental abundances in high-mass stars. Effects predicted from theoretical evolution models that include rotationally induced mixing could not be confirmed at that time.

The particular information referred to in connection with Fig 8.6 is from P. Harmanec, *Bull. Astron. Inst. Czech.*, Vol. 21, p113, 1970; while the papers cited with regard to the specifics of classical Algol evolution models are those of S. Refsdal & A. Weigert in *Astron. Astrophys.*, Vol. 1, p167, 1969; and *Astron. Astrophys.*, Vol. 6, p426, 1970.

The approach to summarizing the behaviour classical Algols in Section 8.3.1 develops from that of N. S. Awadalla & E. Budding in *Binary and Multiple Stars as Tracers of Stellar Evolution: Proc. 69 IAU Coll.*, (Reidel), p239, 1982. It also draws on E. Budding's review in *Proc. 3rd IAU Asian-Pacific Regional Meeting, (Kyoto)*, eds. M. Kitamura & E. Budding, *Astrophys. Space Sci.*, Vol. 118, p241, 1985; E. Budding, *Space Sci. Rev.*, Vol. 50, 205, 1989; as well as the catalogue of Algol type binary stars in *Astron. Astrophys.*, Vol. 417, p263, 2004, of E. Budding, A. Erdem, C. Çiçek, I. Bulut, F. Soyduğan, E. Soyduğan, V. Bakış, & O. Demircan. Many reference works relating to Algol type binaries will be found in these latter publications.

The relationship of period changes to evolutionary status was explored by J. M. Kreiner and J. Ziółkowski for 18 classical Algols in *Acta Astron.*, Vol. 28, p497, 1978. Our Eqn 8.19 appears as Eqn 20 in that paper. A catalogue of eclipsing binaries that show systematic secular trends in their times of minima was prepared by D. B. Wood & J. E. Forbes in *Astron. J.*, Vol. 68, p257, 1963. Most of the better determined C/P values (cf. Eqn 8.19) in that catalogue are in the range $\sim 10^{-8} - 10^{-9}$ or less. S. H. Negu & S. B. Tessema, in *Astron. Nachr.*, Vol. 339, p709, 2018, have investigated the parameters of Algols, of both Case A and Case B evolution types, in the range of primary and secondary masses between 0.2 and 8 $M_\odot$), and examined their properties in both mass-luminosity and mass-radius planes. Period changes in massive early type binaries were discussed by P. Mayer, *Bull. Astron. Inst. Czech.*, Vol. 38, p58, 1987. A general review of O – C diagrams was given by C. Sterken in *Variable Stars As Essential Astrophysical Tools*, ed. C. Ibanoğlu, (NATO Sci. Ser. C), Springer, p529, 1999.

The idea that TX Cnc could exemplify a primordial contact binary was pursued by J. A. J. Whelan, S. P. Worden & S. W. Mochnacki in *Astrophys. J.*, Vol. 183, p133, 1973. S. W. Mochnacki later compared the evolutionary status of A and W type EW binaries in *Astrophys. J.*, Vol. 245, p650, 1981. This classification of contact binary types was introduced by L. Binnendijk in *Vistas in Astron.*, Vol. 12, p217, 1970. Peculiarities of the W type were noted by A. P. Linnell, *Astrophys. J.*, Vol. 316, p389, 1987. R. W. Hilditch, D. J. King & T. M. McFarlane in *Mon. Not. Royal Astron. Soc.*, Vol. 231, p341, 1988, compiled parameters for some 31 contact binaries, from which it was deduced that there could be different methods of arriving at a given EW type configuration. Again, the indications were that A type contact systems are more evolved than W types.

The properties of 102 contact binaries were collected and compared with those of 118 chromospherically active binaries by Z. Eker, O. Demircan, S. Bilir & Y. Karataş in *Mon. Not. Royal Astron. Soc.*, Vol. 373, p1483, 2006, to allow inferences on the evolutionary behaviour of contact binaries. Comparable studies were carried out by K. D. Gazeas and

P. G. Niarchos, who collected together the parameters of 52 A-type and 60 W type EW binaries in *Mon. Not. Royal Astron. Soc.*, Vol. 370, L29, 2006. These authors argued for a progression from the W to the A type condition on the basis of their collected mass and angular momentum data. A follow-up discussion on the same sample was carried out by K. D. Gazeas & K. Stepień in *Mon. Not. Royal Astron. Soc.*, Vol. 390, p1577, 2008, referring to evolutionary trends and a likely ultimate coalescence of the components.

L. B. Lucy's original 'thermal relaxation oscillation' (TRO) model for contact binaries was published in *Astrophys. J.*, Vol. 205, p208, 1976, and was supported by the computations of B. P. Flannery published in the same journal and volume on p217. In the same year, the 'discontinuity at the contact surface' (DSC) model was released by F. H. Shu, S. H. Lubow & L. Anderson in *Astrophys. J.*, Vol. 209, p536, 1976. Shu summarized similarities and differences between the TRO and DSC theories in the proceedings of *IAU Symp. 88*, eds. M. J. Plavec, D. M. Popper & R. K. Ulrich, p477, 1980. L. B. Lucy & R. E. Wilson compared observational tests of these theories in *Astrophys. J.*, Vol. 231, p502, 1979. Period changes in contact binaries were also reviewed by A. Kalimeris, H. Rovithis-Livaniou & P. Rovithis, *Astron. Astrophys.*, Vol. 282, p775, 1994.

With regard to observational evidence, the asymmetric shape of certain close binary light curves, particularly among contact and near-contact systems, producing what has become known as 'the O'Connell effect', was drawn attention to by D. J. K. O'Connell, *Mon. Not. Royal Astron. Soc.*, Vol. 111, p642, 1951. O'Connell surmised that mass motions of gas would be involved in the explanation of the effect. This topic was considered in detail by R. F. Webbink in *Astrophys. J.*, Vol. 215, p851, 1977, who expected disturbances to the thermal structure of the common envelope mainly in the region of the inner Lagrangian point L_1. The nature of mass flows affecting surface flux distributions in selected examples was considered by J. Kałużny in *Acta Astron.*, Vol. 35, p313, 1985, and modelled by G. Djurašević, M. Zakirov & S. Erkapić in *Publ. Astron. Obs. Belgrade*, No. 65, Belgrade, p5, 1999.

The work of K. Yakut & P. P. Eggleton on the evolution of contact binaries was published in *Astrophys. J.*, Vol. 629, p1055, 2005. A comparable kind of discussion on the possible progression of a close binary through different categories was given by G. N. Dryomova & M. A. Svechnikov in *Odessa Astron. Publ.*, Vol. 12, p187, 1999. Yakut & Eggleton's review includes over 180 references to work in the field. These are divided into 5 main categories: (1) publications presenting and analyzing observations of individual systems that are tabulated within the paper; (2) articles describing data analysis methods; (3) surveys; (4) theoretical contributions on surface conditions and internal structure; and (5) studies of EW evolution. The modelling of 5550 evolving close binaries, mostly in a Case A process of interactive evolution carried out by C. A. Nelson & P. P. Eggleton, was reported in *Astrophys. J.*, Vol. 552, p664, 2001. Despite this great wealth of research, Yakut & Eggleton conclude with reservations on the difficulty of finding a complete understanding of contact binary properties and behaviour. Even so, they urge attention to what they describe as one of the most outstanding theoretical problems about binary stars. The distributions of period with spectral type for short-period binaries, shown in Figures 8.10-12 were published in *Investigating the Universe*, ed F. D. Kahn, Reidel, p251, 1981.

The possibility of applying a mass transfer equation to summarize evolution in contact was considered by E. Budding in *Astron. Astrophys.*, Vol. 130, p324, 1984; though that paper contains a number of slips and misprints that detract from its clarity. The cited paper of V. Trimble, dealing with the distribution of binary mass ratios, was published in *Mon. Not. Royal Astron. Soc.*, Vol. 242, p79, 1990. Evolution into contact with AML had been discussed by T. Rahunen in *Astron. Astrophys.*, Vol. 102, p82, 1981; O. Vilhu in *Astron. Astrophys*, Vol. 109, p17, 1982; and again by by F. van't Veer in a Strasbourg conference on *Stades avancés dans l'évolution des étoiles binaires serrées*, p1, 1992, and, on the basis of a more complete database of 78 system properties by C. Maceroni & F. van't Veer, in *Astron. Astrophys.*, Vol. 311, p523, 1996. S-B. Qian re-examined the absolute parameters of these contact binaries, studying also the period variations indicated by 59 of them, to deduce a broad measure of support for the TRO explanation combined with AML. The AML scenario was also considered by A. V. Tutukov, G. N. Dremova & M. A. Svechnikov in *Astron. Reports*, Vol. 48, p219, 2004; taking into account the 'blue stragglers' in cluster colour-magnitude diagrams with the hypothesis that these are the result of component mergers in W UMa contact binaries. This interpretation calls for further consideration about the possible reasons for the dispersion in the proportion of blue stragglers in different clusters in the Galaxy.

Even with a plausible mechanism for the creation of EW systems from NCBs by AML, accounting for the high incidence of EW systems remains a question. The possibility of an undetected population of low mass-ratio NCBs is thus relevant, and was referred to by J. S. Shaw, J-P. Caillaut & J. H. M. M. Schmitt, *Astrophys. J.*, Vol. 461, p951, 1996. Their review of near-contact-binaries in the ROSAT All-sky Survey found significant heterogeneity in the X-ray luminosities of early and late type NCBs, which they related to the differences between A and W type contact binaries. K. Stępień's discussion of AML (through magnetic braking) in *Mon. Not. Royal Astron. Soc.*, Vol. 274, p1019, 1995; points out the issue of incidence and age: the present incidence of systems of a given type need to be related to a previous population of parent systems. Stępień's more recent modelling of circulation and energy transport in contact binaries was published in *Mon. Not. Royal Astron. Soc.*, Vol. 397, p857, 2009; observational support for which was presented by S. M. Rucinski in *Astron. J.*, Vol. 149, p49, 2015.

The problems of placing W UMa type binaries within a cosmogonic context were set out by S. M. Rucinski in *The Realm of Interacting Binary Stars*, eds., J. Sahade, G. E. McCluskey, & Y. Kondo, *Astrophys. Space Sci. Lib.*, Vol. 177, p111, 1993. A more recent short review is that of S. Zola, K. Gazeas, J. M. Kreiner, & B. Zakrzewski, in *Astrophys. Space Sci.*, Vol. 304, p109, 2006; and, in the same volume, p321, Rucinski reports on the DDO programme to assemble a uniform database of rv-data for short-period binaries. The role of Case A evolution in leading to a common envelope binary had been noted as long ago as 1969 by J. Ziółkowski in *Astrophys. Space Sci.*, Vol. 3, p14, 1969; see also D. L. Moss in *Mon. Not. Royal Astron. Soc.*, Vol. 153, p41, 1971, who made the point that not all EW systems are explained by the same process. As to their ultimate fate, the likelihood AML processes leading to an eventual merger was proposed by R. F. Webbink, in *Astrophys. J.*, Vol. 209, p829, 1976, and echoed by D. H. Bradstreet, & E. F. Guinan, in *Interacting Binary Stars*, ed. A. W. Shafter, *Astron. Soc. Pacific Conf. Ser.*, Vol. 56, p228, 1994. The role of a limiting mass ratio beyond which merger occurs has been examined by B. Arbutina in *Publ. Astron. Soc. Pacific*, Vol. 121, p1036, 2009. In a study of 46 EW binaries showing a low mass ratio and deep common envelope, Y-G. Yang & S-B. Qian, *Astron. J.*, Vol. 150, p69, 2015; argued for their ultimate coalescence.

The high volume of observational data arising from automated surveys has become noticeable. A pointer in this direction came from Prsa, A., et al. (15 authors), *Astron. J.*, Vol. 141, p83, 2011, who catalogued parameters of 1879 eclipsing binaries after the first data release of the Kepler Mission. A detailed examination of the properties of 469 over-contact binaries by S. Kouzuma, appearing in *Publ. Astron. Soc. Japan*, Vol. 70, p90, 2018, established that mass transfer between the components can proceed in either direction. However, some inconsistency was found in the trends of binaries of high and low mass rates of period change, possibly related to the A/W sub-classification. More recently, O. Latković, A. Čeki & S. Lazarević, have collected data on nearly 700 W UMa stars in *Astrophys. J. Suppl. Ser.*, Vol. 254, p10, 2021, mostly using the results from surveys since 2011. This information will be valuable for statistical demarcation of properties – the authors arguing that candidates with temperatures above 7000 K are not really genuine W UMa type binaries. The catalogue has been applied to parameter correlations by A. Poro et al. (16 authors) *New Astronomy*, (in press) 2021.

8.4 The early idea of an EW type system evolving into a cataclysmic binary (R. P. Kraft, *Astrophys. J.*, Vol. 135, p408, 1962; B. Warner *Mon. Not. Royal Astron. Soc.*, Vol. 167, p61, 1974), while not entirely ruled out, has been largely supplanted by the latter being modelled as descending from a much wider binary, as will be discussed in Chapter 11 (see also O. Vilhu, *Astrophys. Space Sci.*, Vol. 78, p401, 1981). The occurrence of helium-rich white dwarfs is linked to binarity, since this offers the RLOF mechanism for stripping a red giant's envelope and exposing its otherwise unobservable core. An example was discussed in detail by J. Liebert, P. Bergeron, D. Eisenstein, H. C. Harris, S. J. Kleinman, A. Nitta & J. Krzesinski, *Astrophys. J.*, Vol. 606, L147, 2004. The subject was reviewed by L. G. Althaus, F. De Gerónimo, A. Córsico, S. Torres, & E. Garcia-Berro, *Astron. Astrophys.*, Vol. 597, p67, 2017.

The 'blue stragglers', that appear as a group above the majority of stars in colour-magnitude diagrams of clusters have been regarded as evidence for the merging of stellar cores. The case of the contact binary system KIC 9832227, that appeared to show a relatively large decrease of period was discussed by L. A. Molnar et al. (10 authors) *Astrophys. J.*, Vol. 840, p1, 2017. The possibility of binary merger producing an outbust in the rare class of luminous blue objects, specifically η Car, was studied by S. F. Portegies Zwart & E. P. J. van den Heuvel, E. P. J. in *Mon. Not. Royal Astron. Soc.*, Vol. 456, p3401, 2016.

Observations and theory of classical novae were discussed by S. Starrfield, J. W. Truran, W. M. Sparks, P. H. Hauschildt, S. N. Shore, J. Krautter, K. Vanlandingham & G. Schwarz in *Astron. Soc. Pacific Conf. Ser.*, Vol. 137, p352, 1998. The possible role of binarity in relation to the formation of SN 1987a was considered by V. V. Dwarkadas, *Mon. Not. Royal Astron. Soc.*, Vol. 412, p1639, 2011.

9

Binaries in Early Stages

9.1 Formation of Binary and Multiple Systems

The discussion of cloud collapse due to J. Jeans (1902) has played a foundational role in ideas on star formation. The argument is basically that a large cloud of interstellar material can collapse when its internal pressure is insufficient to balance its self-gravitational attraction. For a given density and temperature, there is a *critical mass* beyond which gravitational contraction takes over as the predominating process until local fragmentation or star-formation disrupt the infall or change its character.

Consider a spherical gas cloud of radius R, mass M, and internal sound speed c_s. If the gas is disturbed we can expect that in a time

$$t_s = \frac{R}{c_s}$$

$$\simeq 5 \times 10^5 \text{y} \left(\frac{R}{0.1\text{pc}}\right)\left(\frac{c_s}{0.2\text{kms}^{-1}}\right)^{-1} \tag{9.1}$$

the internal pressure will respond to restore the balance of forces.

On the other hand, self-attraction acts to contract the system in a 'free-fall' time

$$t_{\text{ff}} = \frac{1}{\sqrt{G\rho}}$$

$$\simeq 2\text{Myr} \left(\frac{n}{10^3\text{cm}^{-3}}\right)^{-\frac{1}{2}} \tag{9.2}$$

where $n = \rho/\mu$ is the gas particle number-density. The mean mass per particle can be written as $\mu = 3.9 \times 10^{-24}$ g, assuming molecular hydrogen and 20% helium by number of particles. If now $t_s < t_{\text{ff}}$, pressure acts more quickly to restore the balance than gravity can persist in disturbing it. However, if $t_s > t_{\text{ff}}$, gravity wins and the cloud continues to collapse.

Associated with this process is a *Jeans length*, which is given by

$$\lambda_J = c_s \times \frac{1}{\sqrt{G\rho}}$$

$$\simeq 0.4\text{pc} \left(\frac{c_s}{0.2\text{kms}^{-1}}\right)\left(\frac{n}{10^3\text{cm}^{-3}}\right)^{-\frac{1}{2}} . \tag{9.3}$$

There is also a *Jeans mass*, which is the mass within a sphere of radius $\lambda_J/2$, given by

$$
\begin{aligned}
M_J &= \frac{\pi}{6}\rho\lambda_j^3 \\
&\simeq 2\mathrm{M}_\odot \left(\frac{c_s}{0.2\mathrm{kms}^3}\right)\left(\frac{n}{10^3\mathrm{cm}^{-3}}\right)^{-\frac{1}{2}},
\end{aligned}
\tag{9.4}
$$

The rough numerical estimates of expectable parameters, that result in $t_{\mathrm{ff}} \simeq 2$ My, $\lambda_J \simeq 0.4$ pc and $M_J \simeq 2$ M$_\odot$, allow a broad impression of how diffuse masses of interstellar gas could end up as stars of comparable mass to the Sun.

However, early surveys of the solar neighborhood found that most stars are members of binary and multiple stellar systems. More recent multiplicity surveys of star-forming regions and clusters have established that the majority of stars are members of bound groupings. For solar-type stars the multiplicity of field stars is approximately 50 to 60 per cent, though this may reduce down to ~30 per cent for low mass stars. The existence of close groups of stars raises important basic questions: such as "How do binary stars form?", "How does a companion affect the distribution of circumstellar material?", or "Are most of these systems actually formed as binaries and multiples or do they result from later captures?"

After high angular resolution techniques in the near infra-red were developed at the beginning of the 1990s, such questions were expected to be settled from observational evidence. Unfortunately, obtaining complete answers has been inhibited by the inherent difficulty of spatially resolving binaries with small angular separations. At the distance of nearby regions of star formation (~150 pc), a large proportion of gravitationally bound pairs would have separations of substantially less than 1 arcsecond: so still rather a challenge for ground-based observing techniques.

Young T Tauri-type binary stars have often only been spatially resolved at one wavelength, typically 2.2 μm, in the context of high resolution lunar occultation or speckle imaging surveys that revealed their multiplicity. Finding their full stellar and circumstellar properties (i.e. surface temperature, luminosity, mass and accretion rate) from this one resolved measurement is, of course, not possible. That is aside from the fact that young stars usually show substantial excess emission and line-of-sight extinction. Determining a complete set of properties would require more spatially resolved measurements over a broad range of wavelengths.

A few studies, however, were successful in obtaining such measures for wide binaries, or small samples of close binaries, and the combined results of these studies have been useful in understanding how binary stars form. Although the samples are small, the mass ratio and secondary-mass distributions of T Tauri binary stars, as well as their closely comparable ages, tentatively suggest a *core fragmentation* process that dominates binary star formation, while capture and disk-instability scenarios are relatively unlikely. Since around the year 2000, classical capture and fission scenarios for binary and multiple star formation were no longer given much weight. Rotating cloud cores with initially flat density profiles tend to fragment immediately after a phase of free-fall collapse. This can produce systems with a wide variety of properties that are largely determined by internal accretion processes. In turn, these are strongly dependent on the initial conditions of the cloud cores, particularly their initial distributions of mass and angular momentum.

Certain theoretical predictions, such as that close binary systems are likely to have mass ratios near unity, were indirectly supported by statistical studies of evolved binaries. Given that most protostellar cores are assumed to fragment and form binary and multiple systems, we would expect that most of the angular momentum contained in the collapsed region is transformed into the orbital angular momentum of the resulting stellar binary or multiple stars.

The current general picture thus favours stars originating in small groups. Binary stars come from the fragmentation of particularly close protostellar cores as they are collapsing. A typical core would produce ~2-3 stars in the fragmentation. Computerized dynamical models suggest fragmentation occurs through turbulence, or gravitational instability in a protostellar disk. Individual star formation requires the loss of 99-99.9% of the initial angular momentum carried by the the the cores, and a significant fraction of this angular momentum can be deposited in a multiple star system rather than just dissipated.

Angular momentum could be carried out of a condensing system via jets, outflows,* or the ejection of a wide companion. The formation of the binary star could then occur together with the appearance of single stars. In modelling the formation of a binary, disks can form around individual components as circumprimary or circumsecondary, or indeed circumbinary disks. Dynamical resonance effects produced by the binary system greatly influence the evolution of such disks. These effects are thought to relate to *planet formation*, by forcing gas and dust to aggregate into particular orbits within a protoplanetary disk.

In summary, typical low-mass binaries are thought to come from the rotational fragmentation of relatively small gaseous cores. This mechanism is viable for solar-mass stars creating short-period twins. The majority of multiple stars, however, involve a turbulent core with fragmentation in its surrounding disk. Massive stars and binaries would be formed by prolonged accretion within large clusters. Companions are produced in the circumstellar disks. These may grow in mass, migrate, and merge — increasing the mass of the final primary star.

The typical specific angular momentum J/M in a molecular cloud associated with a star formation region is of the order of 10^{21} (cgs units), while in a typical T Tauri star, soon after its formation, one finds $J/M = 5 \times 10^{17}$. Since angular momentum is approximately conserved during the near free-fall collapse of a protostellar cloud, it has been generally supposed that 'spin' angular momentum is converted into 'orbital' angular momentum through fragmentation. In fact, formation processes involving up to hundreds of core and disk fragments evolving into wide binaries or multiple systems, as well as the formation of circumstellar disks around fragments, all make for a rather complex redistribution of the total angular momentum in real situations.

More recently, detailed magnetohydrodynamical (MHD) simulations were carried out to compare effectiveness in the outflows of mass, linear and angular momentum in the formation of a single star, a tight binary ($a = 2.5$ AU), and also a wide binary star ($a = 45$ AU). The outflows were studied in an attempt to understand the contributions of circumstellar and circumbinary disks on the efficiency and morphology of the flows (see Fig 9.1).

In the single star and tight binary case, a single pair of jets were launched from the system. With the wide binary, two pairs of jets appeared. These simulation results show the role of the circumbinary disk on the flow regime to be greater for the tight binary than the wide one. It was confirmed that a single star is the most efficient at transporting mass and momentum outward from the central condensation, while the wide binary case is typically least efficient. The relative efficiencies (mass, linear momentum, angular momentum) for

---

*There is no strict differentiation between a jet and an outflow: a jet is generally considered to be a high velocity and strongly collimated outflow.

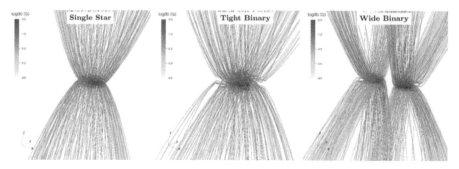

Figure 9.1   3D streamline plots of magnetic field structure around a single star (left), a tight binary (middle) and a wide binary (right). Courtesy of R. L. Kuruwita.

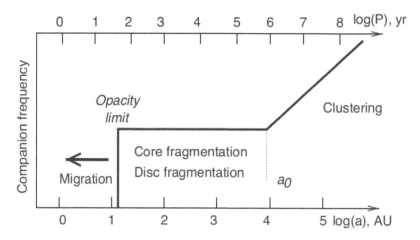

Figure 9.2   Characteristic scales of binary formation. The thick line schematizes the number of binary stars against separation (lower axis) or period (upper axis). Fragmentation is associated with pair separations in the range $10 - 10^4$ AU. Some of these migrate to lower separations, For spatial scales $> 10^4$ AU the original fractal clustering persists. The majority of very wide pairs in the clustering regime are not bound. Courtesy of A. Tokovinin.

the wide binary were in the ratios 50%, 33%, 42%; compared with the single star. The tight binary's efficiency for shedding corresponding quantities falls between the other two cases (71%, 66%, 87%) compared with the single star. By studying the magnetic field structure, it was deduced that outflows from the single star and tight binary are magneto-centrifugally driven; whereas in the wide binary star case, material outflows are driven by a magnetic pressure gradient.

9.2   Binary Separations

A typical (projected) separation s of binaries formed by a fragmenting core is related to the core's initial size (on the order of 10^4 AU) and rotation rate. Observational data give the final separation typically somewhere between 10 and 1000 AU, and this has been matched by theoretical simulations. The smallest separation in this fragmentation scenario, on the order of 10 AU, is set by the so-called *opacity limit*. This is related to the material becoming optically thick, heating and radiating away the infall energy. It has been calculated to happen as a result of dust opacity when the density is greater than about 10^{-13} g cm^{-3}.

Newly formed binaries continue to accumulate gas and at the same time migrate within the condensation, evolving into tighter pairs or even merging (Fig 9.2). Meanwhile, stellar pairs with semi-major axis a of 10^4 AU or wider are found in the field. Typical sizes of condensing cores may be examined against the Jeans criterion (Section 9.1). While size-scales larger than the Jeans length are unstable to gravitational collapse, those on the order of the Jeans length or less should be stable. This points to wide binaries with separations greater than the Jeans length *not* being formed by the fragmentation of a single core. Special wide-binary formation mechanisms have thus been proposed, such as the dissolution of clusters, or the 'unfolding' of more compact dynamically unstable triple systems.

Clearly, the distribution of separations in binary systems should have much to reveal about formation scenarios, and this has been studied in a number of detailed investigations. R. B. Larson discussed the distribution of separations between stars in the Taurus-Auriga star-forming region (SFR) by combining data on large-scale clustering with the statistics of binary companions. He found a break in the power law that approximates the density of companions versus angular separation. At projected separations larger than around 8250 AU, the density of companions c per square degree (Θ) follows the power law:

$$c = 3.4\Theta^{-0.62} \; . \tag{9.5}$$

This corresponds to the logarithmic separation distribution $f(\log s) \propto s^{1.38}$. It can be related to a fractal structure in the stellar clustering with a fractal dimension $D \approx 1.4$. Interestingly, this recalls the observed fractal structure of molecular clouds.

For separations between $\sim$10 and $\sim$8250 AU, the surface density of companions in the Taurus-Auriga SFR is described by: $c = 0.0064\Theta^{-2.15}$, or $f(\log s) \propto s^{-0.15}$. This distribution of the logarithm of separations $\log s$ being fairly flat, the projected separation s then approaches the 'classical' formula deduced by E. Öpik almost 100 years ago:

$$f(s) = ks^{-2} \; . \tag{9.6}$$

The range of separations in question corresponds, physically, to the realm of the binary stars, where the number of systemic companions averages at about one per star. Larson's calculation has been checked by others and referred to the Ophiuchus, and Orion Nebular Cluster (ONC), SFRs. In all SFRs, it appears companion statistics are well represented by a broken power law. The break point for the ONC SFR is, however, found at the relatively low separation of $s \sim 400$ AU. Later studies revised Larson's $D \approx 1.4$ for the Taurus-Auriga SFR to $D \sim 1$.

The transition between clustering and binary regimes occurs at separations on the order of the average projected distance between cluster stars, referred to as the confusion limit. Theoretical simulations have confirmed this result, while noting that the power law at large distance scales can have multiple origins; so this does not necessarily mean fractal clustering with a single index. For a sparse SFR like Taurus-Auriga, it is thought that the break in the separations' power law will correspond to an average spatial separation between stars that is on the order of the Jeans length. However, the transition between clustering and binary regimes is smooth: stars formed in adjacent cores can 'fall' to a common centre, interact dynamically, and produce a bound binary if their relative velocity is small enough. We estimate below the fraction of potentially wide pairs bound in this way.

It has been proposed that the binary and clustering regimes are separated into small regions with a constant density of companions, where primordial clustering has been de-ranged by the motions of the stars. But stars are formed in groups anyway, so the spatial locations of the newly formed stars and their velocities should be inherited from the parent molecular cloud. In fact, recent observational studies have revealed that the interstellar gas is organized in *filamentary structures*. The diameters of these filaments are on the order of 0.1 pc, with a remarkably small dispersion about this typical value. Pre-stellar cores form along these filaments in chain-like linear configurations, with typical separations on the order of the filament diameter. Such linear configurations would correspond to a fractal dimension $D = 1$. The number of stars within a distance R should then grow approximately as R^D.

Gas motions in molecular clouds have been described by 'Larson's law', where the typical rms velocity difference between two fragments σV is roughly proportional to the square root of their apparent separation, so that,

$$\sigma V (s) \approx (s/1\mathrm{pc})^{1/2} \tag{9.7}$$

in km s^{-1}.

On the other hand, the orbital velocity of a binary decreases with its semi-major axis a as $V_{\mathrm{orb}} = 30a^{-1/2}M^{1/2}$ (km s^{-1}), M being the total mass in solar units and a the semi-major axis in AU. This formula describes the velocity in a circular orbit, while the parabolic velocity $V_{\mathrm{par}} = \sqrt{2}V_{\mathrm{orb}}$. A pair of stars separating faster than V_{par} has positive total energy and would not remain gravitationally bound.

Let us suppose that two gas fragments of a collapsing cloud form a binary with $a_0 \approx s$, storing the angular momentum in the orbit. Then the condition $\sigma V \leq V_{\mathrm{par}}$ must hold for these to become a binary system, otherwise the fragments would not stay with each other (Figs 9.2-3). Substituting the foregoing forms for σV and V_{par}, we find

$$s < a_0 = 1.9 \times 10^4 M^{1/2} \ , \tag{9.8}$$

in AU; or $a_0 \approx 0.1$ pc.

The maximum separation of a solar-mass binary a_0 is, in this way, determined from the original gas motions to be on the order of 10^4 AU. This is also the characteristic size of the pre-stellar cores and of the same order as the Jeans length and the typical diameter of filaments in molecular clouds. Note that this scale depends on mass; it is smaller for low-mass cores and larger for more massive cores. Environments with faster gas motions should produce, on average, closer binaries. Cores separated by the distance $s > a_0$ move too quickly, on average, to form a gravitationally bound pair. However, motions in the formative molecular clouds are chaotic, and a certain fraction of well separated adjacent cores may still bind.

The orbital period of a binary is related to its semi-major axis by the third Kepler law, $P = a^{3/2}M^{-1/2}$, where P is in years, a is in AU and M is the mass sum in solar units. The upper scale in Fig 9.2 shows period against separation for a solar-mass binary. The characteristic scale $a_0 = 10^4$ AU corresponds to $P = 1$ My. The time needed for the two cores to fall to the centre of mass, to 'unfold', is then about $P/2$. If star formation and accretion processes last for about 10^6 years, a pair of cores observed to be separated further than a_0 will not have had time to fall into each other, and would not be considered as a binary, even if they may eventually become one.

9.3   Orbital Eccentricities

The changing separations between binary components arising from their orbital eccentricity would be observationally established only if the binary was monitored over at least one full orbit, though this is not possible for the putative early, near-formation, stages discussed above. But it is a reasonable proposition that binary systems would generally form

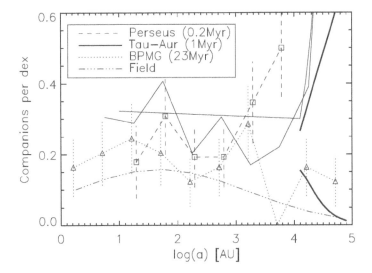

Figure 9.3 Distribution of projected stellar separations in selected sky regions. Courtesy of A. Tokovinin.

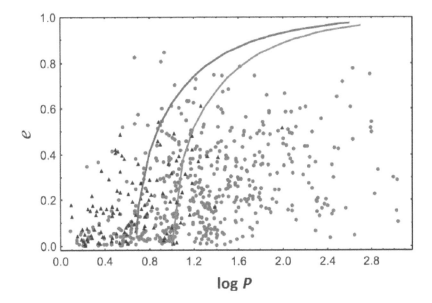

Figure 9.4 $P - e$ diagram, using the data of Bulut & Demircan (2007) (triangles), and Kepler targets from D. P. Kjurkchieva et al. (2017) (circles). The continuous lines describe tracks of constant angular momentum corresponding to circularized periods of 10 (right) and 5 d (left), according to Kjurkchieva et al.

with some eccentricity to their orbit, and orbits that have only been observed with partial completeness can often be fitted with an eccentric Keplerian model to acceptable accuracy.

In this way, large modern surveys, continued over several years, have led to the discovery of many new eccentric binaries. These discoveries have caused a reappraisal of these systems as, not just interesting objects, but important targets for astrophysical research. They became probes for the study of various tidal phenomena: such topics as mechanisms for circularization of orbits and the synchronization of stellar rotation with orbital motion, impermanent mass transfer occurring close to the stars' periastron passage, the 'heartbeat effect', and apsidal motion.

Useful for such research is the catalogue of precise values of eccentricity and periastron angle for 529 detached, eccentric eclipsing binary stars by D. P. Kjurkchieva et al. (2017). The data were presented after modelling long cadence data-sets extracted from the Kepler catalog of eclipsing binary light curves. Some of the included physical parameters were calculated using approximate values from known empirical relations of Main Sequence (MS) stars.

A number of other noteworthy catalogues have been published in this connection, giving details of apsidal motion. Examples are given in the bibliography of this chapter. Among the more recent ones, the list of I. Bulut & O. Demircan (2007) compiles relevant parameters for 274 eclipsing binaries, 150 of them having incomplete photometric observations and unconfirmed eccentricities. Data on eccentric orbits were gathered for such compilations using three large and more general sources, namely: the HIPPARCOS catalog (ESA, 1997), the Atlas of (O – C) diagrams of J. Kreiner et al. (2001), and the 9th Catalogue of Spectroscopic Binary Orbits of D. Pourbaix et al. (2004).

Against this background, the *period-eccentricity relationship* becomes one of the most interesting and debated subjects connected with eccentric binaries. Various attempts were made over the years to explain the observed trend that binaries with longer periods have larger eccentricities. Proposals involved, for example, tidal action, a secular decrease in stellar mass, or the effects of repeated encounters. More recently, it was concluded that tidal effects are especially effective for the circularization of orbits of late stars with convective envelopes.

H. Horrocks, in 1936, derived a theoretical period-eccentricity dependence of the form:

$$PM^3 e^{-6}(1-e^2)^{3/2} = \text{const.} \ (= H, \text{say}) \ , \tag{9.9}$$

that has been supported observationally for binaries with periods $P > 5$ days. Similar results were obtained for evolved MS binaries belonging to the old open cluster M67, but with a cut-off period of around 11 days. At shorter periods, it has been found that the orbital period generally decreases in approximate proportion to e^2. This decay of eccentricity is associated with energy dissipation, together with changes in the orbital axis location.

The $(P - e)$ diagram shown in Fig 9.4 was formed using the catalogue of Kjurkchieva et al. (2017), mentioned above. Most of the collected data (excluding a few score with small periods) fall below the envelope $H = 5$, while most of the members of the catalogue of Bulut & Demircan (2007) fall to the left of this line. Fig 9.4 shows a faint trend of e increasing with the period P. But these tabulations have not confirmed the conjecture that, at the shortest periods, binaries with both components having high temperatures are more likely to be eccentric than those with only one hot, or both cool, systems.

9.4 Cataloguing Detached Systems

It has been estimated that at least $\sim 60\%$ of solar neighbourhood stars are in binary or multiple systems. Binary stars are important for astrophysics, not only because of their high incidence, but because they provide basic stellar parameters as independent observed quantities used in testing theory. Thus, the mass – perhaps the most fundamental physical quantifier of a star – can be determined directly from Kepler's third law for a visual binary component, if the other orbit parameters were fully calibrated.

Reliable stellar masses are also obtained from radial velocity curves, without known distances, given the orbital inclinations (Chapter 6). Resolved double stars (visual + spectroscopic binaries) are thus a special case to provide reliable stellar masses, since the orbital inclinations are available from the apparent orbit, and the absolute orbital sizes known from the radial velocity curves.

As seen in Chapter 7, the light curves of eclipsing binaries provide inclinations as well as radii relative to the semi-major axis of the orbit. If both stars are resolved spectroscopically, accurate radii and masses result from fitting both light and radial velocity curves with appropriate models. In addition to the radii, which are not available from visual binaries, eclipsing spectroscopic binaries provide masses, effective temperatures, and absolute dimensions of the orbit, from which absolute brightness can be calculated. If a binary's parallax is also given, either the physical parameters or the assigned parallax can be mutually tested by comparing photometric and trigonometric distance measures. Moreover, eclipsing binaries are not only recognised to have a relatively high parametrization accuracy, they show a larger mass range, notably towards larger masses, compared with visual binaries, even after including HIPPARCOS or more recently catalogued systems.

In this way, compilations of stellar absolute dimensions with increasing quantity and quality have been critically compiled over the last century. With careful precautions in place, stellar masses and radii from detached, double-lined eclipsing systems, having uncertainties sometimes less than 2%, were published by the end of the last century. Pushing the limits for reliable quantification of individual close binary systems continues into the 21st century.

Being satisfied with a 10% general accuracy measure for the parametrization of MS stars, L. A. Hillenbrand & R. J. White (2004) provided a statistical study of observational constraints on pre-Main Sequence (PMS) evolutionary tracks. Of their 148 stars, 88 are MS systems, 27 PMS, and 33 are on the post-Main-Sequence (OMS). On the other hand, whilst some 6330 eclipsing binaries were listed in the *Catalogue of Eclipsing Binary Stars* of O. Y. Malkov et al. (2006), only 114 binaries are double-lined detached systems. This list is expected to yield masses, radii, logarithmic effective temperatures and luminosities to within typically 10%, 3%, 0.03, and 0.1 of their respective values. Of course, improvements in observing and analysis techniques continue, implying ongoing refinements to our knowledge of stellar properties. Thus, G. Torres et al. (2010) updated the list of detached, double-lined eclipsing binaries that give accurate masses and radii. That work contains information on 190 stars (94 eclipsing binaries and α Cen).

More recently, Z. Eker et al. (2018) increased the number of reference stars with accurately known parameters by 67%, allowing greater coverage of parameter values. We will give particular attention to this compilation in the following discussion. Compared to other field star surveys, this binary-based sample is advantageous for studying stellar properties through its provision of accurate photometric distances together with physical parameters. The ranges of effective temperatures, masses, and radii, in this catalogue, are $2750 < T_{\text{eff}} < 43000$ K, $0.18 < M < 33$ M$_\odot$, and $0.2 < R < 21.2$, R$_\odot$ respectively. The stars sampled are mostly located within 1 kpc of the Sun, but extend to 4.6 kpc within the two local Galactic arms. The number of stars with both mass and radius measurements estimated to be better than 1% uncertainty is 93; better than 3%, 311; and better than 5%, 388. Derived parameter distributions are shown in Figs (9.5-8).

For a given mass, greater periods mean larger sized orbits. Larger orbital size, however, decreases the probability of visible eclipses. Since this sample contains only eclipsing binaries, the decrease towards longer periods in Fig 9.5 reflects that declining occurrence probability. Observer preference might also bias the number of known systems towards an increasing frequency at low P. However, according to Fig 9.5, starting from $\log P_{\text{orb}} \approx 0.4$ (P in days), the number of systems decreases rather quickly towards short periods. The peaking of the selection around $P \sim 2.5$ d might be explained by decreasing orbital period implying decreasing orbital size. Relatively small orbits increase the scale of proximity effects, however, which is something to be avoided in the interests of reliably accurate modelling of

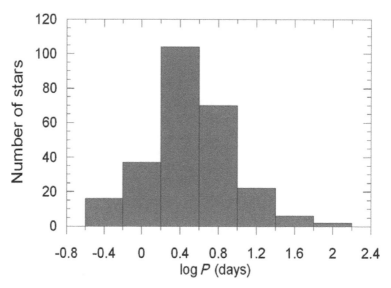

Figure 9.5 Orbital period distribution of 257 systems, after Eker et al. (2014).

fundamental stellar properties.

The shortest orbital periods (0.30 – 0.63 d) were generally found among F and later spectral types. Such short orbital periods are missing for close binaries with spectral types O, B, and A. The pair of normal unevolved stars with the currently shortest known orbital period (P_{orb} = 0.30 d: DV Psc) has components of spectral types K4V + M1V. Among the O-type binaries, the shortest orbital period is 1.62 d, belonging to V1182 Aql, while the longest orbital period in our catalogue selection is, perhaps surprisingly, only 4.24 d for V1292 Sco.

By contrast, the longest period B-type binary included has P = 99.76 d (V379 Cep). Normally, larger radii would increase the probability of eclipses, so one might expect to see more O-type systems with the median orbital period among the earliest spectral types. In fact, the median $\log P_{orb}$ decreases from G to O-type. This increasing incidence of high periods at later spectral type, selection effects notwithstanding, is probably influenced by the increasing incidence of later type (lower mass) stars in general.

The distributions of masses and radii of 586 stars from the catalogue of Z. Eker et al. (2018) are shown in Fig 9.6. The similar appearance of both distributions is not a coincidence, but arises from the compilation of reliable stellar parameters coming mainly from detached MS binaries. The similarity of the trends, in effect, reflects the well-known MS mass-radius relation. The uncertainty range of up to 20%, in this compilation, includes all the mass estimates, while the radii account for 98% of the sample stars. This means there are just nine stars whose radius uncertainty is expected to be relatively large.

9.4.1 Mass Ratios and Spectral Types

With regard to binary mass ratios, the median value ($M_2/M_1 = q$), is almost constant (median $q \approx 0.92$) across the later spectral types (G, K, M), but it decreases from F-type to O-type, to become as low as median $q \approx 0.71$, rendering the most massive primaries likely to have a distinctly less massive companion. Combining this with the faster rotations expected for hot young stars, and therefore increased rotational line-broadening, the chance of detecting the secondary spectrum decreases. Thus, the faster rotation in O-type binaries could contribute to their reduced detections as binaries and low orbital periods.

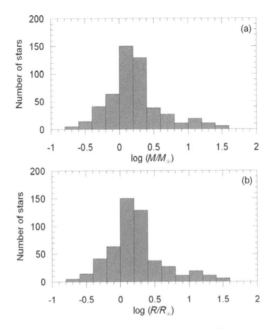

Figure 9.6 Distribution of masses and radii, after Eker et al. (2014).

The (spectroscopic) mass ratio q of the components is one of the basic parameters of binary systems. It is conventionally defined as mass of the secondary divided by that of the primary. Normally, we expect $q < 1$, but in the light-curve-based approach followed in the compilation, the primary is defined as the star eclipsed during the deeper minimum. The mass ratios determined spectroscopically from radial velocity amplitudes as K_1/K_2, correspond to this assignation, so, in fact, K_1 need not always be less than K_2. The mass ratio distribution of the sample stars is displayed in Fig 9.7 for the range $0.2 < q < 1.2$. The range $q < 1$ goes with young unevolved binaries on the MS. On the other hand, the range $q > 1$ relates to somewhat evolved pairs, where the more massive component has cooled to a lower surface temperature than its companion.

Even without evolved pairs, a few empirical mass ratios above unity are expected due to some failing precision with q near unity. According to Fig 9.7, there are 21 systems with $q > 1$. Examining their positions on the H-R diagram, seven of them (TZ For, V1130 Tau, AI Hya, GZ Leo, RT CrB, V2368 Oph, V885 Cyg) have definitely evolved cooler components. The others have some ambiguity, because, with close component similarity, one cannot unambiguously identify primary and secondary stars.

Using published spectral types and counting each component as a single star, the spectral-type distribution of 514 stars (257 pairs) is shown in Fig 9.8. Except for the O-types, there are statistically useful numbers of stars at all spectral types. The most represented spectral types are F, A, and B. Starting with the earliest type, after an abrupt increase at type B, the number of stars in the catalogue increases with spectral type until F, when there is a sudden and significant decline. From spectral type G there is gradual decrease towards M.

It should be recognized that our sources of spectral types use different classification methods, from detailed analytical procedures to coarse visual estimates or photometric colour indices. Also, the width of the spectral type ranges against corresponding temperature intervals is not uniform. The hottest star in the sample – the O-type primary of V1182 Aql – has an estimated effective temperature of some 43000 K. The coolest has $T_{\mathrm{eff}} = 2750$ K: this is the late M-type secondary of KIC10935310

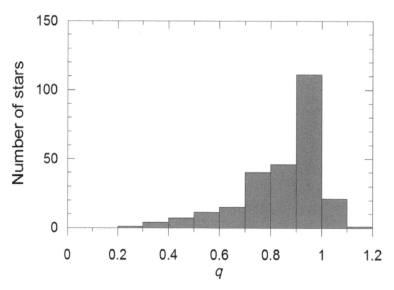

Figure 9.7   Distribution of catalogued mass ratios, after Eker et al. (2014).

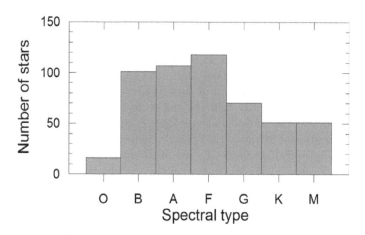

Figure 9.8   Distribution of catalogued spectral types, after Eker et al. (2014).

9.4.2 Proximity Effects

Components of close binary systems are distorted to varying degrees by rotation and tides (Chapter 2). As well, there is the radiative interaction (also known as the reflection effect: Chapter 7). These effects may be relatively small in the case of detached binaries, where single star evolution could be applied to each component to a fair approximation, but it would be useful to know how closely the observational data from binary stars should be compared with classical theoretical models of non-rotating single stars.

Of relevance, in this connection, are the projected rotational velocities ($v_{\mathrm{rot}} \sin i$), listed in the cited catalogue of Eker et al., from which one may determine whether tidal synchronization was achieved. The rotation rate and Roche lobe 'filling factor' are two indicators of the scale of deviation from sphericity. For completeness, a measure of stellar distortion was included using the corresponding *inner* Roche critical surfaces (see Chapter 4) for each binary. As a practical indication of closeness in a binary system we might write $r_1 + r_2 \gtrsim 0.1$; r_1 and r_2 being the fractional radii of the components in terms of their mean separation. Non-interacting binaries are thus taken, in a general way, to satisfy $r_1 + r_2 \lesssim 0.1$, i.e. being well-contained within their 'Roche lobes' the shapes of such stars are effectively spherical (Fig 9.9).

Unfortunately, the number of such systems is only 24 in our present sample, and much less in previous ones. In order to increase the sample size the condition $r_1 + r_2 \lesssim 0.1$ could be relaxed. The catalogue's estimates of the Roche lobe interior filling factor f allow investigators to check to what extent surface distortion may be distorting comparisons with structural and evolutionary models.

Fig 9.10 shows how these fs distribute among the catalogued stars. Accordingly, 89% of them appear spherical to within 1% of their radius. The rest are deformed more than that, but remain distinctly detached, so that all of the selected stars are from binaries in which no significant mass transfer has occurred. Hence, they would all be expected to have evolved as effectively single; though their rotations are specified in the compilation.

9.4.3 Distribution on the Sky

The distributions of detached binaries in equatorial and Galactic coordinates, as catalogued by Eker et al. (2018), are displayed in Fig 9.11. Although the 257 systems appear, at first glance, to be homogeneously distributed, a significant asymmetry between northern and southern hemispheres can be seen from the incidence towards the lower right corner of Fig 9.11a. While there are 168 binaries with positive declinations, we find only 89 binaries in the southern hemisphere. Without going into detail, the excess in the northern hemisphere can be naturally explained by the historic concentration of telescopes and astronomers.

But the real concentration of stars towards the Galactic plane is also noticeable in Fig 9.11a, where the Galactic Centre is marked by a star symbol. Although a north-south asymmetry is again indicated by the numbers at the lower left corner, a nearly symmetric distribution with respect to the Galactic plane shows up in Fig 9.11b. It is also noticeable that there are regions on the Galactic plane where stars appear to be grouping; in particular, towards Galactic longitudes 30, 70 and 110 deg. These directions are associated with the Galaxy's local arm structure. On the other hand, a less populated region towards $l = 250$, $b = -45$ deg is also seen in Fig 9.11b.

Fig 9.12 displays space distribution of the sample in the solar neighbourhood on the Galactic plane ($X - Y$), where X is towards the Galactic centre and Y is in the direction of

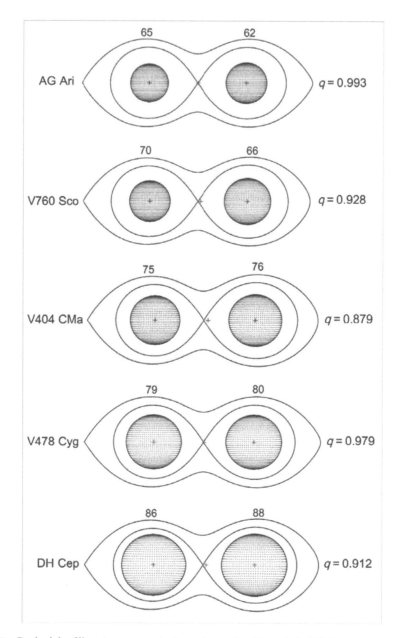

Figure 9.9   Roche lobe filling in response to r_1 and r_2, after Eker et al. (2014).

positive Galactic rotational motion. With a median distance of ~1500 pc, O-type binaries are the most distant objects. The nearest one is SZ Cam at some 870 pc. The most distant is DH Cep at an estimated 2770 pc.

Fig 9.12b shows the distribution perpendicular to the Galactic plane, where the scale height of the thin Disk is taken to be $H = 220$ pc. Distant systems, mostly O-B type binaries in the Galactic plane, are distinguishable from the concentrated central region. It is easily seen that all local binaries ($d \lesssim 300$ pc) are contained within the Thin Disk scale-height, except for just three.

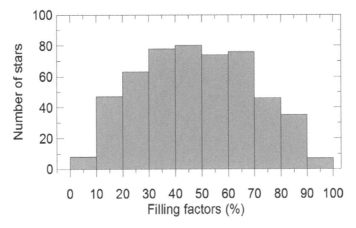

Figure 9.10   Roche lobe filling factor distribution, after Eker et al. (2014).

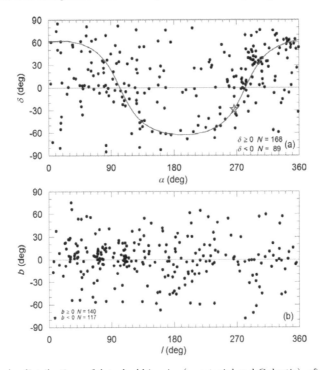

Figure 9.11   The sky distributions of detached binaries (equatorial and Galactic), after Eker et al. (2014).

There are 10 detached binaries with an O-type primary in the catalogue. The distant stars in Fig 9.12a, mostly O-B type binaries close to the $X - Y$ plane, suggest the location of one of the local Galactic arms. Indeed, after re-plotting the O-type binaries in Fig 9.13, their position in the Galactic plane, relative to the Sun, becomes clear. The present sample is mostly located within 1 kpc of the Sun, positioned between the Carina-Sagittarius arm and the Orion Spur. The Orion Spur itself is situated between the Perseus and Carina-Sagittarius galactic arms.

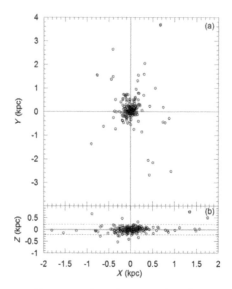

Figure 9.12   Space distribution of binaries, after Eker et al. (2014).

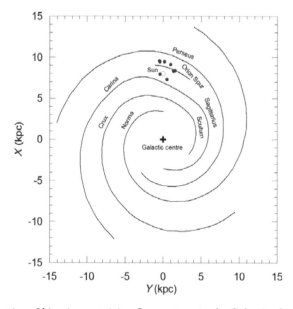

Figure 9.13   Distribution of binaries containing O-type stars in the Galactic plane, as shown by Eker et al. (2014). It is assumed that the Sun is distant 8kpc from the Galactic centre.

9.4.4   Constructing the Herzsprung-Russell Diagram

The Herzsprung-Russell (H-R) diagram is a primary tool to demonstrate and study stellar evolution. Using the most accurate stellar parameters, evolution theory is checked from the observed locations of stars in this diagram. Taking effective temperatures ($\log T_{\text{eff}}$) and luminosities ($\log L$) from the cited catalogue of Eker et al., the positions of 472 stars (236 detached binaries) on the H-R diagram are shown in Fig 9.14, where theoretical lines marking the Zero Age (ZAMS) and Terminal Age Main Sequence (TAMS) are drawn. Almost the whole sample is within the MS band.

There are no supergiants in this collection, though the selection criteria do not exclude them. Considering the fact that large (evolved) stars can be observed at larger distances

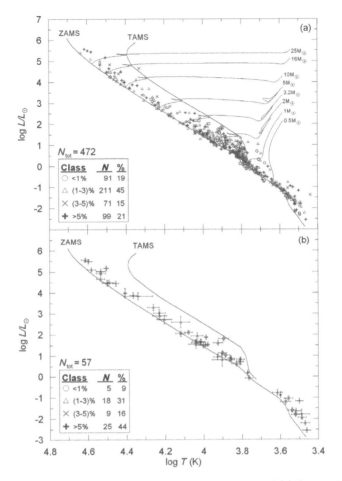

Figure 9.14 Detached binaries and the H-R diagram. In the upper panel (a) data on 472 stars have been compiled from the catalogues of Z. Eker et al. (2014, 2018). In the lower panel (b) the least accurate 57 stars in the sample are displayed with their error-bars. The implication is that most of the errors for the stars in (a) would be comparable to or smaller than the symbols used to mark their positions.

than their MS counterparts, while the probability of observing eclipses is higher if component radii are larger at a given separation, the absence of giants and supergiants in the samples calls for some explanation. The lower proportion of evolved stars – a consequence of the generally shorter durations of evolved phases – and low probability of having both components simultaneously evolved to comparable sizes, permitting the easier detection of eclipses, must be more effective than eclipse-favouring conditions. And while dwarf companions to giants and supergiants should exist, the large difference of visible luminosity between the components means that they would very rarely be picked up as double lined eclipsing binaries.

Although there are 586 stars (293 binaries) altogether in the cited group, we can give only 472 stars (236 binaries) effective temperatures. Indeed, the catalogue contains 21 systems without published temperatures. When analysing light and radial velocity curves, some authors pay scant regard to actual temperature values, being satisfied to estimate temperature ratios only.

Our sample stars were displayed on the H-R diagram as four sub-samples, according to their parameter accuracy estimates. These sub-samples are (for both mass and radius): (1) up to 1% accuracy (91 stars); (2) between 1% and 3% (211 stars); (3) between 3% and 5% (71 stars); and (4) worse than 5% (99 stars). These groups are shown in Fig 9.14a with different symbols, in order to investigate if there are any preferred positions for different classes of accuracy. Except for the most accurate sub-sample, which covers the spectral range A0 to M3, corresponding to a temperature range $\sim 10000 - 3000$ K, the other sub-samples have full ranges, covering the spectral range O5 to M3 and corresponding temperatures from 43000 K down. No preferred locations of the sub-samples is apparent, except for the most accurate one. All other groups are evenly distributed along the MS band.

In order to assess precision of the positions on the H-R diagram, the published uncertainties of radii and temperatures have been propagated to estimate those of the luminosities. Of the 472 stars (236 binaries) plotted on Fig 9.14a, seven systems do not have published uncertainties for their temperatures and seven have temperature error estimates only for their secondaries. Leaving out the systems without temperature uncertainty, and assigning the same uncertainty to the primary for those having only a secondary error measure, we can calculate the relative uncertainties of luminosities for the 229 remaining binary components. All these uncertainties are less than 50%, except for one star: the secondary of V1292 Sco. The 19% temperature error and 52% uncertainty of the radius propagate to a $\sim$130% error for the luminosity. Among the 458 stars then, 87.6% (401 stars) have relative errors less than 20%. The remaining stars were placed on the H-R diagram with error bars (Fig 9.14b) to show any preferred locations for stars with high error bars, though this turns out to be inconspicuous.

The determination of reliable effective temperatures from observations remains the biggest obstacle to using this kind of data in studies of stellar evolution. It will be recalled that some authors have published only internal formal errors, which often look unrealistically small (a few examples are given in the bibliography), so the error bars in our present sample are probably optimistic. Even so, this sample is among the most complete and up-to-date collection of stellar positions on the H-R diagram from simultaneous solutions of light and radial velocity curves of detached eclipsing binaries presently available.

9.5   The Empirical Main Sequence

9.5.1   Correlations with Mass

One of the most fundamental discoveries in stellar astronomy is the mass-luminosity relation (MLR), brought to light in the early 20th century independently by Ejnar Hertzsprung and Henry Norris Russell through use of their data on visual binaries. Parameters from eclipsing binary systems were added later. Information from 13 eclipsing binaries, together with 29 visual binaries and 5 Cepheids were used by A. S. Eddington, while D. McLaughlin included 41 eclipsing binaries in his plots.

Investigations of the MLR continued as the quantity and quality of data increased Especially noteworthy were the critical compilations of absolute dimensions of binary components by D. M. Popper (1967, 1980). After the mid 20th century, the testing of theoretical stellar models using the empirical stellar mass-radius relation (MRR) also grew apace. By the 1990s, the range had extended to more than 100 stars, including more OB type close and massive systems. At the low mass end, parameters from visual/speckle binary systems have been combined with those of eclipsing binaries to probe the connection of stellar mass to bolometric magnitude and radius.

For practical applications, there were attempts to represent relations such as the MLR and MRR by polynomials over appropriate ranges of mass. Such approaches were extended to multivariate compilations, for example, in studying the roles of metallicity and age to the MLR for MS stars of intermediate spectral type.

Mass and chemical composition are two independent basic parameters from which stellar evolution models are often constructed. Radius R, luminosity L, and effective temperature

T_{eff} are the primary products of these models, and represent the basic parameters by which stellar evolution can be tracked. From the observational point of view, M and R are first-rank derivables. In the last few decades, the number of accurately determined M and R values has increased by hundreds. Furthermore, the accuracy with which they are specified has reached the level needed to test stellar evolution modelling, even within the MS band. Chemical composition (or the *metallicity* index [Fe/H]) and T_{eff} are, however, not as accurately known as M and R, as far as present data on binary stars allows, except perhaps in a few special cases.

Unlike M and R, L is a second-rank outcome, obtained from either observed magnitudes and known distance, or from the relation $L = 4\pi\sigma R^2 T_{\text{eff}}^4$. However, both these routes to luminosity are problematic, because they both require accurate determinations of effective temperatures. Such temperatures are rarely accurate, as they are usually inferred indirectly. We will argue below that a revised MS MLR can provide an easy and effective estimate of stellar effective temperatures, directly from known masses and radii.

Recent literature in this field has tended not to address revising the classical MLRs and MRRs, on the basis that the scatter in the mass-luminosity diagram is not due to observational errors, but more likely abundance and evolutionary effects. Instead of writing a single $L(M)$ function, $\log L$ may be related to $\log M$ through simple polynomials with T_{eff}, $\log g$ and [Fe/H] as parameters, rather than the classical approach of writing $\log L$ as a simple proportionality to $\log M$. A similar approach was followed with regard to the MRR.

It is always possible to define a function to optimally represent a given band of data through a least squares method. Moreover, the classical MLR and MRR forms have been generally proven useful and are still widely applied. That being said, revisions of the classical MLR and MRR with modern and more accurate data are will be increasingly necessary for high-accuracy applications.

Most light curve solutions for eclipsing binaries require an effective temperature for at least one component as an input parameter; however, in many cases values were not spelled out in the source material. With this in mind, some authors avoid citing rough estimates, and publish fitting parameters without reference to absolute temperatures, providing only temperature ratios. But since detached double-lined eclipsing binaries yield reliable values for M and R, we can apply the empirical MLR and, making use of $L = 4\pi\sigma R^2 T_{\text{eff}}^4$, stellar temperatures become determinable.

The number of stars with dependable parameters was so limited just a quarter of a century ago, that models of contact and semi-detached eclipsing systems could hardly constrain parameters well enough to help with MLR and MRR definition. With the growth of modern detectors and observational techniques including space-based or high-altitude observatories and high-speed computers, the number of accurate parameters from detached eclipsing spectroscopic binaries has increased rapidly.

The removal of non-MS stars in our stellar samples was completed using the MRR. This provides a much more reliable basis for stellar parametrization than the MLR, as well as an accurate diagnostic tool for analyzing stellar evolution. Although metallicity data are not provided in the cited catalogue of Eker et al. (2014), Thin-Disk field stars in the solar neighbourhood are generally taken to have solar metallicity on average, with upper and lower limits of ± 0.5 dex in actual specification. Since Thin-Disk stars in the solar neighbourhood could be diluted by $\sim 7\%$ due to Thick-Disk and Halo stars, TAMS tracks with solar metallicity should not be regarded as providing sharp borders to observed MS trends. A small number of non-MS stars are also expected to remain in the calibration sample, though generally with a negligible effect on statistical inferences.

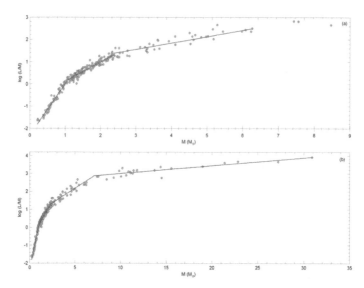

Figure 9.15 Luminosity/mass ratio for Main Sequence stars, after Eker et al. (2015). The diagrams indicate 4 separate regions (lower panel) where fair linear approximations can be made. The upper panel (a) shows the three lower mass regions on a greater scale.

9.5.2   Luminosities and Temperatures

The MLR relation can be expressed in different ways; based on the relation between M and L, or between M and M_{bol}, for example. A fundamental form would be $L \propto M^{\alpha}$, with α the slope on the logarithmic $M - L$ diagram. With early data, it was reasonable to express the MLR for all MS stars with a single power law in this way. However, as the quantity, quality, and range of data have increased, it has become increasingly regular practice to express the MLR in separate power-law forms over limited mass ranges. The ratio L/M, understandable in physical terms as the efficiency of the stellar furnace, is plotted against mass in Fig 9.15. The distribution appears linear between the break points at $M = 1.05, 2.4$ and $7M_{\odot}$.

The reciprocal form M/L, the 'mass-to-light ratio', is a parameter commonly used in extragalactic astronomy. The mean M/L ratio relates to a variety of galactic stellar compositions, and depends on the relative numbers of stars of different types. It is used in the context of spiral galaxy rotation-curve decomposition, and the band-pass dependent slope of the galaxy-magnitude – rotation-velocity trend, also known as the Tully-Fisher relation. In this context, the M/L ratio has been used in efforts to locate dark matter. Although the use of M/L in extragalactic astronomy is a different setting to applications of L/M in stellar astronomy, the same two quantities are, of course, involved. So it is noteworthy that updated information on the energy generation efficiency of varying stellar types should be useful for improving models in future extragalactic studies.

Newly improved data, that we can review here, permit a classical approach to luminosity determination, though within separate mass domains. The linear distributions of $\log L$ within four such domains are displayed in the four panels of Fig 9.16(b) – (e) below the main panel. The MLRs ($L \propto M^{\alpha}$) within each of the four mass domains have been determined by least-squares fittings to separate linear forms. The break point between the very low and low-mass domains is not too clear in Fig 9.16 due to insufficient data. In revising the classical MLR in this way, the linear fits found within different mass domains can be considered as physically real, if not yet fully understood, divisions.

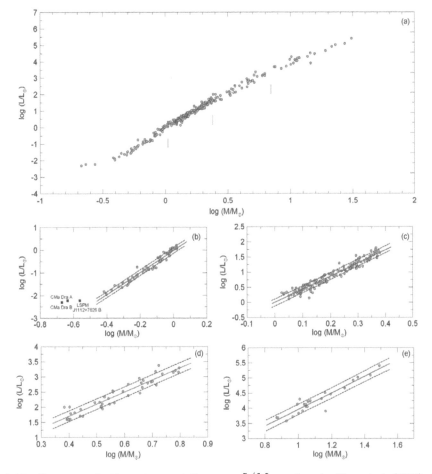

Figure 9.16 Energy production rate per stellar mass L/M, as given by Eker et al. (2015). Separate linear distributions in $\log L$ vs. M are clear.

Corresponding MRR and MTR forms are plotted in Fig 9.17. For masses $M < 1M_\odot$ there is a narrow distribution of radii, while we find a broad band of radii for stars with $M > 1M_\odot$, so the use of a single function to express the MRR seems inappropriate. The continuous line in Fig 9.17a indicates the theoretical ZAMS, although temperature evolution within the MS band is not obvious on the $M - T_{\rm eff}$ diagram. At first glance, it resembles the MLR shown in Fig 9.14(a). We could therefore suppose that an MTR would be expressible as a polynomial over various sections, as with the MLRs of the four mass domains.

9.5.3 Surface Temperature Evaluation

Stellar temperatures are determined by several methods, including stellar bolometry, the use of photometric colors, atmosphere modelling, spectral fitting to selected spectral lines or regions, or spectral line depth ratios. When applied to binary stars, however, those methods often face difficulties, because the colors and spectra obtained are usually for the system, not for the components separately. Methods such as cross-correlation function (CCF) fitting and spectral analysis may give more accurate temperatures, though such techniques are still not fully established.

In many previous studies, including light curve analyses of eclipsing binaries, the temperatures of the primary components were adopted according to roughly estimated spectral types and colors determined from de-reddened $UBVRI$ photometry. Fig 9.18 shows

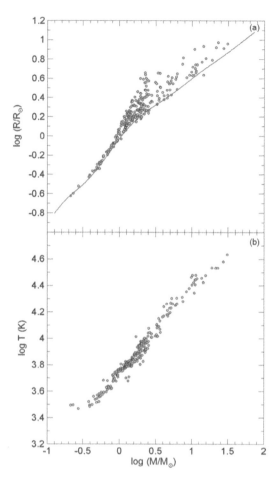

Figure 9.17   MRR and MTR relations, after Eker et al. (2015).

calculated (vertical) and published (horizontal) temperatures for 268 stars in a sample that we may consider, taken from Eker et al. (2015).

Temperatures calculated through use of the MLR are in close agreement with published temperatures obtained by other methods, most of which are based on optical photometry (color or brightness temperatures), with some involving spectroscopic techniques. However, indications are that the formal errors *published* with individual temperature calculations are about three times smaller than the scatter in plotted results; implying that a significant fraction of published temperature values have underestimated the scale of their uncertainties.

Temperature estimates in more recent studies, by comparison, are more rigorously derived. In the calibration sample for Fig 18, 46% of temperatures were taken from papers published in the last seven years. The better agreement between calculated temperatures and those published recently, compared with older sources, is apparent in Fig 9.18 (lower).

For a better comparison between the calculated and published temperatures, the photometric distances were also computed for a limited number (20) of binaries with reliable HIPPARCOS distances. These were chosen to be within 100 parsecs in order to avoid interstellar reddening. The calculation method can be summarized thus: first, the luminosity of each component was determined using its radius and effective temperature. The luminosities were transformed to bolometric absolute magnitudes, which were then converted to visual absolute magnitudes with bolometric corrections corresponding to the temperatures used.

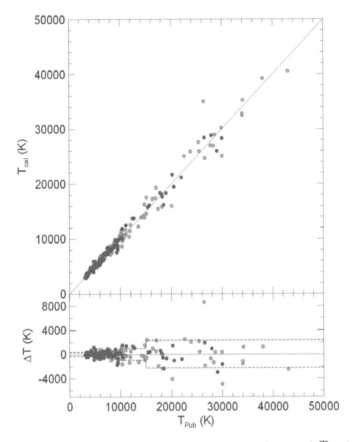

Figure 9.18 A comparison of calculated (ordinates) and published (abscissae) T_e values (upper panel) together with their differences $\Delta T = T_{\text{cal}} - T_{\text{pub}}$, after Eker et al. (2015). Dashed lines show the standard deviations in selected groups. Filled, and empty, circles correspond to values published within the last 7 years, and previously.

The visual absolute magnitudes of the two components were combined to find the absolute visual magnitude of the binary system.

The standard deviations of differences from the diagonal line in Fig 9.18 indicate that the calculated temperatures for these stars are slightly less accurate than published values. Heterogeneity of the published temperatures combined with homogeneity of the calculated temperatures could be a reason for this. The published temperatures appear mostly contained within the MS band for stars with $M > 1 M_\odot$, whereas, for the same stars, calculated temperatures would become more scattered, with some outside the MS. Yet, with that in mind, homogeneity of the temperatures, even with their slightly larger uncertainties, renders such calculated values applicable to wider usage contexts than published ones (Fig 9.19).

9.5.4 Evolutionary Effects in MS Correlations

The classical MLR is a thin, well-defined distribution, whereas MS evolution forms a band. Calculating effective temperatures using the MLR thus propagates the half-width of the band as a dominant uncertainty, greater than $\sim 6\%$. Hence, there are more points outside the theoretical ZAMS and TAMS lines for stars with $M \gtrsim 1 M_\odot$. By comparison, for stars with $M \lesssim 1 M_\odot$, MS lifetimes are greater than the age of the Galaxy. Evolutionary

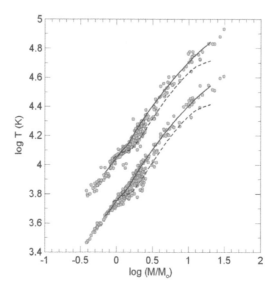

Figure 9.19 MTR data, after Eker et al. (2015). The upper data-set corresponds to the empirical M-T relation shown in Fig 9.17(b), but shifted up by 0.3 dex to enable direct comparison with the lower correlation calculated from the M-L and M-R correlations. ZAMS and TAMS lines are also displayed, as derived from recent theoretical models. The somewhat large errors of the calculated temperatures corresponds to a thicker Main Sequence, though this becomes better at low mass, due to the thinning of the MRL relation for MS stars with $M < 1$ $M_\odot$.

effects will not impact on low mass stars as much as higher mass ones. A thin, well-defined mass-luminosity relationship is thus successful in characterizing low-mass stars.

We can notice that the scatter in calculated temperatures is much narrower, compared with that in published temperatures, for stars with $M < 1 M_\odot$, as shown in Fig 9.19. However, it is also known that low-mass stars in eclipsing binaries show a discrepancy of temperatures and radii with respect to models, perhaps related to their magnetic activity. This may partly explain the larger spread of published temperatures in the low-mass region. Apparently, radii of the catalogue-selected sample in the same low-mass regime were not so disturbed, as indicated on the $M-R$ diagram (Fig 9.17a). A smooth and narrow distribution for calculated temperatures is produced in the same low-mass region of stars in Fig 9.19.

A well-defined MLR would provide a single L for a given M, but then metallicity and evolution information contained in the $M-L$ diagram would be lost with this single value of L. While this may appear a drawback, information lost by constructing an MLR can be re-introduced into the $M-T_{\text{eff}}$ diagram by calculating effective temperatures using the MLR and R values. Accurately determined radii constrain the effects of metallicity and evolution, because R is one of the primary products of evolution theory that uses M and metallicity as free initial parameters. We need not bother about age and chemical composition in the calculation, because their effects naturally propagate to the calculated T_{eff}. Since the same applies to all the stars used in the calibration sample, the calculated temperatures will homogeneously reflect the evolution and metallicity information contained in the $M-R$ distribution. In addition to this propagated effect of evolution, further widening of the distribution for the stars for which $M \gtrsim 1 M_\odot$ comes from real observational errors. In this way, the difference between the observed ZAMS and TAMS reaches to 0.2 dex (in $\log T_{\text{eff}}$), rather than the theoretical 0.15 dex separation of the ZAMS and TAMS lines.

This method of calculating effective temperatures has been applied to a larger sample (371 stars) from the same source catalogue containing less accurate M and R values. This

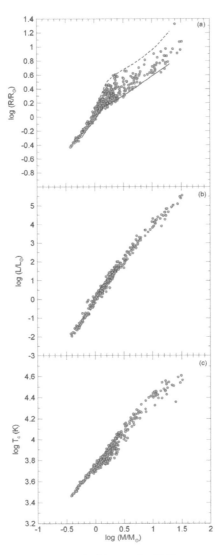

Figure 9.20 (a) Here we show a larger sample (N = 371) of MS stars on the $M - T_e$ diagram. (b) The $M - L$ correlation for the same sample of stars; and (c) the $M - T_e$ diagram using calculated effective temperatures, after Eker et al., 2015.

larger sample satisfied two selection rules: (1) both M and R could have errors up to 6%; and (2) both components had to be MS stars. Unlike for the calibration stars, there is now no limit on the precision of the luminosities. There are 408 catalogue stars with M and R errors less than or equal to 6%. That number reduces to 371 after removing non-MS stars, with the reduced list still containing the original calibration sample.

These calculated effective temperatures are listed in Table 9.1. In this table mass ranges are sorted into selected bins, where the number of stars in each bin N is given in the second column. The next four columns display the mean values for the masses, radii, effective temperatures and surface gravities of the stars, using the usual symbols, grouped according to the mass ranges of the first column. The mean values are calculated as simple arithmetic averages, i.e. sum of masses divided by number N. Corresponding spectral types are listed in column six. Bolometric magnitudes, mass/luminosity ratios and their inverses fill out the remaining columns.

Table 9.1 Mean trend of Main-Sequence stars (after Eker et al. 2018).

Mass Range	N	M ($M_\odot$)	R ($R_\odot$)	T_{eff} (K)	$\log g$ (cgs)	Spec. type	M_{bol} (mag)	M/L ($\odot$)	L/M (cgs)
$0.16 < M < 0.25$	6	0.217	0.238	3105	5.020	M4.5	10.55	45.7713	0.044
$0.25 < M < 0.35$	5	0.281	0.302	3167	4.926	M3.5	9.95	34.0746	0.057
$0.35 < M < 0.45$	11	0.398	0.395	3282	4.844	M3	9.21	24.4342	0.08
$0.45 < M < 0.55$	12	0.503	0.489	3547	4.761	M1.5	8.41	14.8043	0.13
$0.55 < M < 0.65$	11	0.602	0.612	3924	4.644	K9	7.49	7.5421	0.26
$0.65 < M < 0.75$	14	0.690	0.675	4289	4.618	K4.5	6.89	4.9844	0.39
$0.75 < M < 0.85$	14	0.802	0.798	4663	4.539	K3.5	6.16	2.9687	0.65
$0.85 < M < 0.95$	19	0.907	0.935	5279	4.454	K0	5.28	1.4882	1.30
$0.95 < M < 1.05$	18	1.000	1.092	5498	4.362	G5	4.76	1.0222	1.89
$1.05 < M < 1.15$	21	1.093	1.176	5922	4.336	G3	4.28	0.7156	2.70
$1.15 < M < 1.25$	22	1.209	1.337	6181	4.269	F8	3.82	0.5162	3.74
$1.25 < M < 1.35$	43	1.300	1.482	6396	4.211	F6	3.44	0.3941	4.91
$1.35 < M < 1.45$	27	1.398	1.688	6495	4.129	F4.5	3.09	0.3070	6.30
$1.45 < M < 1.55$	23	1.502	1.823	6737	4.093	F3	2.77	0.2443	7.91
$1.55 < M < 1.65$	25	1.596	1.865	7110	4.100	F1	2.49	0.2001	9.66
$1.65 < M < 1.75$	12	1.702	1.818	7794	4.150	A8	2.14	0.1554	12
$1.75 < M < 1.85$	22	1.793	2.230	7560	3.995	A6	1.83	0.1230	16
$1.85 < M < 1.95$	18	1.894	2.035	8153	4.099	A5	1.70	0.1153	17
$1.95 < M < 2.05$	19	1.994	2.059	8606	4.111	A4	1.44	0.0955	20
$2.05 < M < 2.2$	17	2.139	2.353	8722	4.025	A3	1.09	0.0744	26
$2.2 < M < 2.4$	26	2.299	2.683	9154	3.943	A2	0.60	0.0507	38
$2.4 < M < 2.8$	20	2.573	2.653	10030	4.001	A0	0.22	0.0402	48
$2.8 < M < 3.2$	10	2.993	2.610	10999	4.081	B9	-0.14	0.0335	58
$3.2 < M < 3.6$	12	3.362	2.657	12239	4.116	B8	-0.64	0.0236	82
$3.6 < M < 4.0$	8	3.769	3.497	12588	3.927	B7.5	-1.36	0.0137	141
$4.0 < M < 4.6$	8	4.310	2.911	15372	4.145	B5.5	-1.83	0.0101	190
$4.6 < M < 5.2$	9	4.916	3.287	16576	4.096	B4.5	-2.42	0.0067	288
$5.2 < M < 6.0$	8	5.587	3.797	17677	4.027	B3.5	-3.01	0.0044	437
$6.0 < M < 8.0$	7	6.716	4.460	19729	3.967	B2.5	-3.84	0.0025	779
$8.0 < M < 10$	7	9.083	4.488	25057	4.092	B2	-4.89	0.0013	1516
$10 < M < 12$	11	11.143	5.497	26685	4.005	B1	-5.61	0.0008	2386
$12 < M < 15$	8	13.702	6.186	28583	3.992	B0.5	-6.16	0.0006	3235
$15 < M < 18$	8	16.886	6.825	31599	3.998	O9.5	-6.81	0.0004	4761
$18 < M < 24$	5	20.163	8.454	33220	3.889	O8	-7.49	0.0003	7492
$24 < M < 32$	3	27.835	9.037	39067	3.971	O6	-8.34	0.0002	11860

Table 9.2 was produced using also the Sejong Open cluster Survey (SOS) data of Sung et al. (2013), which is a project dedicated to providing homogeneous photometry of a large number of open clusters in the SAAO Johnson-Cousins' $UBVI$ system. Much attention has been given to the use of standard stars in order to properly calibrate the system. A large number of open clusters of different ages were studied. These data were combined with absolute parameters from the MLR, MRR and MTR functions discussed above. The columns of Table 9.2 indicate spectral types, logarithm of $T_{\rm eff}$, $B-V$, $U-B$, and bolometric corrections in the first five columns following the SOS format. The rest of the columns are aligned as in Table 9.1. Table 9.2 should have general applicability to determine the absolute properties of Main Sequence stars when given basic observational data such as spectral type and colours.

Fig 9.20(a) shows the $M-R$ diagram for the 371 MS stars of the wider sample. Fig 9.20(b) shows their distribution in the $M-L$ plane. Finally, Fig(9.20(c) illustrates the $M-T_{\rm eff}$ diagram with effective temperatures computed. All the temperatures have been homogenized as discussed above, and despite having uncertainties $\Delta T/T$ up to 8%, stellar evolution within the MS is noticeable on the $M-T_{\rm eff}$ diagrams in the same way as in Fig 9.17.

Fig 9.20 shows why the usefulness of the MLR for MS stars is universally recognized, but the same has not been the case for the $M-R$ and $M-T_{\rm eff}$ relations. If we neglect mass loss during the MS lifetime, radial evolution pulls the star upward in the $M-R$ plane, while corresponding temperature evolution is downward on the $M-T_{\rm eff}$ diagram. This is what we see in Fig 9.11, where a single value of L for given M is found from the MLRs. A single value of L means that we are neglecting evolutionary and metallicity effects. In fact, evolutionary and metallicity effects are the main contributing factors to the uncertainty of L. As explained above, observational uncertainties of R and T, which

Table 9.2 Spectral types, temperatures, colours and related data for MS stars, using the MLR, MRR and MTR relations of Eker et al. (2018).

Spec. type	$\log T_{eff}$ (K)	$B - V$ (mag)	$U - B$ (mag)	BC (mag)	M_V (mag)	M ($M_\odot$)	R ($R_\odot$)	$\log g$ (cgs)	M/L ($\odot$)	L/M (cgs)
O2	4.720	-0.33	-1.22	-4.52	-6.44	63.980	16.734	3.80	0.00003	57628
O3	4.672	-0.33	-1.21	-4.19	-5.62	44.260	12.312	3.90	0.00007	28981
O4	4.636	-0.33	-1.20	-3.94	-5.13	34.809	10.301	3.95	0.00010	18514
O5	4.610	-0.33	-1.19	-3.77	-4.80	29.669	9.236	3.98	0.00014	13743
O6	4.583	-0.33	-1.18	-3.58	-4.50	25.380	8.362	4.00	0.00019	10270
O7	4.554	-0.33	-1.17	-3.39	-4.20	21.660	7.616	4.01	0.00025	7642
O8	4.531	-0.32	-1.15	-3.23	-3.99	19.215	7.131	4.02	0.00032	6111
O9	4.508	-0.32	-1.13	-3.03	-3.83	17.123	6.721	4.02	0.00039	4929
B0	4.470	-0.31	-1.08	-2.84	-3.45	14.277	6.171	4.01	0.00055	3511
B1	4.400	-0.28	-0.98	-2.40	-2.93	10.459	5.454	3.98	0.00098	1965
B2	4.325	-0.24	-0.87	-2.02	-2.35	7.699	4.967	3.93	0.00174	1110
B3	4.265	-0.21	-0.75	-1.62	-1.68	6.123	3.989	4.02	0.00373	518
B5	4.180	-0.17	-0.58	-1.22	-0.76	4.516	3.214	4.08	0.00928	208
B6	4.145	-0.15	-0.50	-1.02	-0.44	4.007	2.974	4.09	0.01327	146
B7	4.115	-0.13	-0.43	-0.85	-0.18	3.625	2.797	4.10	0.01790	108
B8	4.080	-0.11	-0.35	-0.66	0.13	3.234	2.617	4.11	0.02518	77
B9	4.028	-0.07	-0.19	-0.39	0.57	2.743	2.394	4.12	0.04121	47
A0	3.995	-0.01	-0.01	-0.24	0.86	2.478	2.274	4.12	0.05587	35
A1	3.974	0.02	0.03	-0.15	0.90	2.325	2.362	4.06	0.05897	33
A2	3.958	0.05	0.06	-0.08	1.06	2.216	2.292	4.06	0.06919	28
A3	3.942	0.08	0.08	-0.03	1.23	2.113	2.226	4.07	0.08107	24
A5	3.915	0.15	0.10	0.00	1.57	1.952	2.123	4.08	0.10557	18
A6	3.902	0.18	0.10	0.01	1.74	1.879	2.077	4.08	0.11971	16
A7	3.889	0.21	0.09	0.02	1.91	1.810	2.033	4.08	0.13563	14
A8	3.877	0.25	0.08	0.02	2.07	1.749	1.994	4.08	0.15209	13
F0	3.855	0.31	0.05	0.01	2.37	1.643	1.928	4.08	0.18726	10
F1	3.843	0.34	0.02	0.01	2.53	1.588	1.893	4.08	0.20956	9.22
F2	3.832	0.37	0.00	0.00	2.69	1.540	1.863	4.09	0.23219	8.33
F3	3.822	0.40	-0.01	0.00	2.82	1.498	1.838	4.09	0.25448	7.60
F5	3.806	0.45	-0.02	-0.01	3.30	1.354	1.588	4.17	0.35692	5.42
F6	3.800	0.48	-0.01	-0.02	3.49	1.305	1.508	4.20	0.40325	4.79
F7	3.794	0.50	0.00	-0.02	3.65	1.259	1.434	4.23	0.45450	4.25
F8	3.789	0.53	0.02	-0.03	3.80	1.222	1.377	4.25	0.50125	3.86
G0	3.780	0.59	0.07	-0.04	4.06	1.161	1.283	4.29	0.59553	3.25
G1	3.775	0.61	0.09	-0.04	4.19	1.128	1.236	4.31	0.65401	2.96
G2	3.770	0.63	0.13	-0.05	4.33	1.098	1.191	4.33	0.71723	2.70
G3	3.767	0.65	0.15	-0.06	4.42	1.080	1.165	4.34	0.75756	2.55
G5	3.759	0.68	0.21	-0.07	4.64	1.031	1.097	4.37	0.87824	2.20
G6	3.755	0.70	0.23	-0.08	4.72	1.019	1.081	4.38	0.92834	2.08
G7	3.752	0.72	0.26	-0.09	4.78	1.011	1.069	4.39	0.96765	2.00
G8	3.745	0.74	0.30	-0.10	4.92	0.990	1.041	4.40	1.06553	1.81
K0	3.720	0.81	0.45	-0.18	5.45	0.922	0.951	4.45	1.49639	1.29
K1	3.705	0.86	0.54	-0.24	5.77	0.884	0.903	4.47	1.82840	1.06
K2	3.690	0.91	0.65	-0.32	6.11	0.848	0.858	4.50	2.22864	0.867
K3	3.675	0.96	0.77	-0.41	6.46	0.813	0.817	4.52	2.71007	0.713
K5	3.638	1.15	1.06	-0.65	7.32	0.736	0.727	4.58	4.34779	0.445
M0	3.580	1.40	1.23	-1.18	9.07	0.558	0.541	4.72	10.16323	0.190
M1	3.562	1.47	1.21	-1.39	9.60	0.524	0.508	4.75	12.75337	0.152
M2	3.544	1.49	1.18	-1.64	10.15	0.492	0.479	4.77	15.91749	0.121
M3	3.525	1.53	1.15	-2.02	10.85	0.462	0.452	4.79	20.00404	0.097
M4	3.498	1.56	1.14	-2.55	12.29	0.323	0.338	4.89	32.12216	0.060
M5	3.477	1.61	1.19	-3.05	13.37	0.249	0.284	4.93	42.55068	0.045

propagate to L, are negligible compared with these evolutionary and metallicity effects. So the upward evolution seen for stars with $M \gtrsim 1M_\odot$ on $M - R$ diagrams becomes downward evolution on the $M - T_{\text{eff}}$ diagram, in consequence of luminosities fixed by the MLR.

One should still keep in mind that the evolutionary and metallicity information contained in $M - R$ diagrams must be also in the single thin distribution, of the MLR. Squeezing the three different contributory factors into the one set of values has the effect of making the relative width of the scatter in L appear small. The organization of stars into a narrow band on the $M - L$ diagram demonstrates the remarkable fundamentality of the MLR relationship. This has been refined for stars in the solar neighbourhood to a continuous sequence of four tight linear correlations.

Looking at stellar evolution theory in a general way, for stars with $M \gtrsim 1M_\odot$ the luminosity increases, but in such a way that surface temperature drops a little with the expanding radius. This is true except in the final stages of MS evolution when the central convection zone starts to shrink, due to the depletion of hydrogen in the core. In this final stage, surface effective temperatures increase somewhat, such that a star becomes brighter, though it remains cooler at TAMS than it was at ZAMS. On the other hand, for stars with $M \lesssim 1M_\odot$, the temperature rises initially, but later falls, while L is continuously increasing. The changes of slope of the theoretical ZAMS and TAMS lines in Fig 9.20 at about $M = 1.05M_\odot$ reflect this point.

MS evolution, visible on the $M - R$ diagram, but not seen on the $M - T_{\text{eff}}$ one with published temperatures, becomes clearly visible with empirical effective temperatures calculated from the MLR. Stellar temperatures based on this method are obtained directly from the absolute stellar properties (M, R). On the other hand, published temperatures are determined from properties such as colours or spectral lines, or using parallax and the definition of apparent magnitude. Temperatures estimated in that way involve, in principle, a bolometric correction procedure. Stellar temperatures calculated by the eclipse method for detached double-lined binaries may have apparently larger uncertainties than literature ones, but they have the potential to calibrate the bolometric corrections needed to estimate the effective temperatures of MS stars in general.

In summary: empirical MLRs are useful to provide an easy and independent way of testing the absolute brightness or parallax of an MS star, once its mass has been estimated. Alternatively, the MLR can be useful to estimate masses of single MS stars from observationally determined luminosities, colours or spectral types. Furthermore, an accurate MLR has practical applications in extragalactic research, since it can be used along with the stellar content of a galaxy to estimate the stellar mass of a galaxy.

9.6 Bibliography

9.1 The basic ideas introducing Section 9.1 trace back to the well-known paper of J. Jeans in the *Phil. Trans. Royal Soc. London*, Vol. 199, pp1–53, 1902. Early surveys of the solar neighborhood concluded that most stars are in bound systems of low multiplicity, cf. A. H. Batten, *Binary and Multiple Systems of Stars*, Pergamon Press, 1973; H. A. Abt & S. G. Levy, *Astrophys. J. Suppl. Ser.*, Vol. 30, p273, 1976; a result that was confirmed in later work, for example, A. Duquennoy & M. Mayor, *Astron. Astrophys.*, Vol. 248, p485, 1991; D. A. Fischer & G. W. Marcy, *Astrophys. J.*, Vol. 396, p178, 1992; M. Simon et al., *Astrophys. J.*, Vol. 443, p625, 1995; R. D. Mathieu et al., *Protostars and Planets IV*, eds V. Mannings, A. P. Boss, S. S. Russell, Univ. Arizona Press; p703, 2000; D. Raghavan et al., *Astrophys. J. Suppl. Ser.*, Vol. 190, pp1–42, 2010.

High resolution techniques to determine the properties of newly formed stars were pursued by e.g. A. M. Ghez et al. *Astron. J.*, Vol. 106, p2005, 1993; Ch. Leinert et al., *Astron.*

Astrophys., Vol. 278, p129, 1993; and M. Simon et al. 1995 (op. cit.). Combining results from observations across the spectrum has allowed more detailed discussion on binary formation, e.g. M. Cohen & L. V. Kuhi, *Astrophys. J. Letts.*,, Vol. 227, L105, 1979; P. Hartigan et al., *Astrophys. J.*, Vol. 427, p961, 1994; C. D. Koresko, *Astrophys. J.*, Vol. 440, p764, 1995; W. Brandner & H. Zinnecker, *Astron. Astrophys.*, Vol. 321, p220, 1997; A. M. Ghez, R. J. White, & M. Simon, *Astrophys. J.*, Vol. 490, p353, 1997; J. Woitas, Ch. Leinert, R. Köhler, *Astron. Astrophys.*, Vol. 376, p982, 2001. See also the collection of papers in *The Origin of Stars and Planetary Systems*, eds. C. J. Lada and N. D. Kylafis, NATO Sci. Ser. C, 540, Springer, 1999.

The fragmentation of rotating cloud cores in dependence on initial distributions of mass and angular momentum were modelled by M. R. Bate & I. A. Bonnell *Mon. Not. Royal Astron. Soc.*, Vol. 285, p33, 1997. The subject was reviewed by P. Bodenheimer et al., *Protostars and Planets IV* (op. cit.), p675, 2000; and J. E. Tohline, *Ann. Rev. Astron. Astrophys.*, Vol. 40, p349, 2002. That close binary systems should have mass ratios near unity, for example (see M. R. Bate, *Mon. Not. Royal Astron. Soc.*, Vol. 314, p33, 2000), was supported from the data analysis of J. L. Halbwachs et al., *Astron. Astrophys.*, Vol. 397, p159, 2003. The redistribution of angular momentum was further discussed by B. Reipurth et al., *Protostars and Planets VI*, eds. H. Beuther, R. S. Klessen, C. P. Dullemond, T. Henning, Univ. Arizona Press, p267, 2014. The average number of stars per core was estimated at 2-3 by S. P. Goodwin & P. Kroupa, *Astr. and Astrophys.*, Vol. 439, p565, 2005. For other considerations bearing on the fragmentation scenario see e.g.: R. S. Klessen et al., *Astrophys. J.*, Vol. 501, L205, 1998; S. S. R. Offner et al., *Astrophys. J.*, Vol. 725, p1485, 2010, and *Protostars and Planets VI*, (op. cit.), p53, 2014; K. M. Kratter et al., *Astrophys. J.*, Vol. 708, p1585, 2010.

The multiplicity of solar-type field stars stands at 50-60% according to A. Duquennoy & M. Mayor, 1991, (op. cit.), and A. L. Kraus et al., *Astrophys. J.*, Vol. 731(8), 2011. Lower mass stars have a generally lower multiplicity according to C. J. Lada, *Astrophys. J.*, Vol. 640, L63, 2006; G. Basri & A. Reiners, *Astron. J.*, Vol. 132, p663, 2006; and M. Ahmic et al., *Astrophys. J.*, Vol. 671, p2074, 2007.

The cited application of detailed MHD effects in the modelling is that of R. L. Kuruwita, C. Federrath & M. Ireland, *Mon. Not. Royal Astron. Soc.*, Vol. 470, p1626, 2017. The role of circumstellar disks in stellar formation processes was considered by A. Tokovinin in *Mon. Not. Royal Astron. Soc.*, Vol. 468, p3461, 2017.

9.2 R. B. Larson's review of binary separations (Section 9.2) in *Mon. Not. Royal Astron. Soc.*, Vol. 272, p213, 1995; addressed such data as produced by M. Gomez et al. *Astron. J.*, Vol. 105, p1927, 1993. The power-law distribution of star separations, as presented in the work of E. Öpik in *Tartu Obs. Publ.*, Vol. 25(6), 1924, underlies many such considerations. With the input of much new observational data, discussion of the separations of binary fragments could be placed on a firmer basis, e.g. by M. F. Sterzik et al. *Astron. Astrophys.*, Vol. 411, p91, 2003; O. Lomax et al. *Mon. Not. Royal Astron. Soc.*, Vol. 447, p1550, 2015. The lowest separations are associated with an *opacity limit* (cf. S. P. Goodwin et al., *Protostars and Planets V*, eds. B. Reipurth, D. Jewitt, K. Keil, Univ. Arizona Press, p133, 2007).

Further consideration of binary separations were given by V. V. Makarov et al., *Astrophys. J.*, Vol. 687, p566, 2008. As the typical size of cores is comparable to the Jeans length, observed wide binaries cannot come from the single core fragmentation. So special wide-binary formation mechanisms, such as cluster dissolution, were proposed; see, for example, M. B. N. Kouwenhoven et al. *Mon. Not. Royal Astron. Soc.*, Vol. 404, p1835, 2010; N. Moeckel & M. R. Bate *Mon. Not. Royal Astron. Soc.*, Vol. 404, p721, 2010. Another idea is the 'unfolding' of dynamically unstable triples (B. Reipurth & S. Mikkola, *Nature*, Vol. 492, p221, 2012). A. L. Kraus & L. A. Hillenbrand, *Astrophys. J.*, Vol. 686, L111, 2008;

suggested that the binary and clustering regimes are separated by a small region with a constant density of companions, where primordial clustering becomes lost in the motions of the stars (see also I. Joncour et al. *Astron. Astrophys.*, Vol. 599, p14, 2017). Properties of the filamentary structures associated with the condensation of interstellar gas were set out by P. André et al. *Protostars and Planets VI*, eds. H. Beuther et al., Univ. Arizona Press, p27, 2014.

9.3 Regarding Section 9.3, A. Duquennoy & M. Mayor's survey (1991, op. cit.) of a few dozen spectroscopic binaries with well-determined parameters showed the expectable tendency to low eccentricities (reflecting orbit circularization) for short period systems, with indications of a trend to a linear distribution with e for wide binaries, also in keeping with general theory.* Orbital eccentricity has a natural connection with the subject of rotation of the orbit's major axis — *apsidal motion*. Binary stars showing this have been collected by T. E. Sterne, *Mon. Not. Royal Astron. Soc.*, Vol. 99, p662, 1939; Z. Kopal & M. B. Shapley, *Ann. Jodrell Bank*, Vol. 1, p141, 1956; M. Schwarzschild, *Structure & Evolution of the Stars*, Princeton Univ. Press, (Table 18.1), 1958. I. Semeniuk, *Acta Astron.*, Vol. 18, p1, 1968; A. H. Batten, op. cit., 1973, (Table 9); A. F. Petty, *Astrophys. Space Sci.*, Vol. 21, p189, 1973; C. S. Jeffery, *Mon. Not. Royal Astron. Soc.*, Vol. 210, p731, 1984; T. Hegedüs, *Bull. d'Info. Centre de Données Stellaires*, Vol. 36, p23, 1989; A. Claret & A. Gimenez, *Astron. Astrophys.*, Vol. 277, p487, 1993; A. Gimenez & D. L. Crawford, *Experimental Astron.*, Vol. 5, p91, 1994; A. V. Petrova & V. V. Orlov, *Astron. J.*, Vol. 117, p587, 1999; I. Bulut & O. Demircan , *Mon. Not. Royal Astron. Soc.*, Vol. 378, p179, 2007.

The parameters listed by Bulut & Demircan (op. cit.) were sourced from: 1. the HIPPARCOS catalogue (ESA, 1997), 2. *An Atlas of (O – C) Diagrams of Eclipsing Binary Stars*, by J. M. Kreiner, C-H Kim, & I-S Nha, Pedagog. Univ., Kraków, (2001), and 3. the 9th catalogue of spectroscopic binary orbits, from D. Pourbaix et al., *Astron. Astrophys.*, Vol. 424, p727, 2004.

The circularization of orbits through tidal effects were considered by M. Lecar et al., *Astrophys. J.*, Vol. 205, p556, 1976; and J-P., Zahn *Astron. Astrophys.*, Vol. 500, p121, 1977; see also M. E. Alexander, *Astrophys. Space Sci.*, Vol. 23, p459, 1973.

The role of periodic episodes of periastron mass transfer was studied by J. F. Sepinsky et al., *Astrophys. J.*, Vol. 667, p1170, 2007; also C-P. Lajoie & A. Sills, *Astrophys. J.*, Vol. 726, p67, 2011. 'Heartbeat' binaries were discussed by K. Hambleton in *Astron. Soc. Pacific Conf. Ser.*, Vol. 496, p162, eds. S. M. Rucinski, G. Torres, M. Zejda, San Francisco, 2015. D. P. Kjurkchieva et al. (see VizieR On-line Data Catalog: J/other/RMxAA/53.235; originally published in *Rev. Mex. Astron. Astrofis.*, Vol. 53, p235, 2017) included e and ϖ values for 529 systems from the Kepler Eclipsing Binary catalogue, used in preparing Fig 9.2.

Discussions of a possible period-eccentricity relationship include those of H. N. Russell, *Astrophys. J.*, Vol. 31, p185, 1910; J. H. Jeans, *Mon. Not. Royal Astron. Soc.*, Vol. 85, p2, 1924; M. H. H. Walters, *Mon. Not. Royal Astron. Soc.*, Vol. 92, p786; and Vol. 93, p28, 1932; V. A. Ambartsumian, *Astron. Zh.*, Vol. 14, p207, 1937. The analysis of H. Horrocks, *Mon. Not. Royal Astron. Soc.*, Vol. 96, p534, 1936; was supported, to some extent, by M. Mayor & J. C. Mermilliod in *IAU Symp. No. 105*, eds. A. Maeder & A. Renzini; Reidel Publishing Company, 1984; Mathieu et al., *Astron. J.*, Vol. 100, p1859, 1990; and Raghavan et al. (2010, op. cit.). Duquennoy & Mayor (op. cit.) used the work of Zahn (op. cit.) and T. Mazeh & J. Shaham, *Astron. Astrophys.*, Vol. 77, p145, 1979; to support their findings.

---

*By contrast, irregularity of the longitude of periastron distribution (Barr effect) was noted in the Chapter 6.

V. Van Eylen et al. published their temperature-related comparisons in *Astrophys. J.*, Vol. 824, p 15, 2016.

9.4 Issues raised in *Binaries as Tracers of Stellar Formation*, eds. A. Duquennoy & M. Mayor, CUP, 1993, bear on the subject of Section (9.4): the cataloguing of stellar parameters, as supported by combinatorial analyses of binary stars (see also O. Y. Malkov, A. V. Tutukov & D. A., Kovaleva, *Proc. Special Session IAU XXIV General Assembly, Manchester, 2000*; ed. A. H. Batten, *Astron. Soc. Pacific*, p291, 2001). Notable examples include: D. M. Popper, *Ann. Rev. Astron. Astrophys.*, Vol. 18, p115, 1980; P. Harmanec, *Bull. Astron. Inst. Czech.*, Vol. 39, p329, 1988; O. Demircan & G. Kahraman, *Astrophys. Space Sci.*, Vol. 181, p313, 1991; J. Andersen, *Astron. Astrophys. Rev.*, Vol. 3, p91, 1991; O. Y. Malkov, *Bull. d'Inf. Centre Donn. Stellaires*, Vol. 42, p2, 1993. The spiral structure of the Galaxy shown in Fig 9.13 is reviewed by Y. Xu, M. A. Voronkov, J. D. Pandian, J. J. Li, A. M. Sobolev, A. Brunthaler, B. Ritter & K. M. Menten in *Astron. Astrophys.*, Vol. 507, p1117, 2009.

A general and expectable trend towards increasing definition and reliability of listed parameters over the years can be seen, for example, with the work of Z. P. Kraicheva, E. I. Popva, A. V. Tutukov & L. R. Yungel'son, *Soviet Astron.*, Vol. 22, p670, 1978; S. Y. Gorda & M. A. Svechnikov, *Astron. Reps.*, Vol. 42, p793, 1998; D. A. Kovaleva, *Astron. Reps.*, Vol. 45, p972, 2001; E. Lastennet & D. Valls-Gabaud, *Astron. Astrophys.*, Vol. 396, p551, 2002; L. A. Hillenbrand & R. J. White, *Astrophys. J.*, Vol. 604, p741, 2004; and others. The sources of Hillenbrand and White's data included J. Andersen (1991, op. cit.); Ribas et al., *NATO Sci. Ser. (Math & Phys)*, Vol. 544, p659, 2000, (ed. C. Ibanoğlu), Kluwer; and Delfosse et al., *Astron. Astrophys.*, Vol. 364, p217, 2000.

More recent contributions have been Ibanoğlu et al., *Mon. Not. Royal Astron. Soc.*, Vol. 373, p435, 2006; O. Y. Malkov et al., *Astron. Astrophys.*, Vol. 446, p785, 2006; C. Jordi & E. Masana, *Lecture Notes & Essays in Astrophys.*, Vol. 3, eds. A Ulla and M Manteiga, p33, 2008; G. Torres, J. Andersen, A. Gimenez, *Astron. Astrophys. Rev.*, Vol. 18, p67, 2010; and Z. Eker, V. Bakış, S. Bilir, F. Soydugan, I. Steer, E. Soydugan, H. Bakış, F. Aliçavuş, G. Aslan & M. Alpsoy, *Mon. Not. Royal Astron. Soc.*, Vol. 479, p5491, 2018; whose data is referred to in much of the later sections of this chapter. Eker et al. (op. cit.) took note of improved values for solar constants (see E. M. Standish, *Highlights of Astronomy*, ed. I. Appenzeller, Kluwer, 1998; and M. Haberreiter et al., *Astrophys. J.*, Vol. 675, p53, 2008). The other works of Eker et al. referred to in this connection are in *Publ. Astron. Soc. Austral.*, Vol. 31, p24 (8 authors), 2014; and *Astron. J.*, Vol. 149, p131, (9 authors), 2015.

The reference to SZ Cam is from E. Tamajo et al., *Astron. Astrophys.*, Vol. 539, p139, 2012; that to DH Cep is from R. W. Hilditch et al, *Astron. Astrophys.*, Vol. 314, p165, 1996. A wide range of theoretical stellar models, for comparison purposes were tabulated by O. R. Pols et al, *Mon. Not. Royal Astron. Soc.*, Vol. 298, p525, 1998. The reference to AE For is from M. Rozyczka et al., *Mon. Not. Royal Astron. Soc.*, Vol. 429, p1840, 2013; to XY UMa, see T. Pribulla et al., *Astron. Astrophys.*, Vol. 371, p997, 2001; and to DV Psc, see X. B. Zhang & R. X. Zhang, *Mon. Not. Royal Astron. Soc.*, Vol. 382, p1133, 2007.

9.5 With regard to Section (9.5): the origin of the mass-luminosity relation (MLR), is associated with the papers of E. Hertzsprung, *Bull. Astron. Inst. Netherlands*, Vol. 2, p15, 1923; and H. N. Russell, W. S. Adams, A.H. Joy, *Publ. Astron. Soc. Pacific*, Vol. 35, p189, 1923; together with data from visual binaries. A. S. Eddington added eclipsing binary and Cepheid parameters in *The Internal Constitution of the Stars*, Cambridge University Press, 1926. More data was added by D. B. McLaughlin in the *Astron. J.*, Vol. 38, p21, 1927. The development of the MLR with increasing quantity and quality of data can be traced through the papers of G. P. Kuiper, *Astrophys. J.*, Vol. 88, p472, 1938; R. M. Petrie, *Publ. Dom. Astrophys. Obs.*, Vol. 8, p341, 1950; and *Astron. J.*, Vol. 55, p180, 1950; K. A. Strand & R. G. Hall *Astrophys. J.*, Vol. 120, p322, 1954; O. J. Eggen, *Astron. J.*, Vol. 61, p360, 1956; B.

Cester et al., *Astrophys. Space Sci.*, Vol. 96, p125, 1983; T. J. Henry & D. W. McCarthy Jr., *Astron. J.*, Vol. 106, p773, 1993, who used the data of C. E. Worley & W. D. Heintz, *Publ. United States Naval Obs.*; 2nd ser., Vol. 24, pt. 7, Washington, 1983. D. M. Popper was referred to in connection with his publication in *Astron. J.*, Vol. 72, p316, 1967; and also his 1980 paper, op. cit.

Studies of the mass-radius relation MRR continued through W. H. McCrea, *Physics of the Sun and Stars*, Hutchinson's Univ. Library, 1950; L. Plaut, *Publ. Kapteyn Astron. Lab. Groningen*, Vol. 55, p1, 1953; S. S. Huang & O. Struve, *Astron. J.*, Vol. 61, p300, 1956; C. H. Lacy, *Astrophys. J. Suppl.*, Vol. 34, p479, 1977; & *Astrophys. J.*, Vol. 228, p817, 1979. Z. Kopal's book *Dynamics of Close Binary Systems*, *Astrophys. Space Sci. Lib.*, Vol. 68, Dordrecht, Reidel, 1978 refers to the MRR, and further related studies include those of J. Patterson, *Astrophys. J. Suppl. Ser.*, Vol. 54, p443, 1984, (low mass high energy binaries); A. Gimenez & J. Zamorano, *Astrophys. Space Sci.*, Vol. 114, p259, 1985; P. Harmanec, (1988, op. cit.); and Demircan & Kahraman (1991, op. cit.). T. S. Metcalfe, R. D. Mathieu, D. W. Latham & G. Torres, *Astrophys. J.*, Vol. 456, p356, 1996 (lower Main Sequence). The MLR and MRR of Demircan & Kahraman (1991, op. cit.) have been referred to by L. Li et al., *Mon. Not. Royal Astron. Soc.*, Vol. 360, p272, 2005; I. Hunter et al., *Astron. Astrophys.*, Vol. 479, p541, 2008; D. Jiang et al., *Mon. Not. Royal Astron. Soc.*, Vol. 396, p217; 2009; a A. L. Kraus & L. A. Hillenbrand, *Astrophys. J.*, Vol. 704, p531, 2009; and others.

Whether the parameters of stars in binary systems conform to exactly to expectations from theoretical studies of single stars was questioned by O. Y. Malkov in *Astron. Astrophys.*, Vol. 402, p1055, 2003. The revised MLR and MRR relations of O. Y. Malkov in *Mon. Not. Royal Astron. Soc.*, Vol. 382, p1073, 2007; were considered by Zasche & Wolf, *Astron. Astrophys.*, Vol 527, p43, 2011; R. Gafeira et al., *Astrophys. Space Sci.*, Vol. 341, p405, 2012; A. Fumel & T. Böhm, *Astron. Astrophys.*, Vol. 540, p108, 2012; and others.

Applications of the mass luminosity and radius correlations to the Tully-Fisher relation (R. B. Tully & J. R. Fisher, *Astron. Astrophys.*, Vol. 54, p661, 1977) in galactic research are found in the papers of S. M. Faber & J. S. Gallagher, *Ann. Rev. Astron. Astrophys.*, Vol. 17, p135, 1979; E. F. Bell & R. S. de Jong, *Astrophys. J.*, Vol. 550, p212, 2001; and M. Girardi, et al., *Astrophys. J.*, Vol. 569, p720, 2002. Tables 9.1 & 2 are based on the work of Eker et al. (see above) and the SOS project (Table 9.2) is discussed by H. Sung, B. Lim, M. S. Bessell, J. S. Kim, H. Hur, M-Y. Chun & B-G. Park in arXiv:1306.0309 [astro-ph.SR], 2013. Ongoing revisions of such MLR and MRR relations as discussed above, using modern and more accurate data, can be continually expected.

10

Moderately Evolved Systems

10.1 RS CVn Binaries

Table (7-6) in Kopal's (1959) book listed 18 eclipsing binary systems characterized by light curves whose deeper (primary) minimum was that of a normal Main-Sequence-like star, while the secondary minimum corresponded to the eclipse of a cooler star of comparable, if not greater, size, yet significantly smaller than its surrounding Roche lobe. Such binaries, having normal primaries and swollen secondaries, sometimes called subgiants, were originally regarded as within the broad spectrum of Algol-like systems. Kopal referred to this particular subgrouping as the 'undersize Algols'. In some cases, a spectroscopically determined mass ratio q was available, and that was often close to unity.

Spectra of the cool companions, in this kind of binary, were usually found to show emission features: particularly the Fraunhofer H and K lines of ionized calcium. The showing up of these lines in the spectra of various cool binaries had been pointed out by Otto Struve in 1946. Emission in the Balmer H_α line was also sometimes observed at relatively weak levels. A prominent member of the group was the system RS CVn, whose light curves reveal peculiar additional variations, not attributable to regular proximity effects, but with some tendency to have a cyclic pattern of recurrence on a time-scale much longer than the orbital period of $\sim$4.8 d. Other members of this group showed comparable effects, notably AR Lac, the vagaries of whose light curves were discussed already in 1947 by the pioneer photometrist G. E. Kron.

With later, more concentrated, studies of these binaries, particularly by D. S. Hall and his associates at the Vanderbilt University, a defining set of characteristics became clear, as well as a theoretical scenario that could provide a satisfactory general explanation of their occurrence and properties. These stars, when referred to in the context of moderately evolved close binary systems, are nowadays usually styled *RS CVn* binaries. They are closely related to a somewhat wider grouping known as *chromospericaly active* stars. They are believed to be strongly affected by the presence of large surface regions of concentrated magnetic field — *starspots* — alternatively termed *maculation*. It is that property that gives rise to the peculiar additional variations, that can reach to the order of 0.1 mag in their light curves (see Figs 10.1-2).

RS CVn systems have been divided into a few subgroups, with the orbital period being a key discriminant, thus: (1) Short period: cool detached MS-like stars with $P < 1$ d; (2) Regular: cool stars, MS secondary, primary somewhat evolved, but no Roche Lobe overflow, 1 d $< P < 14$ d; (3) Long period: both stars evolved to subgiant or giant stage, $P > 14$ d.

DOI: 10.1201/b22228-10

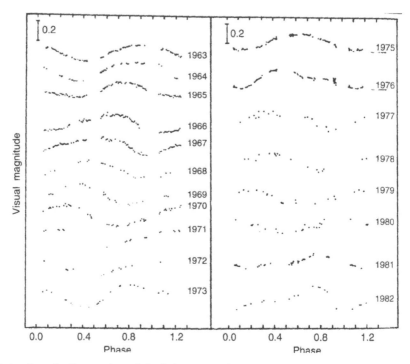

Figure 10.1   Out-of-eclipse regions of the light curves of RS CVn demonstrate the distinct 'photometric wave' effects associated with RS CVn type binaries. The diagram, presented by C. Blanco et al. in *Proc. 71st IAU Coll.*, Vol. 102, p387, 1983, eds. P. B. Byrne & M Rodonò; ©D. Reidel Publ. Co.; shows a compete cycle through the years 1963-82 obtained mostly at the Catania Observatory, but including two light curves from E. W. Ludington (1978) for 1975-76.

Those cool dwarfs that from time to time have sudden episodes of increased brightness, especially noticeable in UV spectral ranges – *flare stars* – are also associated with RS CVn systems, certainly within the context of chromospherically active objects. Flare stars are generally low mass, late spectral type objects, having a rapid rotation that energizes their strong magnetic fields. Binarity helps maintain the fast rotation through tidal engagement, though some flare stars appear to be single, their rapid rotation then a consequence of relative youth. A cool magnetically active star may also be in a binary system with a star that already evolved through to the white dwarf stage. These 'V471 Tau type' binaries share some properties with the classical RS CVns.

The linking theme in all these groupings is clearly that of magnetic activity. This is understood to be fundamentally similar to that of the Sun, complete with photospheric dark regions (starspots), that undergo long-term cyclic behaviour, chromospheric emissions, and occasional dramatic releases of energy — flares — detected across the spectrum from gamma rays to radio. The generation of these strong magnetic fields is associated mainly with (a) rapid rotation and (b) extensive sub-photospheric convection regions. These conditions thus connect aspects of binary evolution to studies of stellar magnetic activity, fostering a natural interest in the latter within the context of the former.

Discussions of the Sun's magnetic field generally refer to a *dynamo* model. This entails the presence of an electrically conductive fluid medium and an ongoing supply of rotational energy with convective motions within the fluid. The dynamic nature of the magnetic field is seen in its pattern of large-scale cyclic changes over a period of typically a few years. The '$\alpha - \Omega$ dynamo' proposes the operation of two complementary processes, in which the

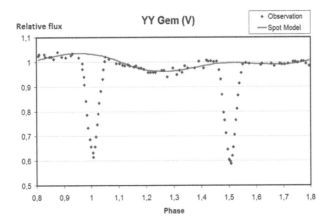

Figure 10.2   Starspot-affected light curve of the M-type short period eclipsing system YY Gem, as presented by C. J. Butler et al. (11 authors) in *Mon. Not. Royal Astron. Soc.*, Vol. 446, p4205, 2015. The starspot model for the maculation wave is shown as the continuous curve through the out-of-eclipse points.

overall magnetic field distribution varies between predominantly poloidal and toroidal arrangements. These processes are confirmed observationally, though theoretical explanations have not been perfectly aligned with observations in all respects.

In short, the Ω process is reflected in the way that the *differential rotation* of the Sun stretches out an originally 'north-south' or poloidal field line embedded in the solar plasma, steadily dragging it into a toroidal concentration. The α process involves the regeneration of the poloidal from the toroidal field structures, and it implies a current being formed along the general direction of the toroidal field. An indication of the α-process in action is that up-welling toroidal lines of force become twisted towards the meridional direction. The rhythmic succession of these two processes on the Sun makes up its $\sim$22 yr magnetic activity cycle.* Around sunspot maximum the magnetism becomes evident as toroidal loops break through the photosphere: at sunspot minimum the global field appearance concentrates towards relatively large polar caps. It is naturally of great interest to see how far this $\alpha - \Omega$ model may be applied to RS CVn and related stars.

Large darkened surface regions have been regularly mapped for RS CVn stars, using the technique of *Doppler Imaging* (DI), or its developed form that utilizes the field-related line-splitting Zeeman effect, i.e. *Zeeman Doppler Imaging* (ZDI) to demonstrate the presence of cooling surface magnetic field concentrations. These effects appear to parallel those observed directly in the Sun's active regions, but on a much grander scale. Fig 10.3, derived from K. G. Strassmeier (2009); is aimed at bringing out principles in Zeeman Doppler Imaging. The profiles of absorption lines in the spectrum of a rotating spotted star contain information about the distribution of the spots over the stellar surface, as mentioned in Section 6.4. Spectral line profiles are presented below the stellar disk cartoons in the four Stokes parameter forms (I, V, Q, U). Line features in I are observed to migrate across the profile in successive spectrograms, allowing parametrization of spot positions and possible variation of rotation with latitude (as in regular Doppler Imaging). With the additional data sets relating to polarization effects (V, Q, U – shown here increased by 25 with respect to I), the full geometry of the magnetic field can be reconstructed. The figure illustrates

---

*The period for the full cycle of magnetic activity, in which field polarities reverse half way through, is double that of the familiar cycle in the numbers of surface maculae.

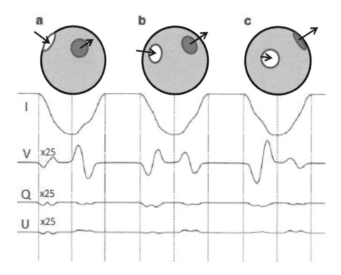

Figure 10.3 A rotating star with regions of concentrated magnetic surface field – 'starspots' – can be mapped with the use of Zeeman Doppler Imaging techniques, as discussed by K. G. Strassmeier in *Astron. Astrophys. Rev.*, Vol. 17, p251, 2009 (see text).

three rotational phases (a, b, c) of a model with two spots at different latitudes that are cooler than the $\sim$ 5500 K surrounding photosphere by some 1270K. The two large spots are characterized by mainly radial surface magnetic fields of 1.5 kG acting in opposite directions. The three phases are separated by $\sim$0.05, with the inclination of the rotation axis set at $\sim$60°. The variation of the profiles, particularly the relationship of the circular polarization V to the projected field strength (note the changes in the second row of profiles), demonstrate the respective sensitivity to the field geometry and their modulation with the stellar rotation. The Q and U parameters give information on non-radial components in the field, but the relatively low signal/noise ratio implies correspondingly high uncertainties to the modelling.

It is by now generally understood that the evolution of RS CVn stars is characterized by at least one of the component stars evolving off the Main Sequence, as in normal single star evolution, expanding as it crosses the Herzsprung gap region, but not to the point of full-blown Roche Lobe overflow (Fig 10.4). During the time of this transition to the Red Giant Branch the more evolved star's outer regions cool and develop a deepening sub-photospheric convective region. At the same time, tidal locking maintains the high rotation velocity: the combined circumstances strongly associated with the generation of the observed intense magnetic fields. The measured parameters of some binaries associated with the group also point to modest amounts of mass transfer, perhaps related to stellar winds.

In the 9th catalogue of spectroscopic binaries of Pourbaix et al. (2004), about 5% of entries could be associated with the RS CVn type. This is roughly of the right order of magnitude, though a little high, compared with the ratio of times spent ascending the Red Giant Branch to Main Sequence lifetimes, i.e. $\sim$1%. But perhaps the additional factors of emission lines and photometric irregularities, renders the RS CVn type to be preferentially selected in observational identification and follow-up studies. Loss of systemic angular momentum through winds may also bias the incidence of these binaries towards shorter orbital periods, where they become more discoverable.

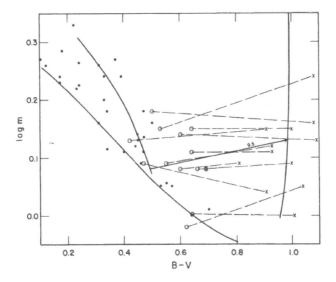

Figure 10.4 A basic theoretical interpretation of RS CVn stars was proposed by D. M. Popper and R. K. Ulrich in *Astrophys. J.*, Vol. 212, L131, 1977; ©AAS and illustrated in the above diagram. The systems contain a post-Main Sequence primary (x), heading towards, or close to, the Red Giant Branch on the right, and a secondary (circle) still in the vicinity of Main Sequence on the left. Regular detached Main Sequence pairs are shown (dots) for comparison. Observational selection effects tend to enhance the chances of the RS CVn combination being picked up as eclipsing binaries.

The extent to which a star's magnetic field structure affects its evolution remains unsettled, at least as regards discriminative observational testing. However, calculations have shown that the outcome of the initial core collapse in star-forming clouds can be significantly affected, i.e. as single or multiple condensation products, by the strength, structure and extent of an included magnetic field. In effect, strong fields tend to inhibit cloud fragmentation. The initial mass function and extent of binarity in star-forming regions thus relates to the local distribution of the galactic field. We can expect stellar magnetic fields naturally to continue to influence binary system behaviour through and after the original formation.

10.1.1 Spotted Stars

A broadbrush picture of RS CVn stars involves processes basically similar to the magnetic activity of the Sun, with activity parameters increased typically by a couple of orders of magnitude. Instead of a spotted surface region having an angular extent of one or two degrees as in the familiar sunspots, a macula on an RS CVn star could spread over an area ∼100 times bigger, thus angular extents of ∼10-30 degrees are reported in evaluations of the photometric effects (see Figs 10.3, 10.5)

Typical temperature differences between a starspot and the surrounding photosphere have been deduced, by monitoring colour variations through the maculation waves, to be ∼1000K, i.e. comparable to those over sunspots with a substantial umbral component. Surface magnetic fields of order several thousand gauss are required for this scale of cooling. The measurable maculation effects allow the activity to be monitored, as with comparable parameters of solar activity. Many studies have thus examined the links between alternative tracers of stellar activity across the spectrum. It transpires that care is needed in drawing analogies. The surface locations of starspots, their migration behaviour and relationship to

AB Doradus, Decembre 1996

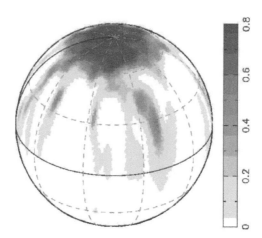

Figure 10.5 Surface field distribution model for the active cool dwarf AB Dor, derived from a ZDI analysis procedure. (Courtesy of J-P. Donati in *Mon. Not. Royal Astron. Soc.*, Vol. 302, p437, 1999.)

underlying magnetic field structures and energy transport mechanisms require significant differences in modelling, even for superficially similar phenomena.

The value of statistical analysis is naturally enhanced with an extensive database of coverage. There has thus been ongoing interest in the application of automated patrol detectors to active cool stars. Such devices are increasingly associated with other kinds of astronomical monitoring across the electromagnetic spectrum, where space-based observations are of particular importance. The surveillance of spotted stars provided an early persuasive case for automated photometry.

With enhanced observational techniques, including ZDI, more detailed information on starspot properties is elucidated. The modelling of dynamo processes active throughout the convective envelope involving magnetism on the scale of several kG has been thus enabled. With increasing rotation rates and convection zone depths, the field distributions become stronger and allow finer structural details to be revealed. Long-term cycles in such field distributions, typically on the order of decades, may be reflected in recorded small changes in the orbital parameters of RS CVn binaries.

10.1.2   RS CVn Light Curves and Inferences

There is a long historical background to the explanation of stellar light variation in terms of the photometric effects of starspots. As recalled in Chapter 1, Goodricke in his 1783 report on Algol, had considered surface maculation, as well as eclipses, as possible causes of the observed effects. In the 19th century, a kind of 'minimization of hypothesis' principle was implicit in the maculation idea, given the range of feasible placings of spots on a self-luminous sphere. The subject was discussed in some detail by H. N. Russell in 1906. He argued that, in general, the variability of a light curve that was to be accounted for by the rotation of spots on a stellar surface could be done in an infinite variety of ways. This property would render indeterminate the inverse problem of finding a unique mapping of the flux distribution over the surface from the light curve. Russell noted, however, that light curves interpreted from the more restricted hypothesis of an eclipse model do allow limited parametrization. The determinacy of the inverse problem thus depends on the conditions imposed by the explanatory hypothesis.

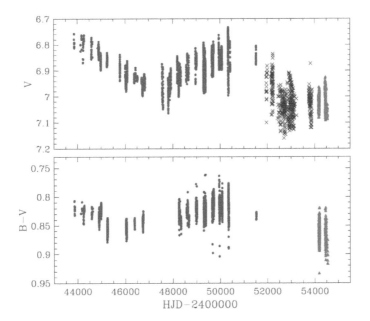

Figure 10.6 Regular photometric monitoring of the young dwarf multiple star AB Dor, that shares phenomenological similarities with RS CVn systems, shows a long-term cyclic trend in the maculation. Colour variations – when available – (lower panel) support the interpretation that the effects correspond to cool areas on the stellar surface. The illustration is from J. Innis et al. *Inf. Bull. Variable Stars*, No. 5832, 2008.

Maculation scenarios might have been postulated for variable stars in a general way in the absence of spectroscopic evidence, but, with more channels of information coming into the picture, inadequacy of the model to explain all related facts becomes apparent. But with the increasing support for magnetic activity in RS CVn stars in the second half of the 20th century through a range of interrelated phenomena, neglect of surface maculation appears as a modelling insufficiency, at least for stars with properties similar to those of RS CVn.

Concerning light curves, *photometric resolution* needs to be taken into account. This refers to a feature to be identified photometrically having an effect greater than the inherent scatter of the measurements. At the time when systematic photometric programmes to monitor the photometric behaviour of RS CVn stars over long periods were getting underway, this meant about 1% of the mean intensity, corresponding to a size of macula of the order of 0.1 radians, i.e. a radius of ∼6 deg of stellar surface as the minimum size of a discernible feature.

The effect of a dark spot on the photosphere of a rotating star can be treated formally along parallel lines to that of eclipses, as presented in Chapter 7. The formulation is simplified by adopting the basic spot model to be of circular outline; so with one size parameter and fixed shape having projected area A (Figure 10.6). As with the previously defined α-integrals (Section 7.3), we can write

$$\pi \sigma_n^m = \iint_A x^m z^n dx dy, \tag{10.1}$$

where the z-axis coincides with the line of sight to the centre of the star, presumed spherical, and the x-axis points in the direction from the centre of the stellar disk to that of the elliptical outline of the spot (Fig 10.7). These σ-integrals turn out to be a simpler form of surface

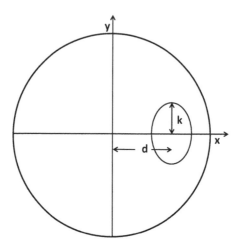

Figure 10.7   Schematic representation of a starspot of radius k whose elliptical outline is centered about a point distant d from the centre of the star.

area integral than the α-integrals of Chapter 7. The σ_n^m formulae are just combinations of elementary and trigonometric functions, and do not involve elliptic integrals.

The properties and application of σ-integrals to maculation problems have been described elsewhere, but an example was shown in Fig 10.2, with the light curve of YY Gem. The diagram illustrates *separation* of the effects of maculation and eclipses. This resembles the perturbation approach to non-linear problems, validated when disturbances from a simple preliminary description are small. Directly summing maculation and eclipsing effects would not necessarily account for the all data shown in Fig 10.2, since the spot wave applies to an uneclipsed surface. In effect, there could be some region within an eclipse minimum where the spot is eclipsed, so the adopted model for the eclipse of a smoothly illuminated stellar surface becomes inappropriate. The eclipse minimum is then affected in a non-linear way by the two separate descriptions.

This issue has wider implications, for example in connection with stars whose surfaces become distorted by binary proximity. It relates to the historic discussion about 'rectification' and 'synthesis' approaches to binary light curves. In the former, a separation of the proximity effects was performed that resulted in an idealized curve with clearly identifiable eclipse tangency points, from which an explicit 'solution' for the sought main parameters was directly obtained. Since light curve synthesis, by itself, doesn't give an explicit formulation for solution parameters, it was open to question as to whether they were uniquely definable.

In practical applications, these differences are more apparent than real, given the underlying perturbations methodology. In a general way, an eclipse effect, produced by the integration of an emergent flux over a phase-dependent boundary, can be regarded as a cross-correlation of aperture f and surface flux distribution g functions $f * g$, say, where f and g might well become complicated in a general case. The perturbations approach amounts to setting $f = f_0 + \Delta f$ and $g = g_0 + \Delta g$, where f_0 and g_0 are relatively simple forms having the known result $f_0 * g_0$. The distributive property of the cross-correlation entails that the additive (first order) forms $f_0 * \Delta g$ and $g_0 * \Delta f$ are matched to the residuals from the $f_0 * g_0$ model. Outside the eclipse phase ranges the cross-correlation $f_0 * g_0$ reverts to zero when the boundaries do not overlap. For eclipsing binaries this is tantamount to the discussion presented in Section (7.3.7), with $f_0 * \Delta g$ representing the integrated perturbed flux and $g_0 * \Delta f$ the boundary perturbations, respectively.

Maculation analysis using the circular spot approximation is, in method, a variation of the same procedure. The spot can be associated with a different aperture function, h, say, playing a separate but parallel role to f, except if there is a (small) phase range where the eclipsing star's boundary overlaps with that of the spot. This produces a small deviation inside the eclipse minimum, but a maculation model can be derived from the phases outside eclipses.

The question arises, since there are two stars in close binary systems, how do we know which one is affected? For the classical RS CVn systems, it has been shown from spectroscopic information that almost all the magnetic activity comes from the subgiant component with its rapidly rotating convective envelope, allowing plausible models to be developed. For the 'short period group' — essentially fast rotating cool MS stars, when one star is significantly more luminous than the other (e.g. XY UMa, SV Cam, RT And), starspot modelling could be effectively assigned to the properties of those (primary) stars. In other cases, account of the relative amplitudes of given spot waves in different wavelength ranges has been used to associate spots with individual components. Alternatively, it was found possible to assign which spot is probably affecting which star by combining single wavelength photometry with analysis of the colour variations.

Drawbacks due to oversimplification of the real physical situation and uniqueness questions about light-curve modelling are assessed as these relate to the photometric resolution and aims of the investigation. Given non-uniqueness for a general fitting function in the data-matching problem, there remain constrained optimization approaches, for example using circular spots of adjustable location and size, or a set of contiguous darkened pixels of adjustable individual sizes, locations and overall number.

Guided by the solar paradigm, what is generally sought is a representation that adequately fits the data, without loss of significant information or over-interpretation, for use in statistical contexts. It has been argued that concentrations of individual maculae, or 'groups', would be more applicable wording for the representation, though the evidence of colour (temperature) data, together with information on the vertical structure of active regions points to the term 'spot' being more relevant than could have been initially expected. Note that the maculation waves that originally drew attention to the RS CVn stars as a group, concerns the differential component of the flux over a phase cycle. There may well also be a more uniformly distributed component of the maculation that is not discerned from a single light curve.

A minimal set of parameters required for a single spot specification can be set out as: longitude of spot centre λ, corresponding latitude β, inclination of the rotation axis to the line of sight i, and the angular extent of the spot γ. The apparent semi-major axis of the spot k (Fig 10.7) follows as $k = \sin\gamma$. The reference light level for the unspotted state U, would be usually normalized to unity for an individual light curve (cf. Section 7.3). The derivation of the relative flux measures from the source observations involves a magnitude decrement Δm_0 from a calibrated comparison star. Δm_0 may vary from light curve to light curve and it bears on the uniformly distributed component of the maculation. For RS CVn systems like HK Lac (Fig 10.8) the relative luminosity of the spot-bearing star L_1 would be a reasonably large fraction of U. The temperature-dependent ratio of the mean flux over the starspot to the normal photospheric flux, over a spectral window, κ_W say, is a small quantity ($\kappa_W \lesssim 0.1$) in starspot optical light curves. The limb-darkening coefficient u of the spotted star's photosphere also enters this simple characterization.

With certain parameters given from other lines of evidence (e.g. inclination i or limb-darkening u), spot parameters can often be assessed from inspection of the light curve. Thus, λ follows from the orbital phase of the spot minimum. If a coherent spot wave is observed though the time it takes to build up a complete light curve, i.e. a few weeks or months, then it indicates near-synchronism of the rotation to the orbital motion. The orbital phasing

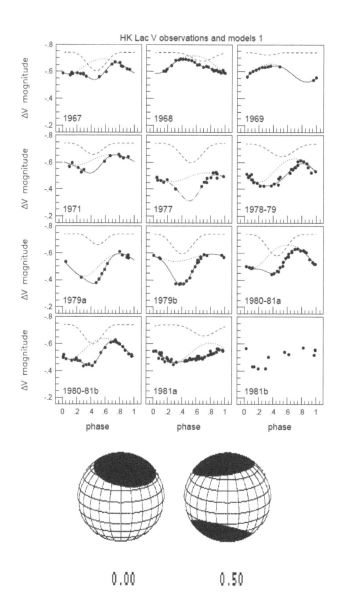

Figure 10.8 Long-term photometry of the RS CVn binary HK Lac. The light of the primary dominates the system and the maculation effects can be modelled using σ-integral type spot modelling. A predominantly poloidal magnetic field structure appears in the global illustrations at the bottom (cf. Fig 10.5). The 'south' polar region visible at phase 0.5, lower right, (phases marked below) corresponds to a relatively shallow effect in a limited phase range. This corresponds with the dashed curves in the upper panels, while the 'north' counterpart is the broader and deeper feature marked by the dotted curves. The diagram is from K. Oláh et al. *Astron. Astrophys.*, Vol. 321, p811, 1997; ©ESO.

thus provides a reference point for longitude assignment. The maculation wave's phase of minimum light, or spot longitude λ, is found to slowly drift in time (Fig 10.1), which is interpreted in terms of differential rotation of the spot-bearing photosphere.

A *Rossby parameter* $Ro = P_{\rm rot}/\tau_{\rm conv}$, where $P_{\rm rot}$ is the rotation period and $\tau_{\rm conv}$ the turn-over time for convection in the stellar envelope, has been used in this connection for statistical studies of RS CVn star activity. $\alpha - \Omega$ type dynamo theories have been successful in explaining observed characteristics, such as the relationship between chromospheric Ca II H-K flux and Ro, although dynamo model efficiency strongly depends on the rotation rate and convective time-scales, $\tau_{\rm conv}$ is not a direct measurable, but polynomial fits relating theoretical estimates of its value at given $B - V$ colours are available.

The size parameter γ is indicated by the depth of the minimum, though the spot latitude β comes into this, while β is also reflected by the relative duration of the wave. If we set ψ equal to the semi-width of the spot minimum on the light curve then we have a loose constraint that $\tan \beta \approx -\cos \psi \tan i$. A short shallow wave can then be associated with a spot in the lower (or 'southern') hemisphere (more remote from the line of sight). Its counterpart in the upper hemisphere would affect a range of phases greater than $180°$ as well as being deeper. In this way, a trend for a generally prevailing poloidal field arrangement is indicated for the system HK Lac shown in Fig 10.8.

This kind of parametrization for RS CVn type light curves has been used to study how dark spots change or migrate, seeking parallels with the well-known solar case. Lifetimes of individual maculae may range between a few months to several years. These durations have been reconciled with the sizes of emerging spots and their disruption from the shearing of differential rotation. Spots are reported to emerge relatively quickly in preferred longitudes. The positions of spots are linked to differential rotation of the underlying photosphere through a formula such as $\omega(\beta) = \omega_0(1 - k \sin^2 \beta)$, as used for the Sun; ω being the angular rotation velocity. The differential rotation coefficient k in this expression is found to be typically an order of magnitude less in active RS CVn stars than the solar case, for which $k \approx 0.2$. In fact, a trend towards more rigid-body type rotation $k \to 0$ is observed for fast rotators and those primaries close to filling their Roche lobes.

Long-term cycles of mean brightness have been deduced, with the Sun's well-known ~ 11 year sunspot (Schwabe) cycle influencing ideas. Various RS CVn stars that have been monitored closely appear to show cycles of comparable duration. These concern primarily the changes in mean brightness Δm_0, but also relate to latitude drift effects and small changes of orbital period. While the time-scales for monitoring behaviour hitherto are not extensive enough to be conclusive, it has been posited that orbital revolution synchronizes with the rotation speed at about $30°$ latitude, so a downward drift in the phase of spot minimum (lag) would be associated with a high latitude spot, due to the differential rotation.

Multiwavelength observations

Multiwavelength observations gathered from different observatories simultaneously monitoring have been fruitful in studies of RS CVn stars. As with the Sun, the radio domain has the potential to probe an active region in its vertical dimension thus complementing the 2D picture from optical techniques. Large plasma loop models were successfully applied to microwave data, especially regarding the very fast rise times of stellar radio flares. This has given support to the 'large monolith' picture of the active regions, a modelling approach supported also by UV and X-ray observations.

The short term (minutes to hours) impulsive component in microwave data associated directly with flares is distinguished from a more long-lasting quiescent component. The latter exhibits undulations on time-scales comparable to, or less than, those of the stellar rotation. Source sizes comparable to the stellar radius were inferred. VLBI measurements have found bright core regions surrounded by 'halo' structures that are relatively large compared with the sizes of the stars. Brightness temperatures, for areas comparable to the stellar disk, are in the range 10^8 to several times 10^{10} K, often with some circular polarization. Large halos would not be associated with wavelike variations on the order of the orbital period or less, but they could form a slowly decaying background after an initial large flare.

The non-flaring radio emission has often been interpreted in terms of gyrosynchrotron process from mildly relativistic electrons. If we consider the temperatures of the quiescent component corresponding to active regions smaller than the stellar disk, the Lorentz beaming factor ($\gamma \gtrsim 10$) implies a fully relativistic regime in which emission concentrates into a narrow cone (of semi-angle $\sim 1/\gamma$) about the direction of instantaneous motion of the electron. From this viewpoint, if the beaming factor and maximum intensity could be better constrained, microwave observations could provide a useful guide to the latitude of active regions.

In a general way, non-optical activity markers are found to correlate with the mean rotation rate at a given spectral type, and are in keeping with the magnetic activity of RS CVn systems being driven by a stellar dynamo.

10.1.3   Angular Momentum Changes and RS CVns

Observational evidence of a decaying rotation rate for stars with spectral types later than F on evolutionary time-scales was published by A. Skumanich after studying the projected equatorial speeds ($v \sin i$) of late-type stars in open clusters of varying ages. This decay in stellar rotation is explained in terms of an angular momentum loss (AML) process, acting through magnetically driven stellar winds, also called magnetic braking. For tidally locked binaries with late type components, such an AML can come from the reservoir of orbital angular momentum (OAM). In this way, AML from the components of spin-orbit coupled binaries causes their orbits to shrink. Spin-orbit coupling and a shrinking orbit then imposes a spin-up effect on the rotation of binary components. This is clearly different from single stars, where AML would slow the rotation rate.

This mechanism has been considered as a way to form common envelope binaries from initially detached systems. However, different authors have given different accounts of the period evolution and time-scale for forming contact binaries. In one approach, the AML of a component star was computed directly from Skumanich's formula ($V_{\mathrm{rot}} \sim t^{-1/2}$). This was derived from relatively slowly rotating stars ($V_{\mathrm{rot}} \lesssim 17$ km s^{-1}). AML from the two components, thus derived, was equated to the change of OAM. The corresponding orbits evolved initially with almost constant periods but finally shrank sharply to form contact binaries. In another approach, a comparison of young cluster and field distributions of G type binary star rotations was used to construct a period-evolution function for the braking law of orbits.

A common element in these researches is that spin-orbit coupling is set by synchronization. In consequence, the proposed mechanisms do not apply for asynchronous binaries. With this in mind, we may restrict attention to the OAM distribution of RS CVn systems with $P_{\mathrm{orb}} < 10$ d, for an observational estimate of the rate of orbital AML, and also deduce a braking law from the upper boundary of the OAM distribution.

The outcomes from this kind of test are not decisive, however. The upper period boundary implies that the study will not be representative for all systems. Mass loss and AML may have a complex dependence on age, given the role of magnetic activity. For a better understanding of the orbital period evolution of detached active binaries, systems with different ages should be considered closely. While the ages of well-observed binaries are estimated using detailed evolutionary isochrones, the ages of stellar groupings can sometimes also be estimated kinematically.

In a recent study, a sample of 237 systems were divided into two groups: possible moving group (MG) members (95), and older field binaries (142). A comparison of the total mass, orbital period, mass ratio and orbital eccentricity of these two groups showed clear observational evidence that chromospherically active binaries (including RS CVn systems) lose mass and angular momentum, so that their orbits shrink and circularize. Being independent of input from physical models, O – C diagrams have been frequently checked in the study of orbital properties of binaries.

At the present time, mean time-of-minimum deviations as small as 10 s are detectable for relevant binaries with well-defined eclipses. Orbital period decreases of individual systems, unfortunately, are not yet sufficiently well established from the commonly used O – C diagrams (Section 8.4) constructed from eclipse minimum timings. The following reasons can be given for this: (1) insufficient time-spans covered by existing O – C data (at most 100 y) in comparison with durations implied by the predicted kinematical ages (which are of the order of 10^9 y); (2) a large scatter of unevenly distributed time of minimum estimates (especially visual and photographic observations); (3) the co-existence of complicated, large amplitude, short time-scale fluctuations, caused by various effects both physical in the source and affecting the way the timings are measured.

It may be possible to determine rates of change of mass loss and AML statistically from the new and more accurate space-based data, taking into account better known absolute dimensions and kinematics. Such a more-informed approach will be advantageous in detecting changes appropriate to evolutionary effects where O – C diagrams, by themselves, have been insufficient hitherto.

AML scenario

The rotational or spin angular momentum J_s of a star with mass M, radius R, and rotation period P_s is

$$J_s = k^2 R^2 M P_s^{-1} \tag{10.2}$$

where k is the gyration constant, varying typically between ~ 0.07 and 0.15 in dependence on the density distribution inside the star.

By comparison, the orbital angular momentum J_o of an interacting binary system, with component masses $M_{1,2}$ and orbital period P_o, is given by

$$J_o = \left(\frac{G^2 P_o}{2\pi}\right)^{1/3} \frac{M_1 M_2}{(M_1 + M_2)^{1/3}} \tag{10.3}$$

where G is the universal gravitation constant.

Our basic premise is that a star's angular momentum is lost by stellar winds. However, for cool stars with convective envelopes ($M \leq 1.6 M_\odot$) this mass loss occurs not from the photosphere itself, but from the tops of magnetic loops, located at an *Alfven radius* R_A. The angular momentum loss for such a star may then be considered in terms of the mass loss at the Alfven radius or the spin period change P_s at the photosphere and we can write:

$$\frac{\mathrm{d}J}{\mathrm{d}t} = k^2 R_A^2 P_s^{-1} \frac{\mathrm{d}M}{\mathrm{d}t} = -k^2 R^2 M P_s^{-2} \frac{\mathrm{d}P_s}{\mathrm{d}t} \tag{10.4}$$

This will allow, for the magnetic braking of single cool stars,

$$\frac{\mathrm{d}P_s}{\mathrm{d}t} = -\left(\frac{R_A}{R}\right)^2 \frac{P_s}{M} \frac{\mathrm{d}M}{\mathrm{d}t}. \tag{10.5}$$

The key point here is that R_A is an order of magnitude larger than R, and so a fairly low rate of mass loss may spin down cool stars appreciably in a relatively short time. This argument has been supported by numerous observations of cool star rotation rates.

As we have seen, RS CVn type binaries are characterized by strong surface magnetic fields. These binaries are expected to undergo secular angular momentum loss from their component stars, primarily via a braking process like that just considered. This angular momentum loss can be fed from the reservoir of orbital angular momentum J_o through tidal interaction. The result is the spiralling in of the binary, that must eventually result in a contact system, unless some other process intervenes.

To follow through this scenario, a semi-empirical mass loss rate has been constructed using relevant observational data on the period changes of a selection of RS CVn binaries, thus:

$$\frac{\mathrm{d}M}{\mathrm{d}t} = 0.068\frac{M_1 M_2}{(M_1 + M_2)^{1/3}} P^{-2/3} \frac{\mathrm{d}P}{\mathrm{d}t} \tag{10.6}$$

The masses here are in solar units, with the mass loss rate in solar masses per year, while the orbital period is in days. A time-scale for the RS CVn systems to evolve to contact can be estimated by integrating this equation, thus:

$$t \approx \frac{0.204}{\dot{M}} \frac{M_1 M_2}{(M_1 + M_2)^{1/3}} (P_0 - P_t)^{1/3} \tag{10.7}$$

where P_t is the orbital period at a given age t yr after an initial epoch, when the period is P_0. Note that for $t > 0$, P_t is always smaller than P_0. From this model it is found that a few billion years would be required for a short period RS CVn system like RT And to evolve into a contact configuration. The period changes are small for the first billion or so years, but for a feasible range of initial periods (say 1 to 5 d) coalescence occurs after 2 to 4×10^9 yr.

We would thus generally expect the angular momentum evolution of RS CVn systems to be towards smaller orbits, and for this process to be effective, spin-orbit coupling ($P_s \approx P_{orb}$) is necessary. Relatively long period ($P_0 \geq 10$d) binaries that remain detached are not expected to evolve into shorter period systems over time scales comparable to the age of the Galaxy. On the other hand, with continued orbit shrinkage, the more massive component of a short period binary may reach the point where it fills its surrounding Roche lobe and starts to transfer mass to the other star. Once Roche lobe overflow starts, it would tend to take over as the predominant effect in subsequent angular momentum evolution. In the fast lead up to mass-ratio reversal, the orbital period would decrease, but thereafter it could slowly increase again, on the basis of total angular momentum considerations (Section 8.3.1).

The Applegate mechanism

In 1992, J. H. Applegate raised the possibility of a magnetism-related mechanism that could modulate the orbital periods of close binary systems. Since then, this mechanism has been considered in many contexts where orbital angular momentum changes are evidenced. Studies of the times of eclipse minima occasionally report orbital variations affecting the fifth digit in the binary period specification over time-scales of a few decades. This level of orbital parameter variation can be interpreted in terms of a concomitant internal angular momentum redistribution for the rotating components associated with a magnetic activity cycle. The implied scale of torque variation involves strong magnetic induction fields pervading the outer envelopes of the stars on the order of several kilogauss. This is feasible from some of the available evidence.

In Chapter 3, period changes were associated with apsidal motion or the light-time effect that occurs when a third body is present. Both of these effects should show strictly periodic modulations; however, observations often report period changes showing irregular variations. Some effects associated with putative third bodies point to companions massive enough to allow direct detection, but this is not always confirmed observationally. Similarly, effects attributed to apsidal motion require an eccentric orbit that other evidence does not support. Such being the case, the time scales, quasi-periodicities and supporting phenomena for certain cases of period changes, notably in RS CVn binaries, give credence to a mechanism that involves the interaction of magnetic torques and angular momentum changes.

Applegate showed, in a general way, that a relative variation of orbital period $\Delta P/P$ relates to the ratio of mean magnetic and the relevant gravitational energies E_μ/E_g, so that

$$\frac{\Delta P}{P} = \kappa R^3 B^2 / \left(\frac{GM^2}{R} \cdot \frac{a^2}{R^2} \right) \ , \tag{10.8}$$

where R is the active star's radius and M its mass, B is the mean induction field, a is the orbital radius, and κ is a numerical factor depending on how the interaction of the field and the stellar plasma is modelled. Two possibilities for this were considered: (a) a deformation requiring the magnetic field to contribute to hydrostatic equilibrium in the deformed configuration; in which case $\kappa \approx 6$. (b) a 'transitional' deformation associated with angular momentum exchanges, especially in the outer layers of the star. These layers are effective because of the r^2 factoring of the mass in the inertia tensor, as well as the increased centrifugal action. The angular momentum involved in this transfer builds up slowly during the magnetic cycle and κ becomes approximately the ratio of the period of the modulation cycle to ten times the orbital period, i.e. $\kappa \sim P_{\mathrm{mod}}/10P$.

Inserting relevant numbers then seems to require mean induction fields of prohibitive strength for case (a) (e.g. with $\Delta P/P \sim 10^{-5}$, Applegate estimated that $\bar{B}$ would have to be about 2×10^5 for V471 Tau). However, for the same rate of $\Delta P/P$, κ would increase by a factor of ~ 250 in case (b). For V471 Tau, the corresponding subsurface mean field decreases to the still high, but plausible, value of 11000 gauss.

Applegate also calculated the variations in luminosity that would be expected over the magnetic cycle, associated with the energy changes from the varying angular momentum. The calculation depends on an adopted value of the differential rotation, that is typically in the range 10^{-3} to 10^{-2} and can be checked against observational results. Typical luminosity variations turn out to be on the order of $0.1 \times L_\star$, where $L_\star$ is the mean luminosity of the affected star.

The model was found to be reasonably self-consistent with observational data for 3 of the 4 examples discussed by Applegate, but, for RS CVn itself, the required luminosity variations were higher than the mean value of that of the active component in the system. The model can be adjusted to fit the period modulation in RS CVn, though special assumptions about the energy dissipation mechanism are required and the match-up would still not be as good as for the other binaries. Of course, it should be kept in mind that the timing of minimum light in eclipsing binaries whose light curves are distorted by additional photometric variations in the system will be compromised in ways that may be difficult to calculate in detail, due to complications of the stellar surface light distribution.

10.2 Algol Systems

Most, if not all, the properties under consideration for the RS CVn binaries crossing the Hertzsprung gap in Fig 10.7 (including their magnetic activity) persist through to the classical Algol stage that follows. Added conditions then arise, associated with the originally more massive component reaching its surrounding envelope of limiting stability – its 'Roche Lobe', whereupon the binary attains the 'semi-detached' state.

It was recognized already in the first third of the 20th century that certain eclipsing binary light curves, showing relatively slight proximity effects, had such deep primary minima that they could not be explained as resulting from a pair of Main Sequence stars. In some cases, the eclipse was deep enough and of the appropriate shape to indicate the brighter star being totally eclipsed. But the absence of a significant ellipticity effect also showed this larger occulting 'dark star' to be of relatively low mass. Thus the famous Algol paradox could be deduced, with the benefit of hindsight, from an inspection of the light curve alone. This presents a blatant difference from what standard evolution theory would predict for single stars of corresponding masses.

This classical Algol configuration is associated with an Algol-type (EA) light curve — but not always. There are two types of eclipsing binary with roughly similar light curves: EAD – detached, or EAS – semi-detached (Section 7.2). The physical difference concerns whether there has been interactive evolution with mass loss and/or transfer between the components (Section 8.3). Of particular relevance is the semi-detached (sd) or near-sd condition, where one of the components is close to the structural stability limit (Section 4.2). It is these EAS type binaries that are identified as classical Algols. Apart from the early work of K. Walter and O. Struve, an important recognition of the significance of this sd property for Algols came from F. B. Wood. By the time of B. Paczyński's oft-cited review in 1971, most scientists concurred on the essential solution of Struve's paradox, though the general problem of binary evolution still supports detailed investigation.

So while the evolutionary status of the more familiar kinds of eclipsing binary are now generally agreed on, there are still gaps in knowledge; for example, concerning how many Algols there are and the particular conditions in which we find them. Comparable issues relate to how we systematize knowledge from the accumulation of data on Algol binaries. As with many classification schemes, a central group sharing similar attributes can be identified, together with peripheral candidates whose placing is uncertain. Thus, although many confirmed Algols do have an EAS type light curve, there are cases where photometry alone does not allow a clear decision and other evidence is required. If the orbital axes of close binaries are distributed randomly, about half of them would be found with an inclination between the line of sight and the orbital axis of less than 60°. At least half the sd binaries within the same distance range as the known cases are thus probably missed photometrically.

In order to establish a definite Algol model there should be two clear radial velocity curves to confirm the sd status. Mass ratios and absolute parameters began to be determined in this way for some bright Algols by the late 1970s. This had developed into detailed analysis of high dispersion spectrograms of bright sd binaries before the end of the 20th century, though the familiar single-line radial velocity variation, complicated by 'spurious eccentricity', the so-called 's-wave', or other effects, may remain for some time for fainter objects.

Observational work has extended the available wavelength range; for example into the infra-red, where the subgiant's contribution to the light curve becomes more noticeable, while Algols became targets of interest also in the ultraviolet range. High-energy mechanisms have been revealed through light curves from X-ray wavelengths. Radioastronomy too has seen remarkable advances in both sensitivity and resolution capability, offering fuller geometrical insights into the Algol configuration. Optical astrometry and polarimetry have important roles, as well, in uncovering the properties of Algol systems. In the case of the relatively near and bright prototype β Per itself, the whole picture has been better substantiated by a fuller combination of all such data types and methods of analysis.

10.2.1   Parameters of Interest

Analysis of eclipsing binary light curves yields radii of the component stars $r_{1,2}$ expressed in units of the mean stellar separation. Such radii, when combined with Kepler's third law, give:

$$3 \log R_{1,2} - \log(M_1 + M_2) = 2 \log P + 3 \log r_{1,2} + 1.871 \ , \qquad (10.9)$$

where P is the orbital period in days, $M_{1,2}$ are the masses and $R_{1,2}$ the radii in solar units. The quantities on the right in Eqn. 10.9 are known – the aim for physical studies, is to evaluate the various quantities on the left and various methods of doing this have been used.

The most direct involves two radial velocity curves. If there is only one such curve (that of the primary), then it is commonly assumed that this star is close enough to the Main Sequence to permit standard correlations (Table 9.2) to be used. Applying such a correlation, e.g. $R_1 = M_1^m$, would allow M_2 to be derived from (10.9) and then R_2 from a second use of the equation. Since the ratio R_2/R_1 is known from the photometry, the initially adopted MS character of the primary may be checked and parameters adjusted as appropriate.

Another route, not making MS presuppositions, requires a presumption of the semi-detached configuration, thereby allowing the mass ratio ($q = M_2/M_1$) to be derived from the Roche model, i.e. $q = q(r_2)$ (Table 4.2). The appropriate formula is:

$$\log M_1 = 2 \log(1 + q) - 3 \log q + \log f - 3 \log(\sin i), \qquad (10.10)$$

where f is the mass function obtained from the single radial velocity curve (Eqn 6.17) and i is the inclination. It can be seen by differentiating this equation, that the derived mass will have appreciable errors for relatively small errors of the small quantities q and the radial velocity amplitude K that is used to determine the mass function f. Until recently, these were difficult to estimate with a precision better than several percent.

In fact, double lined sd binaries do indicate that their primaries conform reasonably well to the same mass:luminosity relation as those of reliably determined detached binaries, so a presumption of that relationship should allow the derivation of absolute parameters more confidently than using (10.10) alone. Parameters are more robustly defined, though, by following, where possible, the direct implications of two radial velocity curves, that increased sensitivity and precision of observational facilities are permitting.

10.2.2   Observed Properties

Sky distribution

Fig 10.9 (upper left) shows the distribution on the sky, with declinations vertical against right ascensions horizontal, of 78 well-known Algols that conform to the standard sd config-uration and related properties. The galactic equator is marked as a continuous curve on this projection. Although some clustering towards the galactic equator is apparent, our view is impaired by the incomplete sampling of the southern hemisphere that contains only $\sim 1/4$ of the well-known Algols. The situation is improved by the inclusion of a further 172 sys-tems with EAS light curves (Fig 10.9 upper right), but about which generally less is known, though now almost half are in the southern hemisphere. Fig 10.9 (lower) shows these 250 highly probable Algols distributed in galactic co-ordinates.

The distribution in galactic latitude may be compared with that of field stars in a comparable magnitude range in Fig 10.10. It is seen that the Algols show a noticeably flattened distribution of stars in the Galaxy than the field stars. This is in keeping with reasonable expectations about the masses of Algols' component stars, and the consequences concerning their ages and evolutionary status. Such comparisons have led to an estimate that of the order of 1 star in 1000 is an Algol type binary.

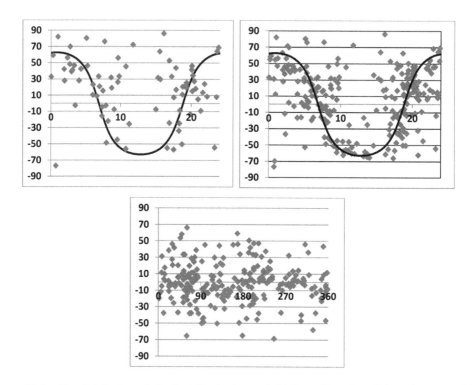

Figure 10.9   Sky distribution of classical Algols: (upper left) 78 confirmed candidates in equatorial co-ordinates; (upper right) a further 172 systems most likely to be Algols added to the set arranged in the same way; and (lower) these 250 Algols distributed in galactic coordinates.

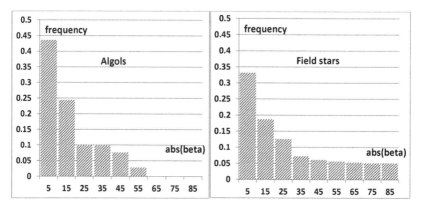

Figure 10.10   Distribution of classical Algols. Histogram of the distribution of 250 Algols in galactic latitude (left) and field stars (right).

Spectral type versus period

The distribution of classical Algols with regard to orbital period and primary spectral type is indicated in Fig 10.11. The maximum of the distribution occurs at primary spectral type A0, and period around 2.5 d. At rather short periods (around 1 d) the proportion of late A or F-type primaries increases). The numbers of Algol systems with primary type earlier than A0 may increase somewhat at periods beyond 2-3 days. In any case, there is a swift decline with earlier spectral type, so that few primaries of the classical Algols are known with type earlier than B5.

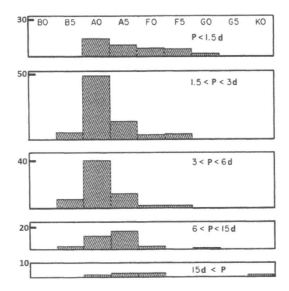

Figure 10.11   Distribution of 250 classical Algols with period and primary spectral type arranged as 2-D sections.

At longer periods, say $\gtrsim 10$ d, the proportion of later type primaries does not decline: it may increase to some extent. The appearance of a few late type primaries at long periods reflects the inclusion of stars like RZ Cnc; systems believed to be essentially similar to classical Algols in their basic properties, except that their presently more massive components already left the Main Sequence and moved towards to the Giant Branch. This grouping would probably include some RS CVn type binaries as well as classical Algols.

Of course, Fig 10.11 refers only to the more directly observed primary spectrum. The spectral type of the secondary, for these photometric Algols, could be derived indirectly from the colour variation during eclipse. From this evidence it appears that there is some correlation between the components, in that the luminosities of secondaries are typically about 1 order of magnitude lower than those of the primary. For the hottest examples, $T_{\mathrm{eff},1} \sim 20000$ K, the luminosity ratio rises to $\sim 1/2$.

Distribution of primary mass

Well-known classical Algols have primary masses that are known with reasonable confidence; their distribution reflects what could be expected from the foregoing spectral type distribution, if the primaries are MS-like stars. The frequencies in uniform intervals of primary mass, and log mass, for the 250 Algols, are shown in Figure 10.12. The maximum occurs at about 2.5 $M_\odot$, in fair agreement with the A0V type in the Main Sequence trend. The mean mass of the sample is 3.3 $M_\odot$; the corresponding MS radius is about 2.5 $R_\odot$. No primary mass has been listed that is less than 1.2 $M_\odot$. This is in keeping with the general tenets of the Roche Lobe overflow theory of Algol evolution, if most of the system's mass is conserved.*

One well-known Algol system with a low overall mass is the relatively bright R CMa, which recent analysis puts at close to $1.9 M_\odot$. About 10% of the 250 likely Algols have a

---

*Potential donors of lower mass than this, at least among local (Population I) stars, would not have had time to evolve off the Main Sequence.

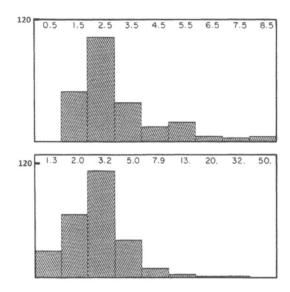

Figure 10.12 Distribution of classical Algol masses: upper, linear with mass in solar units; lower, masses are given in solar units on the horizontal scale but the spacing is in uniform intervals of $\log M$.

total mass less than ~ 2 M$_\odot$. Some of these, like Z Dra and RT Per, appear to be in short period, low mass-ratio configurations like R CMa, and may have odd evolutionary histories.

At the larger mass end, the dispersion looks more protracted. The B3 type primary in RZ Sct must be a massive star: mass values of around 15 M$_\odot$ appearing in the literature. If the giant system BL Tel is essentially similar to an evolved Algol, then a primary mass of around 20 M$_\odot$ could be reasonably expected.

This raises the issue of defining the boundary wherein Algol type binaries are found. Binaries that are recognized as Algols from their EAS light curves have cool secondaries corresponding to the Case B type evolutionary model. Systems larger than a certain mass might evolve in a somewhat different way. For example, for the precursor systems having secondary MS components of type earlier than B5, the role of radiation pressure in the Roche Lobe overflow mechanism may produce some significantly different outcome, such as a Wolf-Rayet star. The initial total mass of such a binary would have to be greater than about 12 M$_\odot$. The existence of RZ Sct would seem to argue against this, but then RZ Sct looks exceptional among the most massive classical Algols. Such systems would spend a proportionally much smaller time in the Algol configuration, but, countering that selection effect, they would more easily be seen out to a greater distance.

In a recent review of semi-detached binaries by O. Malkov, where the potentially important role of selection effects was noted, about 20% showed less of a temperature contrast than the classical Algol type, particularly with higher temperature primaries. These early type systems tend to show a sizeable ellipticity effect in their light curves, which relates to their short periods. V Pup and μ_1 Sco are typical examples.

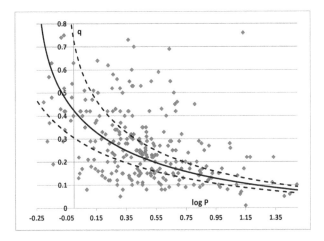

Figure 10.13 Classical Algols: mass ratio versus (log) orbital period (in days). The suggested trends are discussed in the text.

Earlier classifications may have been affected by the difficulty of getting reliable radial velocity curves, but more recent studies have confirmed that the secondaries are hot and essentially filling a surface of limiting stability. But these systems would not be recognized as EAS binaries from their light curves alone. From a classification standpoint, these binaries are regarded as following the 'Case A' model of interactive evolution, unlike the majority of the hot primary, cool secondary classical Algols that evolve according to the 'Case B' paradigm (Section 8.3.1). A complete theory of the Algols should be able to explain, not just the main features of a typical configuration, but also the form of such distributions in mass, spectral type and period as have been presented.

Mass ratio versus period

Potentially significant implications on the mass loss/transfer behaviour of Algols come from the mass ratio q (donor/gainer) versus period P distribution. The general trend of low q at long period is easily noticed in Fig 10.13, and it fits in with the broad expectation of mass being transferred from the secondary to the primary during the course of the Algol condition. The curves in this diagram correspond to the relationship between period and mass ratio when the orbital angular momentum is conserved and the much smaller rotational momentum neglected. The central (continuous) curve goes with a pair of stars of total mass $5M_\odot$ that would just come into contact when their masses were equal with period 0.587 d. We could expect evolution with mass transfer to trend the points in the q - $\log P$ plane downward along a path parallel to this curve. The (dashed) curves on either side are shifted by 0.2 in $\log P$, corresponding to a near 50% range of masses for the same minimum separation of components. The observed distribution is, however, quite different, with the majority of Algols having longer periods than would go with a common envelope minimum separation at q near unity. Also, there are a relatively large number of Algols with low q values, a fair proportion of which have unexpectedly short periods.

A mass transfer process in which a small fraction of the overall mass was lost from the system, but carrying a relatively high proportion of the system's angular momentum, has been found from more detailed studies of the P-q distribution. But there remains in Fig 10-14 the small subgroup of Algols similar to R CMa, whose location in the $P - q$ plane is difficult to explain by the basic mass transfer process, even with moderate angular momentum loss. For such cases, accounting for $\sim$10% of the well-documented Algols, something other than

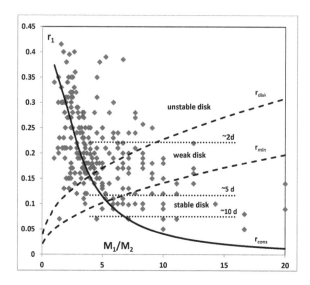

Figure 10.14 Classical Algols: primary relative radius r_1 (derived photometrically) versus mass ratio Q, here in the sense gainer/loser (as used by Lubow & Shu, 1975). The dashed curves r_{disk}, r_{min} correspond to inner and outer limits of a stable accretion structure as calculated by Lubow & Shu. The dotted lines give an indication of the period in days for a system of 3.5 $M_\odot$ evolving conservatively. The continuous curve r_{1cons} follows the trend of a conservatively evolving Algol system whose components are just in contact at $Q = 1$.

the standard picture must apply: perhaps 'third bodies' play a part, or there is some more drastic AML mechanism that might be associated with a common envelope phase in Algol evolution.

Primary radius versus mass ratio

The connection between primary relative radius r_1 and mass ratio has provided a setting for comparative studies of Algol properties since the work of O. Struve, as well as in discussions of particular systems. The distribution of 261 likely Algols from the selection used in Fig 10.10 (with a few additions from the catalogue of M. A. Svechnikov and E. F. Kuznetsova; 1990) in the mass-ratio:r_1 plane is shown in Fig 10.14. The mass ratio is here taken in the sense gainer/donor, Q say, for comparison with data from Lubow & Shu (1975).

Apart from the observed values of Q and r_1, the diagram shows the accretion disk radii r_{disk} and r_{min} values that come from the calculations of Lubow & Shu, referred to in Section 8.3. The outer radius is where quasi-circular particle orbits in the restricted three body problem attain long term stability. Real disks of low density may well reflect this property in their structure. The expansion of a disk beyond this radius entails a likely structural disruption. The inner radius r_{min} is a similar result from modelling. An infalling particle moving with the calculated velocity and direction would not, in the absence of other interactive material, approach the attracting mass centre to within less than this radius.

These radii guide expectations on the accretion disk around the gainer. Algols in Fig 10.14 above the r_{disk} curve have gainer radii bigger than the circular orbit stability limit. They should then not form large and stable accretion disks: there is simply not enough space to get the structure started. On the other hand, points towards the bottom right corner of the diagram have plenty of space about the primary relative to the orbit dimensions in which to accumulate infalling material. Stars lying between the two curves may exhibit intermediate properties regarding accretion disks. From this we would deduce that a majority of observed Algols are unlikely to have formed substantial accretion structures, but a sizeable minority do allow some level of disk accumulation.

The horizontal dotted lines in Fig 10.14 indicate typical periods associated with their corresponding r_1 values, where the primary is a Main Sequence star close to the average mass of the sample (3 $M_\odot$), accompanied by a star of typical secondary mass for a classical Algol (0.5$M_\odot$). Individual systems would not conform to these indications exactly, of course, but the lines suggest that Algols having reasonably stable disk-like structures would be more expected among those with periods greater than ∼5 d.

The curve labelled $r_{1\mathrm{cons}}$ represents the track taken by a primary of constant size in an Algol system, which is conserving its total mass and orbital angular momentum. In this case, its relative radius r_1, neglecting rotational effects, would vary like

$$r_1 = 16r_{1,0}(Q^2/(1+Q)^4) \tag{10.11}$$

where $r_{1,0}$ is its maximum relative radius (∼0.38 in a 'contact' configuration with equal mass components). As a whole, the Algols seem to trend roughly parallel with this track, giving a measure of general support to mass transfer in Algol evolution under conditions not far from AM-conservative. It is notable, however, that with longer periods and lower r_1 values, the Algols trend above the curve — pointing to angular momentum loss during their evolution. Of course, expansion of the primary star would play a part in this trend. But we note, as well, the existence of some systems with both largish r_1 values ($\gtrsim$0.15, say) and large Q values ($\gtrsim$5), supporting a point made before about the $P : q$ diagram. The condition of these stars that have evolved to high $M_{\mathrm{gainer}}/M_{\mathrm{donor}}$ ratios but stayed relatively close is thought-provoking.

10.2.3   Particular Cases

This discussion of the Algols is concluded by giving attention to a pair of systems that exemplify issues arising in studies of individual cases. Loss of parameter resolution, i.e. increase in the probable errors of specification as we increase the number of fitting-function parameters applied to a given data-set has been mentioned. But real physical processes may also be at work that are outside the scope of the modelling discussed hitherto.

R Arae

Variability of the bright component in the visual binary that makes up the R Arae multiple system[*] was first reported by A. W. Roberts in 1894. E. Hertzsprung provided a light curve from photographic observations, but he did not suspect any significant variation of the orbital period over the preceding 50 years. However, spectroscopic observations of R Arae show abnormal radial velocities, peculiar Balmer emission features and other irregularities, that could be associated with relatively strong interactive evolution effects in an Algol binary. This interpretation was borne out in UV and X-ray data. Photometric data on the binary's sd status is ambiguous, though, since the inclusion of third light as an unknown makes a difference to the secondary radius value required to determine the depth of minima.

---

[*]John Herschel noted, in 1833, that R Arae appears as a close visual pair with a separation of about 4 arcsec (∼500 AU perpendicular to the line of sight), the companion contributing about 1/3 to the combined light.

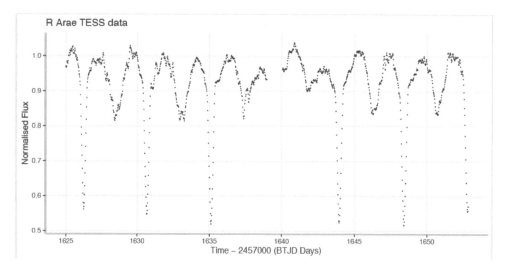

Figure 10.15   Light curves of the peculiar Algol R Arae, recorded by the TESS satellite in 2020, courtesy of the Barbara A. Mikulski Archive for Space Telescopes (MAST). Funding for the TESS mission is provided by the NASA's Science Mission Directorate. The data is from Sector 12. The observations have a 30 minute cadence. The central gap is associated with data communications. The diagram was prepared by T. Banks using Lightkurve, a Python package for Kepler and TESS data analysis (Lightkurve collaboration, NASA, 2018). The clear ongoing irregularities visible in the successive light curves support previous ground-based findings, for example, by D. Blane, M. G. Blackford, E. Budding & P. A. Reed in *Inf. Bull. Variable Stars*, No. 6267, 2019.

The situation is complicated by the inherent photometric irregularities of the binary, which show up in the light curves as ill-defined fluctuations, of order 0.1 mag (Fig 10.15). These irregularities may be related to the other indications of the system's active mass transfer. While these variations are real, from the point of view of light-curve modelling, they can be regarded as large amplitude scatter that would be reflected in correspondingly large parametrization errors. A more detailed model might resolve additional parameters that could remove some of the scatter in the original data, but this should require additional supporting evidence.

Models for cyclic phenomena in R Arae have been advanced that account for at least some of the additional light curve meanderings (Figs 10.15,17), and it may well be that certain period and mass-ratio combinations are propitious in allowing the development of relatively large and unsteady accretion structures. Such considerations are supported by the relatively large rate of orbital period increase that has been observed for R Arae.

V Puppis

A. W. Roberts also provided an initial light curve and analysis of the bright close binary V Pup. The early picture of this B-type binary, such as presented by H. Shapley, was of a pair of similarly sized, massive stars, strongly distorted by tides in a close-to-contact orbit with period about 1.45 days. The binary is about 2.5° away from the complex multiple star γ^2 Vel, and even closer to the open cluster NGC 2547. It appears a likely member of the Vela OB2 association, a circumstance that should constrain modelling.

The light curve is classified as of EB type, the similarity of eclipse depths pointing to similar surface temperatures; so superficially quite different from what one would expect of a classical Algol. The essential Algol paradox about V Pup, however, emerges when we consider the radial velocity curves (Fig 10.18).

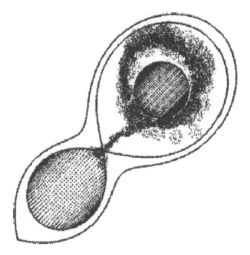

Figure 10.16 Relatively massive and asymmetric accretion structure accounting for light variations in R Arae. The illustration is taken from P. Reed *Proc. IAU Symp.*, Vol. 282, eds. M. T. Richards & I. Hubeny, p325, 2012.; published by Cambridge University Press.

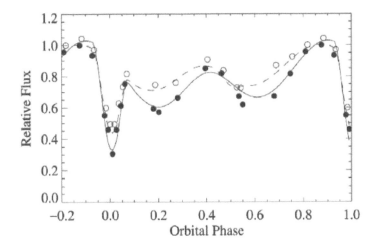

Figure 10.17 Photometric effects of the accretion structure in R Arae From P. Reed — in the same paper as Fig 10.16.

It then becomes clear that, although we have two stars of similar size and luminosity, one of them is only about half the mass of the other. If we ascribe to the slightly larger of the two stars properties similar to a MS star of appropriate size and temperature for its B1V spectral type, the secondary becomes 'paradoxical'. The configuration seems odd, in that the relative separation of the stars is still not much greater than that of a contact binary, yet the donor's mass has dropped significantly since it was the same as that of its partner. The separation should have almost doubled if systemic angular momentum had been approximately conserved during a mass transfer process. Various papers have discussed the physical condition of this binary offering interesting suggestions for further research; for example, with the possibility of it hosting a Black Hole as a wide companion.

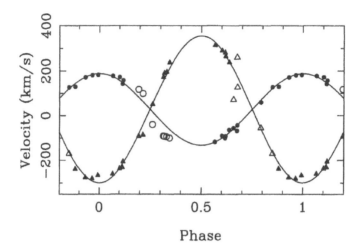

Figure 10.18 Radial velocity curves of V Puppis. The diagram appeared in D. J. Stickland, C. Lloyd, I. Pachoulakis, & R. H. Koch, *The Observatory*, Vol. 118, p356, 1998.

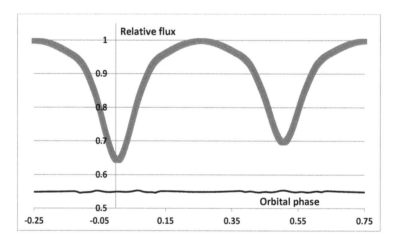

Figure 10.19 WINFITTER model fit to the complete TESS photometric data-set for V Pup, binned by a factor of 10 down to 1513 individual values. The scatter of individual readings is of order 0.0002, but systematic departures of up to an order of magnitude greater than that can be seen in the residuals in the regions of the eclipses when plotted about the mean flux level 0.55 in this diagram. This is mostly accounted for by the fitting function's quadratic approximation being inadequate to represent the stars' photospheric limb-darkening effect with sufficient accuracy.

The stable EB type light curve and well defined eclipses are clear in high precision photometry from the TESS satellite from early 2019 (shown in Fig 10.19). A Case A type Algol model could well be invoked to match these data. However, a closer examination reveals that this is not as straightforward as it may first appear.

At high magnification, short-term low-level fluctuations are seen in the data. These variations can be almost removed by combining and averaging all 14 individual light curves into one complete cycle. The residuals from a standard model are shown in Fig 10.19. Whilst evidently small, there remain small systematic effects, of amplitude ~ 0.002. Such variations over the phase ranges of the eclipses can be reasonably associated with shortcomings of

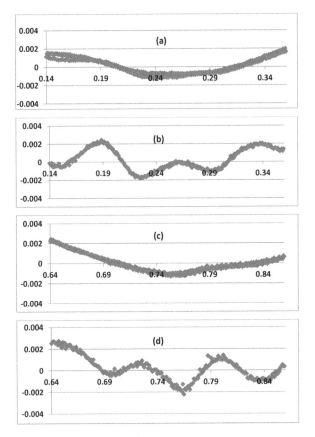

Figure 10.20 Panels (a) and (c) show the light-level of the out-of-eclipse phases resulting from the binning of the 14 individual TESS light curves of V Pup. This has mostly smoothed out the intrinsics β Cep variation visible in individual light curves, shown in panels (b) and (d), although traces of a low-amplitude trend can be seen in the binned data.

the employed quadratic approximation for the limb-darkening of the stars to match the real centre-to-limb light variation. Notwithstanding the eclipse regions, residuals from the fittings of the standard model over the out-of-eclipse phases are reasonably flat.

When we look at the fittings to *individual* light curves, however, we observe a low-level, but definite, intrinsic variation, with a mean period of around 0.075 that of the system, i.e. ≈ 2.6 h (Fig 10.20). This result points to the massive B-type primary being an incipient β Cep variable, with the rather low amplitude of ~ 2 mmag. Although this complicates a standard modelling procedure, the finding of such effects in individual cases attracts attention and fosters deeper insights into the nature of stars that show both intrinsic and interactive peculiarities.

10.3 Bibliography

10.1 The 'undersize Algols' discussed in Kopal's (1959) book had been previously introduced in his paper in *Ann. d'Astrophys.*, Vol. 19, p298, 1956. This group of stars significantly overlaps with the grouping pointed out by D. S. Hall in the *Proc. IAU Colloq. 29; Multiple Periodic Variable Stars*, ed. W. Fitch, Reidel, p287, 1975, where indications of strong chromospheric and related activity provided a key characteristic of these 'RS CVn systems'.

That this connection should be present for the rapidly rotating cool stars in the undersize Algols follows from the physical review of their properties given by D. M. Popper & R. K. Ulrich in *Astrophys. J.*, Vol. 212, L131, 1977.

G. E. Kron's pioneer paper discussing starspots on AR Lac was in the *Publ. Astron. Soc. Pacific*, Vol. 59, p261, 1947. In the proceedings of *IAU Symp. 176: Stellar Surface Structure*, eds. K. G. Strassmeier & J. Linsky, Kluwer, 1996, summaries can be found of the numerous developments in this field over the decades that followed Hall's introduction of the RS CVns and related stars. A review was given by S. V. Berdyugina in *Living Rev. Solar Phys.*, Vol. 2, p8, 2005.

An early qualitative presentation of what has become known as the $\alpha - \Omega$ model for the solar dynamo was that of H. W. Babcock in *Astrophys. J.*, Vol. 133, p572, 1961. For a recent review see P. Charbonneau, *Living Rev. Sol. Phys.*, Vol. 17, p4, 2020; (https://doi.org/10.1007/s41116-020-00025-6).

The light curves of RS CVn shown in Fig 10.1 are from the paper of C. Blanco, S. Catalano, E. Marilli & M. Rodonò in *Proc. IAU Colloq. 71: Activity in Red Dwarf Stars*, eds. P. B. Byrne & M. Rodonò, ©Reidel, p 387, 1983. The two light curves for 1975 and 1976 are from the doctoral thesis of E. W. Ludington (1978), University of Florida at Gainesville, FA, USA. This unique data-set demonstrates not only a distinct maculation effect, but also, it is argued, differential rotation in the source photosphere related to the difference in periods of the wave and the binary rotation. The photometric wave distortion in RS CVn was first discussed by S. Catalano & M. Rodonò in *Mem. Soc. Astron. Ital.*, Vol. 38, p345, 1967. There have been very many papers presenting light curves affected by surface maculation since the mid-twentieth century. *The Sun in time* project was referred to in this connection by J. D. Dorren & E. F. Guinan, *Astrophys. J.*, Vol. 428, p805, 1994.

The photometry shown in Fig 10.2 is discussed in some detail by C. J. Butler et al. (11 authors) in *Mon. Not. Royal Astron. Soc.*, Vol. 446, p4205, 2015. This was part of a multiwavelength observational campaign on YY Gem. A comparable study of II Peg was carried out by P. B. Byrne et al. (17 authors) in *Astron. Astrophys. Suppl.*, Vol. 127, p505, 1998; see also T. Hackman, M. J. Mantere, M. Lindborg, I. O. Ilyin, O. Kochukhov, N. Piskunov, I. Tuominen, *Astron., Astrophys.*, Vol. 538, A126, 2012. For further substantiation of the spot model in the characterization of active star photometry see also G. W. Henry, J. A. Eaton, J. Hamer & D. S. Hall, *Astrophys. J. Suppl. Ser.*, Vol. 97, p513, 1995. Complementary approaches to stellar astrophysics via data on double and multiple stars were brought together in *IAU Coll. 135* (Astron. Soc. Pacific Conf. Ser.), Vol. 32, eds. H. A. McAlister & W. Hartkopf, 1992.

Fig 10.3 is discussed in the review of K. G. Strassmeier in *Astron. Astrophys. Rev.*, Vol. 17, p251, 2009. Other reviews concentrating on ZDI include those of O. Kochukhov, in *Cartography of the Sun and the Stars. Lecture Notes in Physics*, Vol. 914, p177, eds. J-P. Rozelot & C. Neiner, Springer, 2016; M. Semel, *Astron. Astrophys.*, Vol. 225, p456, 1989; and T. A. Carroll, M. Kopf, I. Ilyin & K. G. Strassmeier in *Astron. Nach.*, Vol. 999, p789, 2007.

Fig 10.4 is from the paper discussing the evolutionary condition of the RS CVn stars of D. M. Popper and R. K. Ulrich, cited above. The data used to prepare Fig 10.5 and similar ones comes from J-F. Donati, A. Collier Cameron, G. A. J. Hussain and M. Semel in *Mon. Not. Royal Astron. Soc.*, Vol. 302, p437, 1999. General reviews of Doppler Imaging (DI – the predecessor to ZDI) were given by e.g. V. L. Khokhlova, *Sov. Astron.*, Vol. 19, p576, 1976; S. S. Vogt & G. D. Penrod, *Publ. Astron. Soc. Pacific*, Vol. 95, p565, 1983; J. B. Rice, W. H. Wehlau & V. L. Khokhlova, *Astron. Astrophys.*, Vol. 208, p179, 1989; and A. Collier Cameron, *Lecture Notes Phys.*, Vol. 397, eds. P. B. Byrne & D. J. Mullan, p33, Springer-Verlag, 1992; and others.

Fig. 10.6 derives from data presented by J. L. Innis et al., *Inf. Bull. Variable Stars*, No. 5832, 2008. Comparable figures are shown by A. F. Lanza et al. (8 authors), *Astron.*

Astrophys., Vol. 386, p583, 2002; and Ö. Çakırlı et al. (6 authors), *Astron. Astrophys.*, Vol. 405, p733, 2003; for RT Lac, see also G. Cutispoto & M. Rodonò, *Lecture Notes Phys.*, Vol. 397, eds. P. B. Byrne & D. J. Mullan, p267, Springer-Verlag, 1992.

σ-integrals on spherical surfaces were introduced by E. Budding in *Astrophys. Space Sci.*, Vol. 48, p207, 1977. A critical discussion of their application to the analysis of relevant photometric data was given by Zs. Kővári & J. Bartus in *Astron. Astrophys.*, Vol. 323, p801, 1997. The application to the system HK Lac was presented by K Oláh et al. (6 authors) in *Astron. Astrophys.*, Vol. 321, p811, 1997. A more recent application of the technique to stars showing quasi-sinusoidal light variation is that of A. Erdem et al. (13 authors), in *New Astronomy*, Vol. 14, p109, 2009. The comparable appearance of spots on the giant component of the RS CVn binary XX Tri together with their time evolution, as deduced from Doppler Imaging analysis, was presented by A. Künstler, T. A. Carroll & K. G. Strassmeier in *Astron. Astrophys.*, Vol. 578, A101, 2015. A paper showing the complementarity of photometric starspot analysis and Doppler Imaging was that of E. Budding and M. Zeilik in *Astrophys. Space Sci.*, Vol. 222, p181, 1994. See also the paper of Zs. Kővári et al. (6 authors) in *Astron. Astrophys.*, Vol. 373, p199, 2001; which shows complementarity of photometric and spectroscopic studies applied to σ Gem. The existence of large coherent spots on stellar surfaces was directly confirmed for ζ And by R. M. Roettenbacher et al. (15 authors) in *Nature*, Vol. 533, p217, 2016; using long baseline IR-interferometry.

A classic general review of activity cycles in RS CVn stars is that of S. L. Baliunas & A. H. Vaughan in *Ann. Rev. Astron. Astrophys.*, Vol. 23, p379, 1985; see also S. V. Berdyugina & I. Tuominen, *Astron. Astrophys.*, Vol. 336, L25, 1998 and A. F. Lanza, N. Piluso, M. Rodonõ, S. Messina & G. Cutispoto, *Astron. Astrophys.*, Vol. 455, p595, 2006.

The use of the Rossby number in the context of magnetic activity cycles was discussed by K. Stępień in *Astron. Astrophys.*, Vol. 292, p191, 1994; while differential rotation properties of RS CVn stars were reviewed in the paper of G. W. Henry et al. cited above. N. O. Weiss & S. M. Tobias, in *Space Sci. Rev.*, Vol. 94, p99, 2000, reviewed the applications of dynamo theory to explaining stellar activity phenomena. Data on the connection between a cool star's convective turn-over time and its colour were provided by R. W. Noyes, S. W. Hartmann, S. Baliunas, D. K. Duncan & A. Vaughan, *Astrophys. J,*, Vol. 279, p763, 1984.

Multiwavelength studies of chromospherically active stars have been engaged in by, for example, O. Vilhu et al. (8 authors), *Astron. Astrophys.*, Vol. 278, p467, 1993; R. Dempsey et al. (14 authors), *Astron. Soc. Pacific*, Vol. 154, 1402, 1998; E. Budding, K. L. Jones, O. B. Slee & L. Watson, *Mon. Not. Royal Astron., Soc.*, Vol. 305, p966, 1999; J. C. Pandey, K. P. Singh, R. Sagar & S. A. Drake, *ASP Conf. Ser.*, Vol. 362, eds. Y. W. Kang, H.-W. Lee, K.-C. Leung, & K.-S. Cheng, p74, 2007; and others.

Angular momentum loss (AML) in RS CVn systems was discussed by C. Maceroni in *Türk. J. Phys.*, Vol. 23, p289, 1999. The role of magnetic braking in stellar evolution had been noted by E. Schatzman, in *IAU Symp. 10*, ed. J. L. Greenstein, p129, 1959. Also R. P. Kraft, in *Astrophys. J.*, Vol. 150, p551, 1967; and L. Mestel, in *Mon. Not. Royal Astron. Soc.*, Vol. 138, p359 and Vol. 140, p177, 1968; made influential contributions to the subject. The 'Skumanich law' on rotational braking refers to the work of A. P. Skumanich in *Astrophys. J.*, Vol. 171, p565, 1972. D. R. Soderblom reviewed the subject in *Proc. IAU Symp. 102*, p439, Reidel Dordrecht, 1983; as did O. Vilhu & S. M. Rucinski in *Proc. 71st IAU Coll.*, Dordrecht Reidel, p475, 1983; C. Maceroni & F. van't Veer in *Astron. Astrophys.* Vol. 246, p91, 1991; O. Demircan in *Türk. J. Phys.*, Vol. 23, p425, 1999; and others. The coalescence scenario of Section 10.1.3.1 was discussed by E. F. Guinan and D. H. Bradstreet in *Formation and Evolution of Low Mass Stars; NATO ASI Ser. C*, Vol. 241, eds. A. K. Dupree & M. T. V. T. Lago, p345, 1988.

The 'Applegate mechanism' was presented by J. H. Applegate in the *Astrophys. J.*, Vol. 385, p621, 1992. This was a development of a model from J. J. Matese & D. P. Whitmire in *Astron. Astrophys.*, Vol. 117, L7, 1983. It was elaborated on by A. F. Lanza, M. Rodonò & R. Rosner in *Mon. Not. Royal Astron. Soc.*, Vol. 296, p893, 1998; and further by M. Völschow, D. R. G. Schleicher, V. Perdelwitz & R. Banerjee in *Astron. Astrophys.*, Vol. 587, p34, 2016; who constructed an Internet-based resource to allow on-screen testing of the possible relevance of the mechanism for systems whose details can be sufficiently well specified numerically.

10.2 An early recognition of the 'Algol paradox', raised in Section 10.2, was that of O. Struve in *Ann. d'Astrophys.*, Vol. 11, p117, 1948. But the semi-detached status of classical Algols had already been pointed out by K. Walter in the *Königsberg Veröff.*, No.2, 1931. For early calculations of the Roche model and appreciation of its relevance see also G. P. Kuiper in *Astrophys. J.*, Vol. 93, p133, 1941. Evidence such as that presented by A. H. Joy, *Publ. Astron. Soc. Pacific*, Vol. 54, p35, 1942; and *Publ. Astron. Soc. Pacific*, Vol. 59, p171, 1947; regarding the source of emission lines visible during primary eclipses of the classical Algol RW Tau was influential in the development of ideas about RLOF. This 'Struve revolution' was recounted in Chapter 1 of J. Sahade & F. B. Wood's well documented book *Interacting Binary Stars*, Pergamon Press, 1978. Its observational basis and early history of theoretical studies of the Algol process are also detailed in the later chapters of A. H. Batten's *Binary and Multiple Stars*, Pergamon Press, 1973; and Batten's later review in *Publ. Astron. Soc. Pacific*, Vol. 100, p160, 1988. Relevant papers are also in the proceedings of *Nonstationary Evolution of Close Binaries*, ed. A. N. Żytkow PWN, Warsaw, 1978. The discussion was updated in G. E. McCluskey Jr's paper in *The Realm of Interacting Binary Stars*, eds. J. Sahade, G. E. McCluskey & Y. Kondo; *Astrophys. Space Sci. Library*, Vol. 177, p3, 1992; and summarized in R. W. Hilditch's *An Introduction to Close Binary Stars*, CUP, 2001, (Chapter 6.3).

Accurate parametrization of Algol systems was required to confirm this revolution, and an important indication of this was the work of D. M. Popper on AS Eri in *Astrophys. J.*, Vol. 185, p265, 1973. This was found to show encouraging agreement with theoretical modelling by S. Refsdal, M. L. Roth & A. Weigert *Astron. Astrophys.*, Vol. 36, p113, 1974. The topic is reviewed by M. Plavec and R. S. Polidan in *Structure and Evolution of Close Binary Systems: Proc. IAU Symp. 73*, eds. P. Eggleton, S. Mitton, & J. Whelan, Reidel, p289, 1976.

Much of Section 10.2 is drawn from E. Budding's catalogue *Bull. d'Inf. Centre de Données Stellaires*, Vol. 27, p91, 1984; and related discussion in *Astrophys. Space Sci.*, Vol. 99, p299, 1984; *Pub. Astron. Soc. Pacific*, Vol. 97, p584, 1985; *Proc. 3rd IAU Asian-Pacific Regional Meeting*, eds. M. Kitamura & E. Budding, *Astrophys. Space Sci.*, Vol. 118, p241, 1985; and *Space Sci. Rev.*, Vol. 50, p205, 1989. Fig 10.11 comes from E. Budding's article in *Investigating The Universe*, ed. F. D. Kahn, Reidel, p251, 1981. It may be compared to Fig 2 in the paper of Plavec & Polidan (op. cit., 1976).

Statistical properties of Algols have also been reviewed by R. Stothers, *Astrophys. J.*, Vol. 185, p915, 1973; D. S. Hall, who concentrated on the distinction between Kopal's 'undersize subgiants' and classical Algols in *Acta Astron.*, Vol. 24, p215, 1974. Some Algols were included in the catalogue of well-defined systems of C. H. Lacy in *Astrophys. J.*, Vol. 228, p817, 1979. That work was included in the comprehensive review of D. M. Popper in *Ann. Rev. Astron. Astrophys.*, Vol. 18, p115, 1980; where the general class of Algols was subdivided into a hot subgroup of systems and a subgroup of giant binaries with a semi-detached component as well as the the classical Algols having EAS light curves. While Algol primaries are often regarded as having properties essentially similar to Main Sequence stars of the same mass, matter transferred from the loser should transfer also angular momentum. J. Mukherjee, G. J. Peters, & R. E. Wilson, *Mon. Not. Royal Astron. Soc.*, Vol. 283, p613,

1996, reported this to be generally difficult to confirm observationally. Algol evolution was discussed by W. Packet in *Close Binary Stars: Observations and Interpretation; Proc. IAUS 88*, Reidel, p211, 1980; R. E. Wilson & J. B. Rafert in *Astron. Astrophys. Suppl. Ser.*, Vol. 42, p195, 1980; and G. Giuricin, F. Mardirossian & M. Mezzetti in *Astrophys. J. Suppl. Ser.*, Vol. 52, p35, 1983. A review of evolution-related interaction effects in Algol systems, with particular attention to the case of U CrB, is found in PhD thesis of R. H. van Gent, Utrecht University, 1989.

O. Y. Malkov's recent review of semi-detached binaries is in *Mon. Not. Royal Astron. Soc.*, Vol. 491, p5489, 2020. This refers to the catalogues of O. Y. Malkov, E. Oblak, E. A. Snegireva & J. Torra in *Astron. Astrophys.*, Vol. 446, p785, 2006; E. A. Avvakumova, O. Y. Malkov, A. Y. Kniazev in *Astron. Nach.*, Vol. 334, p860, 2013; and that of M. A. Svechnikov & E. F. Kuznetsova, *Vizier Online Data Catalog*, V/124, 2004.

Malkov's review confirms the peak in observed incidence of primaries around temperatures corresponding to spectral type A0V, as is the cut-off towards spectral type F5V. With more secondary spectral data having appeared in recent years, a large peak in the number of late-G – early-K type secondaries is found. These systems can be associated with the classical Algols having EAS type light curves and probably evolving in the Case B model. However, about 15% of the sample have B type secondaries, associated with the Case A model of interactive evolution. There also appears to be a small group with intermediate type secondaries, perhaps corresponding to the Case AB scenario.

Analyses of data such as shown in Figs 10-14 and 10-15 have suggested that Algol evolution is not conservative: that typically up to half the original mass of the system may be lost, and that the angular momentum has a rather sensitive statistical dependence on mass, i.e. that the mass lost from the system takes a high proportion of the angular momentum. For further background on this see e.g. B. Paczyński & J. Ziółkowski, *Acta Astron.*, Vol. 17, p7, 1967; M. V. Popov, *Perem. Zvezdy*, Vol. 17, p412, 1970; G. Giuricin & F. Mardirossian, *Astrophys. J. Suppl. Ser.*, Vol. 46, p1, 1981; E. Budding, *Astrophys. Space Sci.*, Vol. 99, p299, 1984; R. W. Hilditch, *Space Science Reviews*, Vol. 50, p289, 1989; W. van Rensbergen, C. De Loore, D. Vanbeveren, *Astron. Soc. Pacific Conf. Ser.*, Vol. 367, p379, 2007.

The discovery of the Algol type variability of R Arae was published by by A. W. Roberts in *Astron. J.*, Vol. 14, p113, 1894. E. Hertzsprung's paper on it was in *Bull. Astron. Inst. Netherlands*, Vol. 9, p275, 1942. Figs 10.16 & 10.17 are from P. A. Reed's review of R Arae and some comparable interactive Algols in *Proc. IAU Symp.*, Vol. 282, eds. M. T. Richards & I. Hubeny, p325, 2012. This work relates to the development of Doppler tomography as a tool to study interaction effects in Algol binaries as discussed by M. T. Richards, G. E. Albright & L. M. Bowles, *Astrophys. J.*, Vol. 438, L103, 1996; for more recent applications see also M. T. Richards, M. Agafonov & O. I. Sharova, *Astrophys. J.*, Vol. 760, p1, 2012. In this latter connection, (magnetic interactions in Algols) we may note also the survey of R. T. Stewart, O. B. Slee, G. L. White, E. Budding, D. W. Coates, K. Thompson & J. D. Bunton, *Astrophys. J.*, Vol. 342, p463, 1989.

The early work of A. W. Roberts on V Puppis, who established the period of its light variability, is recorded in *Astron. J.*, Vol. 20, p172, 1900, and *Astrophys. J.*, Vol. 13, p177, 1901. The comprehensive study of H. Shapley, that includes V Pup, is in *Astrophys. J.*, Vol. 38, p158, 1913. The radial velocity curves in Fig 10.18 are from D. J. Stickland, C. Lloyd, I. Pachoulakis, & R. H. Koch, as reported in *The Observatory*, Vol. 118, p356, 1998. They were obtained from observations in the UV range using the IUE satellite. A total of 25 radial velocity observations were used in the study, which was also aimed at quantifying the effects of stellar winds in close binary systems.

The possibility of a black hole forming a triple system with V Pup was advanced by S-B. Qian, W-P. Liao & E. F. Laj'us in *Astrophys. J.*, Vol. 687, p466, 2008. This would be a 'quiescent' black hole, not currently engaged in mass transfer or accretion processes, (cf. C. D. Bailyn, arXiv:1610.09694 [astro-ph.HE]). Figs 10.19 and 10.20, and the identification of a β Cep type light variation associated with the primary component, are from an intensive recent study of V Pup by E. Budding, T. Love, T. S. Banks, & M. J. Rhodes in *Mon. Not. Royal Astron. Soc.*, Vol. 502, p6032, 2021. This included TESS photometry as shown. The high intrinsic accuracy of TESS data shows up the low-amplitude β Cep type variability of the primary. (see J. R. Percy, *J. Royal Astron. Soc. Canada*, Vol. 61, p117, 1967; for an account of this type of variability). The circumstance of eclipsing binarity should allow interesting constraints to be placed on this class of star, if the different sources of light variation in the system can be properly separated. Comparable examples of β Cep type variability in well parametrized close binary systems occur with CW Cep (J. W. Lee & K. Hong, arXiv:2010.03711 [astro-ph.SR]); Spica (A. Tkachenko et al. (19 authors), *Mon. Not. Royal Astron. Soc.*, Vol. 458, p1964, 2016); V453 Cyg A (J. Southworth, D. M. Bowman, A. Tkachenko & K. Pavlovski, *Mon. Not. Royal Astron. Soc.*, Vol. 497, L19, 2020); and probably more, yet to be pointed out.

11

Advanced Evolution

Previous chapters had much ground to cover in setting out the basic properties of pairs of normal, or more generally familiar, types of stars. Particularly since the middle years of the twentieth century, knowledge and understanding of the nature of binary star evolution has grown apace, together with remarkable advances in observational technology across the spectrum. A prodigious amount of new information has been released through the various branches of this development, especially regarding binary evolution products beyond the basic semi-detached arrangement of the classical Algols. In what follows, we trace only the outline of this development, touching on linking elements and common themes where possible. The bibliography section at the end of the chapter should also help point the way forward into these broad and lively avenues of active ongoing research.

11.1 Symbiotic Binaries

The term 'symbiotic star' was used by P. W. Merrill in the 1950s in connection with a group of stars having spectra with some resemblance to those of planetary nebulae but showing an odd and complex combination of high excitation emission lines together with a late type (typically K to M) stellar spectrum. Cyclic changes in the appearance and displacement of both sets of lines were observed, suggesting a binary nature to the sources. A handful of stars were first included in the group, including RW Hya, BF Cyg and Z And.* By the time of the 1984 catalogue of D. Allen over 100 examples had been found, but with a rather inhomogeneous set of characteristics.

11.1.1 White Dwarfs in Binaries

The pointer to a more advanced stage of evolution for these objects is the normal presence of a white dwarf in combination with a cool giant-like companion, the pair engaged in material interactions usually accompanied by cyclic changes of brightness on the order of a magnitude and over a timescale of the order of a year. The separation of the two stars is expected to be usually in the range 1-10 AU. If the cool star's optical spectrum is directly observed the system is referred to as of S (stellar) type. About 80% of symbiotics are of this type. Sometimes, the spectrum appears dominated by opaque dusty material surrounding the source, challenging spectral analysis. These D (dusty) type variants would normally

---

*The group, or a subgroup of it, is sometimes referred to as 'Z And type stars'.

DOI: 10.1201/b22228-11

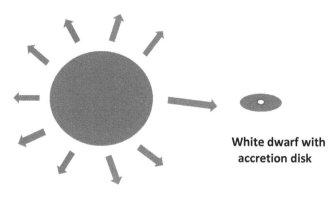

White dwarf with accretion disk

Cool giant with wind

Figure 11.1 Schema (not to scale) of typical arrangement for a symbiotic system. A red giant, with radius of order 50 $R_\odot$ exudes a significant wind-flow of material, some proportion of which is intercepted by the high-temperature accretion structure surrounding a collapsed stellar remnant (at right).

contain a Mira-like variable star as the cool component, and the arrangement associated with longer orbital periods ($\gtrsim 15$ y). Miras are thought able to produce a dust shell that can envelope the binary system.

These S or D type systems are then the 'classical' symbiotic stars, but other possibilities have been countenanced, for example that the non-giant star is still a Main Sequence object (e.g. CI Cyg), or that it may be a neutron star (e.g. V2116 Oph).

Symbiotics move between active and quiescent states of behaviour. During an active stage, the system may show quasi-cyclic brightness increases of a few magnitudes in optical light. Such bursts of brightening can persist through a year or two before the system returns to quiescence at a more or less constant lower brightness level, where it is likely to remain for a decade or so. The transitions between these states appear to have a chaotic character that poses a challenge for detailed modelling. Indeed, some stars, regarded as symbiotic on general observational grounds, have yet to show an active phase.

Models for the interaction process often involve a stellar wind from the cool component, or a Roche lobe overflow process (Fig 11.1). The transferred matter would flow towards the compact component through a surrounding accretion structure. Testing such ideas with models of spectral line formation processes, mass-transfer dynamics and evolution calculations provides the stimulus for investigation.

The parametrization of symbiotics has improved considerably over the last few decades, particularly by taking more detailed account of the binary situation of the component stars. For example, high resolution optical spectra of the eclipsing symbiotic system SY Mus by W. Schmutz et al. allowed the radial velocity curve of the giant M-type component to be drawn up, as well as the determination of its rotation velocity. With the plausible assumption of co-rotation, a value of the giant's radius was calculated to be 86 $R_\odot$. Its effective temperature was then derived from detailed modelling of the spectrum (of M4.5 D type), so that a bolometric luminosity was found to be ~ 1000 $L_\odot$. From this luminosity a mass of 1.3 $M_\odot$ could be determined by means of evolutionary models. The mass of the secondary component was worked out as 0.43 $M_\odot$ using the mass function. After that, Kepler's third law fixes the separation of the stellar components at 1.72 AU.

The geometry of the system in relation to the surface of critical stability was assessed and the distance from the center of the red giant to the inner Lagrange point L_1 turned out to be $\sim$1.05 AU. Since the giant extends to less than 40% of the distance to the L_1 point, the SY Mus binary system must be detached and normal RLOF should not be involved in the system's energetics.

Schmutz et al. did not measure any significant changes in the absorption spectrum of the giant as a function of phase, downplaying the role of any reflection effect. However, among additional spectral effects, the H_α line showed peculiarities, including rapid intensity variations near elongation phases. The rapidity of the variations indicated the emission to come from a relatively small, high-density region close to the red star's surface, facing the hot companion.

Comparable results were found by H. Schild et al., who utilized UV data from the IUE satellite in their study of the symbiotic eclipsing binary RW Hya. Although more than 90% of the light in the UV range around 1270 Å is eclipsed out, the data points are still too sparse to carry out a full eclipse analysis. At optical wavelengths there appears a noticeable reflection effect, though any eclipse would have to be small, given the predominating light of the giant in the optical region. Nevertheless, good radial velocity data were gathered using the 1.4-metre Coudé Auxiliary Telescope (CAT) at ESO's La Silla observatory. This enabled an analysis along essentially similar lines to that of SY Mus. Again, the $\sim$60 $R_\odot$ giant is well-detached from its Roche Lobe, showing that standard RLOF should not be a direct factor in the system's behaviour.

A useful tool for the analysis of data on symbiotic stars was introduced by H. Nussbaumer and his colleagues in 1989, who noted that broad emission bands at around 6830 and 7088 Å are observed in more than 50% of symbiotic stars — and *only* in those systems. It was proposed that these emissions arise from inelastic (Raman) scattering of the high energy O VI resonance doublet at 1032 and 1038 Å by neutral hydrogen atoms. The source photons, scattered by the red giant's wind, show up as the observed strong spectral features. Schmid and Schild went on to monitor these emissions in a selection of symbiotic stars. The accompanying polarization effects confirmed their origin from a Raman scattering process.

The scale of the polarization can be up to 15%, but is typically around 5%, and with significant variations in the degree and angle of the polarization across the line profiles. From their survey, Schmid and Schild found the polarization vector in the blue wings of the lines to be always perpendicular to the line connecting the two binary components, allowing a geometry for the emission scenario to be devised.

These findings were broadly confirmed by T. J. Harries and I. D. Howarth, who went on to reproduce observed effects by modelling the Raman scattering of the O VI lines from a hot source in the extended, expanding atmosphere of a cool companion. While not all details in the data could be fully accounted for, the authors argued that the Raman-line polarization properties would enable insights into the physics of the red-giant wind; particularly when observations have sufficient orbital phase coverage (see also Fig 7.17).

Another procedure involves monitoring of the molecular absorption bands that have been used for estimating the luminosities of cool giant stars, notably the CN bands in selected regions of the infra-red. As a result, the giants in most symbiotics are found to be detached from their surrounding Roche lobes. However, there are some examples (e.g. CI Cyg, RS Oph), where the luminosities are large enough to indicate a semi-detached configuration. The bands of TiO and VO molecules can also be used to assign spectral types to the cool giants in symbiotic stars, though good phase coverage is required. The depths of these bands correlate with the systems' brightness, and that has been interpreted as related to heating from the secondary star, that may, in turn, depend on the rate of mass transfer. Although most S-type symbiotics are not semi-detached, the mass-loss rates appear higher than for basically similar red giants that are single. It is thus supposed that heating from the secondary enhances the wind and consequent matter accretion.

Supporting evidence for this enhanced wind mass transfer mechanism comes from the fact that ellipticity-effect light variations, characteristic of tidally distorted stars, are rarely observed for symbiotic stars. Only a small number of symbiotics have been reported with measurable light variations attributable to a pear-shaped photosphere.

11.1.2 Period Distribution and 'Case D' Mass Transfer

The orbital-period distribution of S-type symbiotics (from $\sim$100 to $\sim$1400 d) cannot be definitely explained by present binary population synthesis (BPS) models. Such modelling has involved both dynamically stable and unstable modes of mass transfer. Mass-transfer that is unstable on a dynamic time-scale leads to the relatively rapid formation of a common envelope, with subsequent spiralling-in of the components. Even with optimistic assumptions regarding the retention of orbital angular momentum, this process is calculated to produce much shorter orbital periods than the observed ones. By contrast, if mass transfer were stable, a widening of the systems would generally be expected, due to angular momentum conservation. From such considerations standard BPS simulations predict a gap in the orbital-period distribution: but this is where most of the S-type symbiotics are actually observed!

Attention was called to this issue by R. Webbink over twenty years ago and a number of proposals to resolve it have been offered. One approach involves a 'quasi-dynamical' mass transfer mechanism that includes elements of both dynamical and Kelvin time-scale mass transfer. If the mass ratio is relatively close to 1, the mass-transfer rate becomes large enough to allow the formation of a common envelope, comparable to those of the over-contact binary models, but without significant spiralling in.

Spiralling in is avoided as long as the envelope remains in co-rotation with the binary, since there is then no frictional angular momentum loss. Typically, the common envelope remains tidally locked to the binary if its size is less than about twice the orbital separation. In this way, most of the mass lost from the giant ultimately becomes shed from the system, mainly through regions around the outer Lagrangian points.

This process could lead to a circumbinary disc that would tidally couple the two stars and influence the evolution of the orbit But when the mass ratio has reversed sufficiently, the envelope will recede below the gainer's critical Roche surface, and subsequent evolution will resemble that of the classical Algols, involving stable, though to some degree non-conservative, mass transfer.

But recent observations of the symbiotic binary Mira (*o* Ceti) have confirmed that the conventional understanding of binary interactions, as outlined in Chapter 8, is incomplete. In the D-type symbiotic system Mira the orbital period is estimated to be larger than 1000 y. Such wide binaries would not have been generally regarded as interacting according to conventional schemes. Nevertheless, X-ray observations by M. Karovska and her colleagues resolved the giant mass-losing (donor) star, producing an image of it apparently filling its surrounding Roche lobe. This would not refer to the stellar photosphere, which is a factor of $\lesssim$10 smaller than this image, but the observation still gives an interesting picture of the slow wind concentration in the system.

Mira winds are associated with pulsations of the dynamically unstable envelope, but are only accelerated to their terminal speeds at $\gtrsim$5 stellar radii, where dust can form. Radiation pressure on the dust then provides the necessary acceleration. If this acceleration region is comparable in size to the radius of the Roche lobe, then the slow wind effusion will respond to the gravitational potential structure of the binary system. This gives rise to the donor star's wind having a form comparable to that of the contact component in a semi-detached binary.

The importance of this type of wind-driven Roche-lobe overflow (WRLOF) is that a large fraction of lost material can end up being transferred to the companion. The mass-transfer rate may exceed the estimate expected for the accretion from a simple spherical outflow regime by up to 2 orders of magnitude. This WRLOF idea has provided a promising mechanism to account for mass transfer effects in wide binaries, and has been sometimes referred to as Case D mass transfer. Since any mass lost from the system is strongly confined to the orbital plane, it produces a disc-like outflow or even a circumbinary structure. This should have implications for the shapes of asymmetric planetary nebulae.

Case D mass transfer would also be important for massive stars. Calculations have shown that massive red supergiants will develop dynamically unstable envelopes and experience Mira-like variability. In many cases, the consequences of WRLOF are comparable to Case C mass transfer, thus, in effect, leading to a significant expansion of the period range for which late phases of mass transfer are important. This has implications for various types of supernova progenitors (such as those of Type II-L, and IIb supernovae), and even some binary gamma-ray bursters, the modelling of which often requires such slow phases of mass transfer. A few symbiotic binaries have been detected as continuous sources of soft X-rays, perhaps associated with interactions between the winds from the cool donor star and the hot compact object. The present time is still early days for exploring all the consequences of Case D mass transfer.

11.2 Cataclysmic Variables

11.2.1 General Properties

The symbiotic stars of the previous section could be regarded as forming a link to, or being towards one edge of, a much more widely encountered grouping known as the cataclysmic variables (CVs). The CVs are a diverse category of objects essentially characterized by occasional and sudden explosive increases in brightness that may reach to the order of 10 magnitudes. The bright stage would usually be relatively short-lived (a few days would be characteristic for the well-known example U Gem; Fig 11.2), compared to the interval between such brightenings, and may be accompanied by rapid variations or *flickering*. This is followed by a slower decline to a quiescent state that would usually be relatively faint compared with, say, typical stars of the Henry Draper catalogue. This point entails that CVs are frequently first picked up by relatively small-scale patrolling facilities, though study of the complex and finer details of CV properties naturally calls for larger flux-collectors and specialized equipment. There is, by now, a large literature on this class of object and our present review touches on only a few of the salient issues.

The total number of known CVs is on the order of $\sim$10% of the number of eclipsing variables. The CV classification is associated with historically known *novae*, originally meaning 'new star', and *supernovae*; relating to those examples that became temporarily visible as naked-eye objects. U Gem does not become so bright as that, its range of brightness variation is more like $\sim$6 magnitudes in the visible range: it is thus distinguished from the classical (or 'old') novae, such as RR Pic, and is described as a dwarf nova. The great majority of CVs are such dwarf systems, although they comprise a half-dozen or more sub-groupings that are not necessarily mutually exclusive: GK Per, for example, has shown a classical nova scale of outburst as well as those of lower amplitude. Generally speaking, dwarf novae are associated with intervals between outbursts in the range of tens to a few hundreds of days. The characteristic amplitude of the outburst correlates positively with this interval.

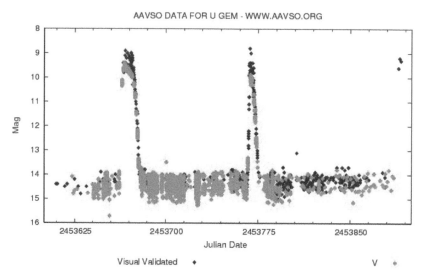

Figure 11.2   Light curve of the archetypal CV U Gem, courtesy of the AAVSO (https://www.aavso.org).

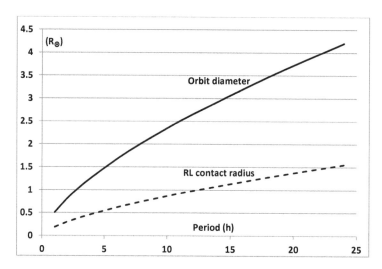

Figure 11.3   Typical sizes (continuous) of CV systems of total mass $\sim 1 M_\odot$ for periods in the range 1 h to 1 d. The approximate size of the Roche lobe for a secondary of comparable mass to the primary is also indicated (dashed).

The relationship of CVs to binary stars became realized following the recognition by Merle Walker in 1954 that the nova DQ Her was made up of two stars: the primary (more massive) being a white dwarf (WD) in a relatively short period orbit (4.6 h) with an M2 type Main Sequence dwarf. Walker's discovery was followed by similar ones from others, notably Robert Kraft, who, over the following decade, established that all observed CVs are in such relatively very close (cf. Fig 11.3) binary relationships involving a white dwarf and a cool companion generally of relatively low mass. A CV system with period 3 h and total mass 1 solar mass would have an orbit diameter of just ~ 1 $R_\odot$. In spite of such a low separation, the white dwarf, of radius only $\sim 1/100$ that of the Sun, would, in its own

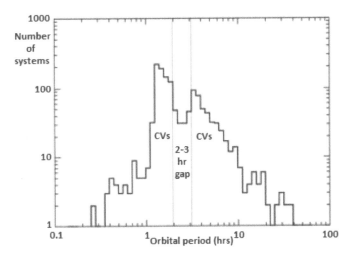

Figure 11.4 Distribution of the periods of CVs according to the catalogue of Ritter & Kolb, 2003. The diagram is from B. Kalomeni et al. in *Astrophys. J.* Vol. 833, p83, 2016 (Fig 14); ©AAS.

terms, still be well separated from its companion. It is widely believed that prior to this arrangement the predecessors of CVs must have co-existed in a common envelope when considerable angular momentum was lost from the system.

The periods of regular CVs are mostly in the range ~1 to 24 hours, but a noticeable drop in numbers, or 'gap', in a short interval around 2.5 h has been frequently pointed out (Fig 11.4). It was deduced early in the study of these binaries that the cool dwarf secondaries (on or near to Main Sequence) would often be engaged in Roche Lobe overflow. Matter, mostly hydrogen, is transferred from the donor towards the white dwarf, falling down an enormous potential energy well in the process. This energy E, for a mass m falling in from a great height, can be assessed by writing

$$E = G m M_{\mathrm{WD}} / R_{\mathrm{WD}} \ , \tag{11.1}$$

where M_{WD} and R_{WD} are the mass and radius of the white dwarf.

If we use accepted values of these quantities for U Gem and, taking m as the mass of a hydrogen atom, regard the thermalized equivalent of its acquired kinetic energy as $\frac{3}{2}kT$, where k is Boltzmann's constant ($\approx 1.4 \times 10^{-9}$ J K^{-1}), we find a temperature of order 10^9 K. This scale of energization – simplistically obtained, but of relevance to the general scenario – is such that CVs could be expected to become sources of high-energy radiation, including hard X-ray emission. Alternatively, we could replace (11.1) with

$$L = \epsilon G \dot{m} M_{\mathrm{WD}} / R_{\mathrm{WD}} \ , \tag{11.2}$$

where L is the luminosity of a thermalization process for a steady transport of mass $\dot{m}$, whose efficiency is ϵ. If ϵ is of order unity while $\dot{m}$ is feasibly of order 10^{-10} M$_\odot$ y^{-1}, then $L \sim 0.7 L_\odot$, i.e. several magnitudes brighter than where a quiescent version of the system would be in comparison to the Sun.

But under normal conditions this transferred matter would retain sufficient angular momentum to prevent direct impact on the surface of the white dwarf. Instead, an *accretion structure* around the white dwarf is maintained by the mass-transferring stream. Explanations of the possible modes of behaviour of CVs are then sought from modelling the physics of this structure. Thus, a classical nova explosion occurs when the density and temperature at the inner boundary of the accretion disk attain appropriate values to trigger a runaway hydrogen to helium fusion process. Less violent instabilities may occur further out in the disk, associated with the more frequent but less dramatic outbursts of the dwarf novae that occur on a variety of timescales. The magnitude and topology of magnetic fields also have an important bearing on the type of phenomena witnessed. Frictional processes in this disk, as well as other effects, entail that the system continually loses angular momentum, so that the components gradually become closer to each other. The white dwarf itself has a stability limit (the Chandrasekhar limit), which may be approached if the accretion process builds up enough mass in the star. The huge energy release of supernova explosions (of Type Ia) have been modelled in this way.

In order to quantify some points about the Roche Lobe overflow process, we refer again to the discussion of S. Lubow and F. Shu (see Fig 11.5). In reviewing the formation process for a steady near-circular disk of accreted material orbiting the receiver component, those authors found that an initial inflow would have sufficient angular momentum to bring it to a minimum separation r_{di}, say, neglecting here the possibility of direct impact on the gainer star. So, for present purposes, r_{di} can be taken to be relatively small compared to the orbit (Fig 11.5a). As the inflow accumulates, the stream should interact with previously transferred material and settle towards a steady form of flow (Fig 11.5b). At this point it is argued that the stream of incoming matter accreted to the outer edge of the disk would satisfy the following condition. The velocity component of the incident stream resolved in the direction tangent to the disk edge at the point of impact equals, the speed of the disk edge at that point (Fig 11.5c). The latter is calculated from classical models of stable orbits of particles within the Roche configuration. The resulting near-circular disk's inner and outer radii r_{di}, r_{do} come close to showing an inverse parabolic dependence on the mass ratio q so that we can write

$$r_d = (\sqrt{b^2 - 4a(c - q)} - b)/2a \ , \tag{11.3}$$

where a, b and c are constants that follow from the numerical data of Lubow and Shu (see Fig 11.6).

About 20% of CVs are characterized by a strong magnetic field of at least several million gauss retained by the white dwarf. This field produces special characteristics in the systems, sufficient to distinguish them into the separate classes of *polar* (prototype AM Her) and *intermediate polar* (prototype DQ Her) categories of CVs. In the former, the WD has synchronized its rotation period to that of the orbit. This is not the case for the latter, but the finding may be more related to the separation of the two stars than the field strength alone.

The different components in CV systems involved in the overall emission processes are then: the component stars, the disc-like accretion structure around the WD component, the boundary layer between the disc and the WD surface, the gas stream from the secondary, the 'hot spot' where the stream impacts the disc, and the hot corona and chromosphere about the disc and around the cool secondary.

The secondary star's main contribution is usually in red and IR regions. The primary WDs have temperatures ranging between 10000 and 50000 K. They are expected to become relatively strong in the UV, but they may be visible also in the optical range if the companion is low enough in luminosity. The accretion disc may contribute to total emission in the whole range between EUV and IR. Calculated output fluxes depend on the choice of disk parameters. The high energy radiation is typically supplied from a zone in the vicinity of the WD, where optically thick material can surround high-temperature sources. The hot inner boundary of the disk adjacent to the WD surface should produce strong X-radiation locally that would emerge from the system in the EUV and short wavelength UV ranges. The optically thin and relatively cool gas stream from the secondary is expected to affect

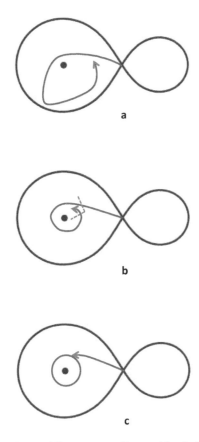

Figure 11.5 Possible stages of the RLOF process as discussed by S. H. Lubow & F. H. Shu in *Astro-phys. J.*, Vol. 198, p383, 1975; ©AAS. The donor star is on the right and the system is revolving in an anticlockwise sense. In the uppermost sketch (a) a particle released from the inner Lagrangian point L_1 with relatively low kinetic energy, subject to mild dissipation and initially deflected upwards in this sketch by the Coriolis force, could approach the gainer's limiting equipotential surface after swinging past the gainer star if there is space enough to perform this orbit. However, dissipation would increase as the returning particle encounters the continuously in-falling stream again. The implied loss in energy leads to the situation (b) where shock fronts are formed (dashed lines) with repeated high-velocity impacting between returning and incoming streams. The flows then tend to relax into alignment. The process matures into a steady state (c), where overflow matter has accumulated into an accretion disk with relatively high density.

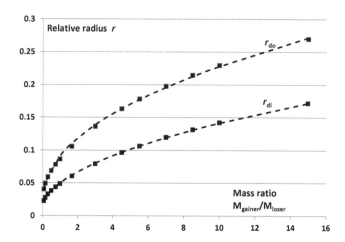

Figure 11.6 Indications of disk relative radii r_{do} and r_{di} (outer and inner), as calculated by Lubow & Shu (1975) ©AAS, referred to in the caption to Fig 11.5. The diagram shows the individual radii from Table 2 of Lubow and Shu (op. cit.) as points, with the optimized parabolic forms Eqn 11.3 as dashed lines.

spectral line formation in the red and IR regions. The hot spot is, for many systems, associated in optical photometry with a periodically recurring 'hump' in the light curves. The visibility of this will depend on the geometry of the system in relation to the observer. The hot corona and chromosphere on both sides of the disk and around the cool secondary, made up of optically thin and hot gas, emit X-ray and UV line emission, and in some cases, microwave radio-emission. There is by now a wealth of data and corresponding models for these different radiation effects and many publications have compared theoretical predictions with observations of CVs across the spectrum.

CV systems with WDs not possessing strong magnetic fields appear to evolve through the following stages: (1) Onset of mass transfer at periods of order half a day; (2) Transition to shorter periods with mass transfer and angular momentum loss affected by frictional processes and magnetic braking. This involves mass-transfer rates on the order of 10^{-8} $M_\odot$ yr^{-1}; (3) Fall-off of mass transfer at periods of about three hours, forming the 'period gap'; (4) Re-establishment of mass transfer at periods of about two hours. This is associated with gravitational wave emission and relatively low mass-transfer rates, on the order of 10^{-10} $M_\odot$ yr^{-1}; (5) Arrival to a period minimum range of around 20-60 minutes, where the mass-transfer rate has dropped significantly from that of earlier stages. A significant part of the CV population may form from systems with low-mass donors that join this sequence directly at stage (5).

In this general picture, the typical progenitor of a CV would probably have been a wide binary, with an intermediate-mass primary evolving to the RGB or AGB stage and producing substantial wind outflow, accompanied by a lower mass star still on the Main Sequence. Such a combination is reckoned to experience a spiral-in process in a common-envelope phase, ending with the WD core of the giant and its MS companion in a close binary arrangement. The foregoing summary conforms to the standard model of CVs, going back at least to the work of B. Paczynski in the 1970s.

This picture faces a number of problems, however. One is that studies of the supposed progenitors of CVs find that the majority of cases ($\gtrsim$75%) should have relatively low-mass primaries, say $\lesssim$0.5 solar masses. The predicted mass distribution of WDs in CVs is then dominated by WDs of low mass. But this runs contrary to the observed trend of WDs in CVs being relatively massive, with a mean mass of $\gtrsim$0.7 $M_\odot$.

In addition, theoretical models generally predict a rather large space density of CVs ($\sim 3 \times 10^{-5}$ pc^{-3}), compared with observational findings ($\sim 6 \times 10^{-7}$ pc^{-3}). It is possible that the pre-CVs with low-mass WDs do not make it to become regular CVs because the first (few) nova outburst(s) remove significant systemic angular momentum. This could result in subsequent unstable mass transfer and merger of the two components. Such a process might significantly decrease the space density of CVs and resolve the discrepancy with their observed incidence.

Most of the known CVs were originally discovered because of their dramatic optical variability, exemplified by the outbursts of dwarf and classical novae. More recently, however, UV and X-ray sky surveys have led to discoveries of many new CVs, including members of subclasses that have less spectacular optical light variations (e.g. the nova-like variables). Analysis of the INTEGRAL/IBIS survey observations has revealed that the rare intermediate polar and asynchronous polar cataclysmic variables (CVs) are consistently found to emit in the 20-100 keV energy band, whereas synchronous polars and the common non-magnetic CVs do so rarely. Follow-up investigations at radio wavelengths have also led to detections of a few well-known, mostly relatively near, CVs as significant radio sources.

11.2.2 Thin Disk Model

It has been mentioned above that many of the properties of CVs are bound up with those of the accretion disk that accommodates, at least for a time, matter transferred from the stellar companion towards the WD. It is appropriate now to take a look at a basic version of this accretion structure. The *thin disk model* adopts that the disk, presumed circular and axisymmetric about the gainer star, has an extent perpendicular to the central plane, of order h, that is small compared to the disk radius r_d. The mass of the disk is sufficiently low that the angular velocity at radius r is taken to be of Keplerian form, i.e.

$$\omega(r) = (GM/r^3)^{1/2} , \tag{11.4}$$

where M is the mass of the central star.

The configuration lends itself to use of the cylindrical co-ordinates r, ϕ, z. The revolution velocity at r, u_ϕ, is then given by $u_\phi = r\omega(r)$, and so proportional to $r^{-1/2}$. As well as u_ϕ, material within the disk is taken to have a small inward drift velocity $u_r(r,t)$, having, in principle, a dependence on time t. The short cylinder of the disk at radius r experiences a viscous torque T that we take to be of Newtonian form, and, in the cylindrical context, satisfies

$$T(r) = 2\pi\nu\Sigma r^3 \frac{d\omega}{dr} , \tag{11.5}$$

where ν is the kinematic viscosity of the disk material (having the dimensions of length $\times$ velocity) and $\Sigma(r,t)$ is the (2-dimensional) surface density of the disk. Mass conservation is expressed through

$$r\frac{\partial\Sigma}{\partial t} + \frac{\partial(r\Sigma u_r)}{\partial r} = 0 . \tag{11.6}$$

In a similar way, the change in angular momentum of disk material at r satisfies

$$r\frac{\partial(\Sigma r^2\omega)}{\partial t} = \frac{1}{2\pi}\frac{\partial T}{\partial r} - \frac{\partial(\Sigma u_r r^3\omega)}{\partial r} . \tag{11.7}$$

It should be remarked that these two equations are supplemented in more detailed discussions by inclusion of the effects of tidal forces on the disc material, and the possibility that transferred matter may still join the disk some way into its structure, rather than the stream being separated from the disk at an actual boundary as in the present idealization.

Eqns (11.4, 6 & 7) can be combined to yield

$$r\frac{\partial \Sigma}{\partial t} = -\frac{1}{2\pi}\frac{\partial}{\partial r}\left(\frac{1}{X}\frac{\partial T}{\partial r}\right) \;, \tag{11.8}$$

where $X = \mathrm{d}(r^2\omega)/\mathrm{d}r$, so that, using (11.4), $1/X = \sqrt{GMr}$. Substituting now (11.5) for T, (11.8) is rewritten as

$$\frac{\partial \Sigma}{\partial t} = \frac{3}{r}\frac{\partial}{\partial r}\left(r^{1/2}\frac{\partial(\nu\Sigma r^{1/2})}{\partial r}\right) \;, \tag{11.9}$$

which forms as the underlying summarizing equation for this Keplerian thin disk model. It can be recognized as of diffusion-equation form, and implies that a circulating ring of material at r_d, in the manner discussed by Lubow and Shu and represented by (11.3), would tend to widen out with time.

A quasi-steady condition can be visualized, with matter being drained from the disk into the stellar sink and a corresponding influx from at an outer limiting surface. Assuming such a steady condition, forces in the radial direction amount to:

$$\frac{1}{\rho}\frac{\partial P}{\partial r} = -\frac{GM}{r^2} + \frac{u_{(\phi)}^2}{r} - \frac{u_r\partial u_r}{\partial r} \tag{11.10}$$

with P the local pressure and ρ the volumetric density. The first two terms on the right of (11.10) are taken to cancel each other out in the Keplerian approximation. Now, using (11.5), the local frictional force on the inner edge of the disk at radius r becomes

$$F = \nu\rho r\frac{\partial\omega}{\partial r} = \alpha P \;, \tag{11.11}$$

so that

$$\frac{1}{\rho}\frac{\partial F}{\partial r} = -\alpha\frac{u_r\partial u_r}{\partial r} \;, \tag{11.12}$$

where we we have assumed that F is some fraction α, of order unity, of the surface normal pressure. A constant value of the kinematic viscosity ν is taken to be representative for disk conditions, at least in a first approximation, and a dimensional argument applied to (11.12) leads to u_r being of order ν/r, where u_r is smaller than the rotation velocity u_ϕ, by the factor $\nu(r_d/GM)^{1/2}$ that turns out to be several orders of magnitude.

This standard thin-disk model adopts the gas to be in hydrostatic equilibrium in the z-direction so that, regarding forces parallel to the disk axis:

$$\frac{1}{\rho}\frac{\partial P}{\partial z} = -\frac{GMz}{r^3} \;. \tag{11.13}$$

Introducing the sound speed as a, and using the Keplerian approximation for u_ϕ, we then have

$$\frac{a^2}{\rho}\frac{\partial\rho}{\partial z} = -\frac{u_\phi^2}{r^2}z \;. \tag{11.14}$$

The Mach ratio u_ϕ/a, is taken to be constant in the z-direction, M_ϕ, say, and so (11.14) may be integrated to show a Gaussian variation of density with height above the central plane:

$$\rho = \rho_0\exp(-z^2/2h^2) \;, \tag{11.15}$$

where the scale height of this variation is $h = r/M_\phi$. The whole character of the thin disk approximation $h \ll r$ is thus tantamount to the disk's rotation speed being highly supersonic $M_\phi \gg 1$.

In the steady state, the left side of (11.7) becomes zero and the right side integrated to yield

$$\Sigma u_r r^3 \omega = \frac{T}{2\pi} + f(r) \ ,$$

(11.16)

where f is an arbitrary function of r. We have set the disk's Keplerian azimuthal velocity structure through (11.4), while (11.16) determines the rate of the angular momentum transfer. Notice that this angular momentum transfer is outward, although the matter is settling inward. Balancing the units, the appropriate expression for f is $\dot{m}(GMr_{di})^{1/2}/(2\pi)$. Writing now for the steady state mass transfer $\dot{m} = -2\pi r \Sigma u_r$, (11.16) reduces to

$$\nu \Sigma = \frac{\dot{m}}{3\pi} \left(1 - (r_{di}/r)^{1/2}\right) \ ,$$

(11.17)

the density vanishing on the inner radius r_{di}, which would be regarded as the radius of the star's equator in a fully relaxed state. As r becomes large compared with r_{di} the surface density Σ tends to the constant $\dot{m}/(3\pi\nu)$ while the inward flow velocity approaches $u_r \approx -3\nu/(2r)$.

The energy balance in the disk means that the work done by friction in reducing orbital kinetic energy, associated with the inflow, transfers into a radiative efflux $l(r)$, say, for which we write

$$l(r) = hFr\frac{\partial\omega}{\partial r} \ ,$$

(11.18)

where $Fh = T/(2\pi r^2)$. Combining (11.5), (11.11) and (11.17) we then have

$$l(r) = \frac{3GM\dot{m}}{4\pi r^3} \left(1 - (r_{di}/r)^{1/2}\right) \ .$$

(11.19)

Integrating $l(r)$ over the disk, we find

$$
\begin{aligned}
L &= \frac{3GM\dot{m}}{2} \int_{r_{di}}^{\infty} r^{-2} \left(1 - (r_{di}/r)^{1/2}\right) \mathrm{d}r \\
&= \frac{GM\dot{m}}{2r_{di}}
\end{aligned}
$$

(11.20)

which may be compared with (11.2). The key role of the inner radius of the disk is apparent.

Although L does not depend explicitly on the viscosity coefficient in (11.20), ν is involved in fixing the surface density distribution to maintain a steady state mass flow through (11.17). A lot of physical interest surrounds the nature of this important parameter that seems to elude direct measurement. The range of values of ν that might be gathered from measures of the flow of gases in laboratory conditions, i.e. $\nu \approx v_T \lambda$, where v_T is the mean velocity of particles at temperature T and λ is the mean free path, that should be comparable to $n^{-1/3}$, can be shown to lead to values of ν that are too small to fit the observed properties of disks. Typical parameters deduced from observations of CV accretion disks, at radius $R \sim 10^{10}$ cm, say, would be of order $T \sim 10^4$ K, $\dot{m} \sim 10^{16}$ g s^{-1} and $n \sim 10^{16}$ cm^{-3}, from which we would derive a kinematic viscosity of order 10.

But this is not consistent with the foregoing estimate of the surface density in the outer regions of the disk, for which a much greater value of ν is called for. Instead of interactions at the level of atoms, turbules, with sizes approaching the thickness of the disk, affect the highly supersonic flow. In an oft-cited paper N. I. Shakura and R. A. Sunyaev thus argued for a kinematic viscosity of order

$$\nu = \alpha a h \ ,$$

(11.21)

where α is a numerical parameter of order unity.

Eqns (11.5) and (11.11) can also be combined to produce an expression for the local effective temperature T_e at radius r, noting that the energy is radiated through both faces of the disk. We write

$$\sigma T_e{}^4 = \frac{9}{8} \frac{GM}{r^3} \Sigma \nu \ . \qquad (11.22)$$

The foregoing equations describing an idealized accretion disk, with suitable boundary conditions applied, may be treated along similar lines to the equations of stellar structure. For given M_{WD}, $\dot{m}$, R_{WD} and α, density, effective temperature, and disk thickness is specified as a function of r. This approach to disk modelling, with α as an adjustable parameter, has proved effective in understanding the general attributes of accretion disks. However, physical details of the viscosity remain obscure, with little independent evidence on the α parametrization. Alternative approaches referring to the role of magnetic forces acting on the disk material allow a fuller theoretical basis to the discussion, though without reducing hypothesis. In fact, observations do indicate more complex fluid mechanics are at play than in the α-disk models. But the latter still have considerable value in allowing insight into key variables of the problem.

The significance of spatially resolved observations to the modelling of accretion disks became clear, notably in the application of the Hubble Space Telescope to the eclipsing nova-like variable UX UMa. The circumstance of eclipse permits interpretation in terms of the structure of the occulted object. The applied modelling allows spectra of the accretion disk that are suitably spaced in time to map disk properties in dependence on the co-ordinates. The transformation from observational to model information has been effected through the use of the *maximum entropy method* (MEM).

This can be summarized as follows. With regard to image construction, many arrangements may be possible that are consistent with the observed effects to within a pre-assigned level of measuring error. But the preferred distribution of image pixels should be that with the highest probability. The use of the word entropy recalls Boltzmann's principle that, of the number of ways in which a macro-state can be achieved, the most frequently occurring arrangement of its micro-states is that most likely to be found at any particular time. Maximizing information entropy implies minimization of superfluous, or inaccessible, detail in the prescription that creates the numerically generated model. A MEM result, given the context of a large range of possible pixel arrangements, is the one that imposes fewest particulars on the source, though there must be additional constraints within the applied algorithm relating to the accepted level of error and the use of an error minimization principle, such as least squares.

In the case of UX UMa, UV spectra of the inner (low r) regions of the disk were found to be characterized by a continuum extending towards the optical blue, having various absorption features. This transitions into an emission spectrum with increasing disk radius. In the azimuthal co-ordinate, a significant asymmetry was found (see Fig 11.7). There appeared to be an absorbing ring of gas at lower temperature, whose density or vertical scale increases with disc radius. The inflowing stream introduces complications, with the spectrum of gas in the stream differing from that of the disc at the same radius. This stream material appears to continue its flow in the same direction, complicating the idea of a simple separation between stream and disk regimes. The form of the Balmer discontinuity points to a significant contribution from optically thin gas to uneclipsed light that could be associated with a vertically extended disk-chromosphere and wind. Circumstellar cool gas may also be introduced to support the interpretation of the data.

Although the general form of the temperature distribution with radius was found to be in agreement with that of formula (11.20), there was evidence of increases in the mass accretion rate that could be related to a contemporary increases of brightness. The estimated rate of mass transfer for UX UMa can vary by $\sim 50\%$ on the time scale of a few months. In short,

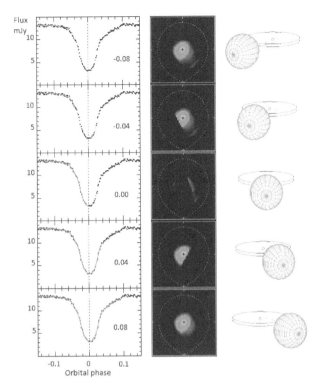

Figure 11.7 Illustration of eclipse mapping of an accretion disk (courtesy of R. Baptista. The illustration appeared in *Lect. Notes in Physics*, Vol. 573, p307, 2001, ©Springer Nature.) The simulated light curve on the left shows what looks like a regular eclipse minimum at first sight. It is distorted however, due to the asymmetric arrangement of luminous material about the accreting subdwarf star. This is seen more directly from the final image in the central panel, when the eclipsing star has almost completed its eclipse of the glowing disk material to the right of the high-intensity central region. The model's account of the light variation is shown by the curve passing through the data points on the left. The curve is completed up to each of the orbital phase values shown in the panels. A 3D-rendering of the underlying geometry of the eclipse is shown in the corresponding representations on the right. The technique has been applied to eclipsing CVs like UX UMa.

eclipse-mapping methods reveal significant additional physical processes beyond those of the standard model presented before.

11.2.3   Dwarf Novae

Regarding dwarf novae, Y. Osaki (1974) called attention to the fact that the luminosity of the hot-spot in dwarf novae light curves was comparable to the luminosity calculated for the complete disk in quiescence. The hot spot luminosity L_s, say, could be reasonably connected with energy released at the outer radius of the disk and to scale with $\dot{m}/r_{do}$. For the disk luminosity, from (11.2), however, we might expect $L_d \sim \dot{m}/R_{WD}$. But R_{WD} would be typically about 2 orders of magnitude smaller than r_{do}!

The mean luminosity of the disk at outburst is indeed two or more orders of magnitude greater than that of the hot spot, but the comparison becomes better if we take into account the amount of time in outburst as a fraction of the complete cycle, as well as the relatively small difference between hot spot and quiescent disk luminosities. These points,

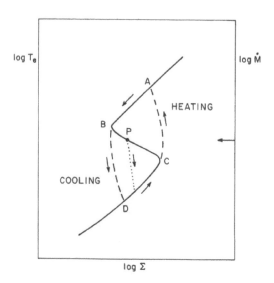

Figure 11.8 The 'S-curve', associated with the DIM model for the outbursts of dwarf novae (after J. Smak, in *Publ. Astron. Soc. Pacific*, Vol. 96, p5, 1984; ©AAS.)

taken together, suggested the basis for what has become the *disk instability model* (DIM) for explaining dwarf novae light curves.

In this model, accretion of matter to the white dwarf is an intermittent process resulting from some instability within an accretion structure that is, for most of the time, at a distance of the order of hundreds of white dwarf radii, beyond the surface of the latter. The instability results in a rather fast inward migration of substantial amounts of matter with corresponding thermalized releases of the acquired kinetic energy. Osaki found, in this way, that the amplitudes of outbursts in U Gem stars, and their period-amplitude relation, as well as the 'standstill' brightness levels of the related Z Cam stars, could all be reasonably accounted for. A more-or-less steady mass transfer from the secondary continues through this process, allowing the cycle of events to be maintained.

Osaki did not elaborate on the details of the instability mechanism, but subsequent investigators have found that the transition to an ionized state for the hydrogen accumulated in the disk is the likely primary agent of change. The upshot is that the relationship between local surface density and temperature passes through something akin to a phase transition between stable 'hot' and 'cool' regimes. The situation is often discussed by reference to the 'S curve' shown in Fig 11.8, as taken from the 1984 review of J. Smak.

The relationship of T_e to Σ is schematized in Fig 11.8 with 'hot' (BA) and 'cool' (DC) phases connected by an unstable branch (CB). Smak (1981) considered an initial point P, relating to a cell of material in the outer region of the disk. An arbitrary slight disturbance at P results in a cooling, downward migration to the stable cool phase represented by the CD branch. But in this regime the inward flow of matter is lower than the mean rate of mass transfer $\dot{m}$. The medium therefore tends to increase in density moving up in the direction D to C. This heating stage at some point entails a transition to the AB branch, whereupon events proceed in the reverse direction. The increase of viscosity now causes a faster than average mass depletion until conditions cool down to the CD branch, whereupon the cycle repeats itself. The hot AB region of the cycle goes with significantly greater brightness of the disk, though it is shortlived compared to the heating DC stage. Crude general estimates for typical disk parameters suggested the bright stage to last $\sim 1/10$ that of the faint one.

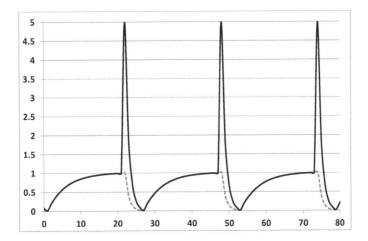

Figure 11.9   Schematic relaxation oscillations with outbursts proportional to the rate of change of discharge (dashed). The graph combines the two solutions (11.25) and (11.26). The observed effect manifests itself as an amplified version of the discharge curve.

Relaxation oscillator outbursts

The behaviour of dwarf novae may thus be compared, intuitively, with that of a *relaxation oscillator*, that can be considered as made up of two modes: 1. a 'charging' mode, in which the number of particles in the structure is steadily built up, with a timescale τ_1 say, in proportion to the remaining available volume, determined by a maximum number n_0; and 2. an 'discharging' mode, in which there is a release of the particles with a timescale τ_2 in proportion to the number remaining in the structure. We can represent these two modes as

$$\frac{\mathrm{d}n}{\mathrm{d}t} = \frac{(n_0 - n)}{\tau_1} \tag{11.23}$$

and

$$\frac{\mathrm{d}n}{\mathrm{d}t} = \frac{-n}{\tau_2} \tag{11.24}$$

Eqn (11.23) has a solution

$$n = n_0(1 - e^{-t/\tau_1}) \tag{11.25}$$

while that of (11.24) is

$$n = n_0 e^{-t/\tau_2} \tag{11.26}$$

The two modes are linked by adding a small constant to the charging number, with the implication that charging recommences when the discharge has dropped below a certain small amount. The result is shown in Fig 11.9, which shows how a relaxation oscillator model simulates, qualitatively, the outbursts of dwarf novae.

The basic disk instability model was applied initially to 'classical' dwarf novae such as U Gem and Z Cam, but it has been developed to account for the more complex behaviours of the separated subgroupings such as the SU UMa and SS Cyg types, and even the low mass binary X-ray (LMXB) transient sources. In such contexts, instead of the original simple model of a thin disk receiving matter at a constant rate with intermittent outbursts due to a thermal-viscous instability, the mass-transfer rate is allowed to vary, and the effects of irradiation are taken into consideration. The role of tides on the disk is also included as well as heating from its hot spot.

SU UMa systems

Although the behaviour of SU UMa is generally similar to that of U Gem, there are a number of special peculiarities. There are periods of accelerated activity when the maxima follow each other at intervals of about a week, rather than the more typical ~16 d between outbursts. 'Supermaxima' occur at intervals of the order of 100 d, when the star brightens increases to a magnitude or so more than during a regular outburst. The supermaxima remain for typically a fortnight, or ~10 times the duration of regular maxima. At minimum light, around $V \approx 15$, the system exhibits a rapid light flickering of order 0.3 mag. As well as the supermaxima, 'superhumps' are seen during superoutbursts. The superhumps drift along the light curve with a period longer by a few percent than the orbital period.

Discussions of the properties of SU UMa type variables, have been thought attributable to sudden mass-transfer changes arising from an instability of the secondary component. Such theory usually refers to an underlying DIM scenario but modified by a tidal resonance effect. This particularly concerns the outer edge of the accretion disk spreading to a radius where its rotation rate is three times that of the stellar orbit. This gives rise to a tidal instability associated with an eccentric accretion disk, whose precession shifts the location of the superhumps. The tidal effect enhances angular momentum transport, which can explain the longer duration and increased brightness of supermaxima. The sequence of normal outbursts builds up the disk until the 3:1 resonance radius is reached, whereupon the tidal torque rapidly develops. This results in a greater shrinkage than normal, and it allows the complete *tidal thermal instability* (TTI) cycle to restart.

The binary system in SU UMa has an orbital period of only 1.83 h, and in fact the geometry of the TTI model is consistent with most of the SU UMa type CVs being short period systems located below the period gap in Fig 11.4.

Polars snd intermediate polars

Localized magnetic fields on normal stars are sometimes found to attain field-strengths of the order of several thousand gauss. With a contraction of area on the order of 10^{-4}, corresponding to the decreased size of white dwarfs, it becomes plausible to expect condensed stars to have fields of order a few tens of millions of gauss. If we place such a field-containing white dwarf into a CV binary, we may anticipate different effects from those of the purely hydrodynamic models considered so far. The driving of gas out from the secondary presumably remains unaffected; but if the magnetic field of the white dwarf is sufficiently strong, plasma shed by the secondary may be completely diverted from the equatorial plane and directed to the magnetic poles of the white dwarf. This is the case for the AM Her type polars, where with surface fields of order 3×10^7 G. Instead of a disk, accretion columns are formed above the magnetic poles of the white dwarf. Such intense fields are confirmed from Zeeman-effect measures in spectral lines. Of the order of 10% of CVs are of this 'AM Her' type.

The increase of kinetic energy in the in-fall remains equivalent to the loss of gravitational potential energy, so the impact at the poles, thermalized through dissipative shocks, corresponds to photon energies in the hard X-ray range (cf. Eqn 11.1). The emergent spectrum, however, depends on the transport of the source shock's heating through surrounding material to an effective photospheric region. Hard X-rays of the impact location are thus generally softened to emission in the spectral region $\lesssim 10$ keV, although there are 'anomalous' intervals when hard X-rays are observed, the soft X-ray component then being correspondingly reduced. The X-ray emission overall comes from a relatively small region above a magnetic pole, that is offset from the rotation axis. The white dwarf's rotation will thus bring the high energy emission zone in and out of view with a quasi-sinusoidal, or sometimes 'square

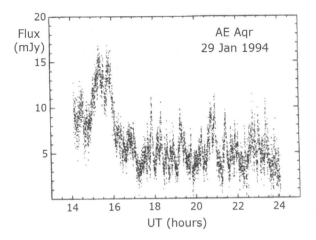

Figure 11.10 Microwave emission from the intermediate polar AE Aqr. The observations, taken with the VLA at 3.6 cm wavelength, cover a complete orbital cycle (9.88 hr) with the high time resolution of ~7 sec. Violent flaring activity can be seen on a variety of time scales that are mostly short compared with the orbital period. (Courtesy of T. Bastian who presented this diagram in *Astrophys. J.*, Vol. 461, p1016, 1996; ©AAS.)

wave' pattern. Geometry of the source in relation to the line of sight is clearly an important factor in determining the observed luminosity and spectrum.

The varying emission patterns have been interpreted in terms of 'blobs' or 'clumps' (inhomogeneities) in the accretion flow. Denser or more frequent blobs are associated with a more submerged initial energy source and a softer X-ray spectrum. The rarer hard X-ray transients are linked to more isolated low density impacts, when the background flux will also be generally less. Above a certain high field strength, magnetism will synchronize orbital and rotational periods for the white dwarf.

The transformation of the accretion structure from a plane to a spheroidal form has a sensitive dependence on a high field strength coupled with a low orbital period, so that the structure becomes either of one or the other types. There are, however, a small group of 'intermediate polars' (IPs), for which DQ Her is the assigned prototype, whose accretion structure combines both a disk-like form in the outer regions and a radial form closer to the white dwarf. The IPs also exhibit strong X-ray emission and high-excitation spectra, with persistent optical and X-ray pulsational effects in their light curves. Their white dwarfs have magnetic moments that are typically in the range 10^{32}-10^{34} G cm^3, so comparable to those of AM Her stars or slightly less. The emission features of the DQ Her stars are thought to have a similar explanation to those of the AM Her type, with softening of the original hard X-ray photons in intervening plasma, though the soft X-ray excess of AM Her stars is not as conspicuous as in the IPs.

Certain of the brighter polars and intermediate polars have been observed to be microwave radio-sources (Fig 11.10). AE Aqr, one of the more persistent of these, is classified as a DQ Herculis-type CV, or IP, comprising a magnetized white dwarf primary and a K5 dwarf secondary. Although AE Aqr is at a distance (84 pc) comparable with that of AM Her (75 pc), its radio flux density is a factor of 4-40 times greater in the 4.9 and 15 GHz bands. The emission usually shows no predominant polarization.

AE Aqr is also considered to be a source of ultra-high-energy (TeV) γ-rays, having quasi-periodicity on the order of the white dwarf's rotation period. This is notably short in the case of AE Aqr. With a period of just over 33 s, it is among the fastest known white

Figure 11.11   Scheme showing the main elements in models of X-ray binary systems. (Courtesy of Robert Hynes, LSU, Baton Rouge, USA.)

dwarf spins. The orbital period, on the other hand, is relatively long at nearly 10 h. The mass transfer rate is estimated to be at an intermediate $\sim 10^{-10}$ $M_\odot$ y^{-1} for this type of CV.

The microwave emission has been modelled in terms of a superposition of flare-type synchrotron emission events. If the mechanism were specified clearly, physical conditions in the accretion region could be well constrained by such observations. Unfortunately, most polars and IPs observed in the microwave region so far have not shown levels of radio-emission with high enough signal to noise ratio to allow detailed modelling. But the polars offer potentially interesting challenges to future radio observations with higher sensitivity.

11.3   X-ray Binaries

X-ray binaries (XRBs) include the brightest X-ray point sources in the sky and constitute a wide and somewhat perplexing class of astronomical object, unified by the release of very high energy radiation often pointing to the existence of a deep well of gravitational potential energy. In turn, this deep well implies the presence of a collapsed object (CO), particularly, in the case of intense or extreme high energy sources, a neutron star (NS) or black hole (BH). Since their accidental discovery (Scorpius X-1 by OSO-3, in 1962) XRBs have been the centre of extensive studies and the driving force of many past and future X-ray missions (e.g. UHURU, EXOSAT, RXTE, Chandra, XMM-Newton, NICER, STROBE-X). The context of highly condensed stars returns us to a subject raised in Chapter 1: namely, the possibility that a mass accumulation may have such an intense gravity that even radiation cannot escape the inward pull.

Michell's presentation on this implies a (Newtonian) constraint between the mass M and radius R of such an object such that $2GM/R = c^2$, where c is the velocity of light. Michell was at liberty to consider M and R as independent quantities. However, early studies of the internal structure of hydrostatically stable gas spheres showed that for a given value of M, R is determined, if a connection between internal pressure P and density ρ of the polytropic form $P = K\rho^\gamma$ can be satisfied, where K is a constant. This led on to the expectation of the Vogt-Russell theorem that the size of a star of specified mass would be constrained by its chemical composition and energy generation process.

But, by the first couple of decades of the 20th century, it was known that certain stars, such as Sirius B and 40 Eri B, would have to have extremely high mean densities compared to normal stars, by a factor of order a million. Sirius B and 40 Eri B became the archetypal *white dwarf* stars, thus identified by W. Luyten, who, in 1922, embarked on an observational programme to study this class of object.

11.3.1 Degenerate Gases and Collapsed Stars

A. S. Eddington's (1926) book on the internal constitution of the stars raised the question of how such white dwarfs night be understood, pointing out an apparent contradiction if a normal gas-law type equation of state was invoked for their structure. A solution to this problem was proposed by R. H. Fowler in 1926, who introduced the idea of a *Fermi gas*, where an equation of state of 'degenerate' form (i.e. with pressure independent of temperature), was applied to white dwarfs. As a consequence, it appeared there might be some finite mass satisfying Michell's conjecture.

But, separate from this argument, the physics of degenerate matter puts a definite value to the constant K in the pressure-density relation for an ensemble of particles, of given type, in compliance with the appropriate Fermi-Dirac statistics. For cold matter γ is found to be $5/3$ and $K \approx 2.2 \times 10^{12}$. However, at the centres of very dense stars, electron velocities may start to become comparable to the speed of light and a relativistic correction is required. S. Chandrasekhar found that, in this case, $\gamma = 4/3$ and $K \approx 3.7 \times 10^{14}$. The ratio P^3/ρ^4 at the centre of a polytropic model for a spherical star can be shown to be of the form $\kappa G^3 M^2$, where κ is a known numerical constant for a given γ. Hence, if $M < (K/G)^{3/2}\kappa^{-1/2}$ then the degeneracy pressure can support the white dwarf against gravity. This limiting mass turns out to be about 1.4 $M_\odot$. Classical white dwarfs are found to have masses below this. Remnants of stellar cores of mass above this value would not remain stable, on the basis of electron degeneracy pressure giving insufficient support against gravitational self-attraction. This situation thus leads to consideration of more extreme forms of CO: neutron stars, black holes, or what other possibilities might be found within the realm of physical conjecture.

11.3.2 High mass X-ray Binaries

High mass X-ray binaries (HMXB) could contain such extreme objects, though they are not very common cosmically. Even so, they are conspicuous because of the high luminosity of the non-collapsed, early type, normal star. It is this observable star that, in practice, associates an X-ray source with either the high, intermediate or low mass category. If the companion is over ∼10 $M_\odot$ it will produce relatively strong winds, some proportion of which interact with the CO, releasing high energy photons as material falls down the deep potential well. Interestingly, HMXBs may lead to the formation of a double CO binary, thus being progenitors of the recently observed gravitational wave sources.

Cygnus X-1, among the best known HMXBs, has received considerable attention since its discovery in 1964. It remains a powerful radiator of X-rays and one of the likeliest candidates for the black hole category, posing exciting challenges to contemporary physics. Rapid variations of the X-ray flux in time testify to the existence of a highly compact source (short light-crossing time), with fluctuations on the order of 1 ms suggesting a Schwarzschild radius less than 300 km. A recent estimate of the compact body's mass is ∼15 $M_\odot$, i.e. significantly more massive than known white dwarfs and neutron stars.

This datum is deduced from the binary context of Cyg X-1, in which the X-ray source is partnered with the O-type supergiant HDE 226868, having an orbital period of about 5.6 d. Cyg X-1 has been construed to be a possible member of the young star association

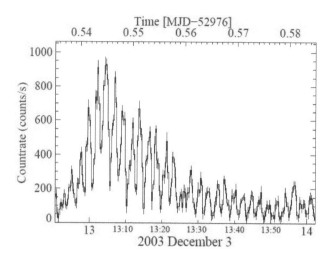

Figure 11.12   Pulsations in the 20-40 keV range light curve of Vela X1 persist through a large flaring event observed with 20 s time resolution (after I. Kreykenbohm et al. in *Astron. Astrophys.*, Vol. 492, p511, 2008; ©ESO).

Cygnus OB3, that enables an assessment of the age of the system and provides a potentially valuable check on the modelling and evolution scenario. Again, the binary system context means that transferred matter has some angular momentum, so we would expect an intermediating accretion disk, paralleling the lower mass examples already discussed. In the present context, the inner disk region could attain temperatures of millions of K, and this high temperature source would generate an observable X-ray flux. Besides the disk, very long baseline interferometry (VLBI) has resolved jets emanating from the central regions of the object that eject highly energized particles.

The configuration of a pair of jets and rapid rotation is reminiscent of a 'pulsar', where jets, associated with high temperature material aligned to the magnetic polar axis, are inclined to the rotation axis. The rotation of the underlying body brings a high temperature near-surface emission region in and out of terrestrial view (the 'lighthouse model'). While sharp, millisecond time-scale, emissions of X-radiation have been detected from Cyg X-1, a more established example of an X-ray pulsar would be the eclipsing HMBX, Vela X-1 (GP Vel), whose X-ray pulses come from a neutron star, rotating in a period of ~283 s. A modulation with half that period can be seen in Fig 11.12 suggesting hot spots at both poles contribute to the emission.

The degeneracy considered before for white dwarfs has a counterpart in the case of neutron stars, but in this case the pressure has intensified so as to make electron and protons to combine into neutrons through the inverse β process. It is presumed that a similar argument would apply concerning the available pressure from neutron degeneracy leading to a limiting mass, but there is still insufficient experience of the properties of matter in the extreme conditions in question to permit a definitive evaluation of this limit. However, observationally determined masses of neutron stars sometimes exceed the Chandrasekhar limit, for example, the NS in the HXMB Vela X-1 has been found to have a mass of 1.9 ± 0.1 $M_\odot$. In moving from the limiting density of electron degeneracy to that of neutrons, inter-particle separation falls by 3 orders of magnitude. There would be a corresponding rescaling of the star's size, assuming that the same general form of structure holds. NSs are then postulated to have sizes on the order of 10^{-5} that of the Sun.

While the detailed properties of NSs are a matter of specialist study, calling for close awareness of nuclear physics and related subjects, some basic premises appear intuitively reasonable; for example, relating to their formation. It was recounted in Chapter 8 that after the exhaustion of hydrogen fusion at the centre of Main Sequence stars a core region is formed with an essential resemblance to a white dwarf. The nature of this core depends strongly on the mass of the original star: higher mass stars being able to carry out fusion with heavier elements at least until iron, whereupon nuclear fusion cannot release further potential energy from the binding of nucleons. This implies that a certain initial size of star will, at some stage, generate a corresponding core that has to be supported by degeneracy pressure alone, though this limits the available core mass, as noted above. If this (Chandrasekhar) limit is exceeded, the core, with further contraction, encounters another restriction associated with the density of nucleons, in particular neutrons, when the increased pressure causes neutronization with the release of an intense flux of neutrinos.

This sudden brake to the infall and consequent energy release produces a *supernova* explosion, wherein the whole stellar envelope above the core is ejected. Incipient NSs are taken to be the remnant of the core after this event, but only for masses less than ~ 3 M$_\odot$. Higher mass remnants fall into the black hole category. Conservation of angular momentum entails that neutron stars are formed with very short rotation periods: on the order of 10^{-3} s. Their magnetic fields are correspondingly highly intensified. They are the natural candidates to explain pulsars. Originally associated with observations at radio frequencies, corresponding pulse phenomena are now known across the whole electromagnetic spectrum. Their interpretation becomes more satisfactory when observations are aligned and related to unifying principles or modelling.

Pulsars

Since their discovery by J. Bell Burnell and A. Hewish in 1967, pulsars have formed the central element in a growing field of intense research that touches on frontier areas of modern physics and technology. Around 5% of the ~ 2000 known pulsars are in binary or multiple systems: a subset that sets up an interesting laboratory in which the physics of matter in extreme conditions can be investigated using appropriate observational methods, with the circumstance of binarity allowing key parameters to be uncovered.

Binarity appears to have a special role in 'spinning-up' pulsation frequencies, accounting for most, if not all, of the observed 'millisecond' pulsars. These are pulsars with periods less than ~ 0.1 s: those with periods greater than that are, for the most part, 'normal' pulsars.

Millisecond pulsars correspond to a later 'recycled' evolutionary stage. It is expected that binary systems containing pulsars are those that must have survived an initial massive star configuration in which one component underwent a supernova explosion. These parent systems are typically made up of a high mass, Main Sequence B or Be star* together with the NS remnant of the other, originally more massive, star. The separations are of the order of an astronomical unit and the orbits are usually quite eccentric. An example is the binary pulsar PSR B1259-63. The NS source of the observed pulsations is thought to have originated with very high magnetic field strength and a short spin period.

At some point, the presently high mass MS star will become large enough to engage in RLOF, or perhaps it may also have a supernova explosion. If the binary survives the

---

*A Be star is an early type star showing hydrogen lines in emission and indications of high rotation speed.

latter process there will be a neutron star binary system, probably in an eccentric orbit. The orbit would subsequently circularize, as angular momentum is steadily shed from the system. This early stage of a binary pulsar's behaviour is termed the ejector phase. Instead of accretion, there is a strong pulsar wind and a gradual 'spinning down' process of angular momentum loss for the neutron star. This has been confirmed by very accurate timing of radio pulsations. The end product may be a pulsar + white dwarf combination when the donor envelope becomes sufficiently depleted. The surrounding plasma builds up around the pulsar, but it is 'propelled' away by the NS's co-rotating magnetosphere, increasing the rate of angular momentum loss.

The NS ultimately slows to spin periods of the order 100-1000 s. The co-rotation radius, defined as the radius at which the Keplerian angular velocity equals the spin angular velocity of the star, would be less than that of the magnetosphere during the propeller phase. Eventually, however, enough rotational momentum is lost to allow the co-rotation radius to exceed that of the magnetosphere, and the pulsar starts to accrete plasma and 'spin-up'.

In a binary system, the companion star provides a natural source of in-falling material, but irradiation of this material from the heated accretion region can reduce the rate of in-fall. There follow cycles of alternating spin-down and spin-up as the accretion rate slows and the magnetospheric radius grows, or *vice versa*. Accreted material accumulates in a column about the pulsar's magnetic poles where it undergoes high heating, and thermal X-rays are scattered by the in-falling particles. Knowledge of subsequent emission processes has become more detailed over the years, but is still under active discussion.

Supergiant XRBs

Studies of HMXBs were boosted in recent years following such technical advances as the INTErnational Gamma-Ray Astrophysics Laboratory (INTEGRAL) observation programme. As well as more standard HXMBs, new variants were discovered, including a curious sub-class of transient X-ray sources having early-type supergiant companions (SgXRBs). Bright, short flaring episodes from such sources (characterized as SFXTs) brought into question the scenario of the wind-CO accretion model. Explanatory efforts tend to invoke the idea of irregularities in the flow or 'clumpiness' of the transferred material.

The SFXT phenomenon may be relatively common in reality, but the detection of sources is hindered by their transient properties and the large Galactic absorption. Given these biases, it is no surprise that only one extragalactic SFXT was clearly confirmed so far. A recent study has shown that SgXRBs, in general, would have systematically higher X-ray absorption and luminosity compared to SFXTs, while the equivalent width of the fluorescent Fe K line is also higher in typical SgXRBs. This is taken to indicate that the SFXTs are an observationally selected subset of SgXRBs.

Be-star XRBs

In XRBs containing Be stars (BeXRBs) the condensed object is expected to encounter in-falling material from the slowly expanding equatorial disk that characterizes its companion. The configuration comes with a range of orbital parameters, but usually there is a CO in an eccentric orbit that is inclined to the equatorial plane of the Be star. When the CO passes through the equatorial ('decretion') disk an episode of flaring X-ray emission may be observed. Typical systems have orbital periods of weeks to months, though this may reach to years. X Per is a well-known example. BeXRBs exhibit several forms of variability besides their X-ray outbursts.

The more typical outbursts are in the range $L_X \sim 10^{36-37}$ erg s^{-1} and are usually correlated with the system's orbital period. Occasionally, BeXRBs may exhibit giant outbursts of a second type with $L_X > 10^{37}$ erg s^{-1}, that have been associated with irregularities in the Be disk. In some cases there must be significant accretion by the CO, according to the strong spin-up effects measured during bright outbursts. Long-term X-ray variability of BeXRBs may be related to slow changes of the binary orbit and the reservoir of available disk material. In about half of the BeXRBs regular X-ray pulsations have been observed. The other half show similar X-ray spectra, suggesting that the COs are indeed neutron stars, though the possibility of black hole mass COs has been sometimes raised.

BeXRBs are also highly variable in the optical band, so that additional information can be gathered from their optical properties. Optical effects in BeXRBs take a variety of forms such as: (i) non-radial pulsations of the Be-star with periods $\sim$ 0.1-2.0 d, (ii) the Be star decretion disk gravitationally interacting with the CO during periastron passage, and (iii) semi-stable distortions of the disk ('warping'), that may also contribute to longer term source variability.

In recent years, the number of identified BeXRBs increased to over 200. The Small Magellanic Cloud appears particularly rich, accounting for about half the known examples. Surveys of these SMC BeXRBs have shown that the majority of the systems show at least one of the associated optical characteristics, with more than 40% having optical variability related to the binary orbit.

Wolf-Rayet and Ultraluminous XRBs

Wolf-Rayet star (WR) X-ray Binaries (WRXBs) are a small group of HMXBs that, apart from Cyg X-3, are located in external galaxies. They are mostly found to have short orbital periods, of order a day. WR stars are characterized by an extremely strong, radiation-driven, stellar wind, with mass-loss rates around 10^{-5} M$_\odot$ y^{-1}. They are massive evolved stars with envelopes severely depleted in hydrogen and are fusing helium in their cores. Wind speeds are typically in the range 2000-5000 km s^{-1}. As the masses of WR-stars in binaries are measured to be at least 8 M$_\odot$, the large luminosity and strong radiation pressure required for driving WR winds have to develop from stars with masses originally well above this lower limit. To produce a helium core larger than 8 M$_\odot$, the WR progenitor must have had a mass of at least 30 M$_\odot$. WRXBs should therefore have had HMXB progenitors with donor masses of at least 30 M$_\odot$. From these general properties and evolution principles it is reasonable to expect that the COs in WRXBs are black holes.

Ultra-luminous X-ray binaries (ULXBs), sometimes referred to as microquasars, have been found as extragalactic X-ray sources with luminosities appearing to exceed the Eddington limit (assuming isotropic emission) for a stellar mass black hole.* It is not clear what powers ULXB radiation. Beamed radiation (rather than isotropic) has been used in modelling, as well as black hole sources radiating at super-Eddington rates. The possibility that such sources are actually background quasars has been shown likely in some cases.

In 2014, a ULXB was linked to a highly magnetised neutron star, and since that initial discovery, a few more ULXB pulsars were detected. Other recent studies have pointed to non-pulsating ULXBs sharing similarities with BeXRBs that host an accreting NS. The powering of these extreme sources clearly still offers challenges to physical understanding.

---

*The Eddington limit comes from the radiation pressure generated in a star reaching the scale of the gravity holding it together. It can be estimated in a simple way, for a spherical body, by requiring that the surface gravity GM/R^2 balance the outward radiation pressure there $L\kappa/(4\pi R^2 c)$, so that $L_{\rm Edd} = 4\pi GMc/\kappa$, where κ is a representative mean opacity. In the ULXB context this implies luminosities of order 10^{39} erg s^{-1} or greater.

11.3.3   Low mass X-ray Binaries

Although some X-ray binaries are classified as intermediate mass (e.g. Her X-1, associated with the optical variable HZ Her), the main general division is between HMXBs and LMXBs, the former more conspicuous in 'hard' X-rays ($\gtrsim$ 15 keV) the latter appearing with 'soft' spectra ($\lesssim$ 10 keV). HXMBs show a pulsational behaviour, but are without the relatively large 'bursts' of LMXBs that makes a distinctive difference between them.

In a general way, the properties of these groups parallel those of early and late type stars, i.e. HMXBs are younger, more massive and concentrated towards the Galactic plane. The LMXBs are distributed like an old population, some notable examples being found in the globular clusters, for example, the 'Rapid Burster' (= MXB 1730-335) in Liller 1. In the cosmic context, we find more LMXBs in old galaxies: the Magellanic Clouds are almost bare of LMXB sources.

X-ray bursters are found in two main types: (i) the more readily observed 'type I's, associated with thermonuclear ignition of accreted matter in the surface regions of the CO, and (ii) the type IIs – linked to instabilities in the accretion process. Parallels are thus drawn between the behaviour of X-ray bursters and CVs, with a neutron star in the former replacing the white dwarf in the latter. Type I bursters then become comparable to the old novae and the type IIs to the dwarf novae. Of course, such parallels should not be overstretched. For example, in an old nova explosion it is hydrogen building up on the white dwarf surface that at, some point, triggers a runaway fusion process that triggers the nova. In an X-ray burster, hydrogen dropping towards the NS in an accretion process is directly converted to helium due to the extreme temperatures and pressures it encounters towards the NS surface.

A clue about the presence of black holes or neutron stars in X-ray binaries thus presents itself, in that the environment that could support surface ignition of accumulated material is not there in the case of a regular black hole. This latter category of CO, within the context of X-ray binaries, might therefore be preferentially sought among HMXBs. On the other hand, analysis of LMXB bursts offers the prospect of determining absolute properties of NSs and their close environments, including their magnetic fields.

Helium accumulates near the surface region of the NS until it reaches a critical mass when an explosive reaction takes place having some resemblance to the 'helium flash' in red giants of a certain mass. The NS attains temperatures of order 10^7 K, expanding and releasing high-energy X-ray photons in a heating process that lasts a few seconds. The heating up involves energies of order 10^{32} J, and is followed by a cooling-down relaxation that takes about 10 times as long. X-ray bursts recur on the time-scale 10-100 days; the interval between bursts correlating with the net energy output of the previous burst.

It is clear that the high energies of photons associated with X-ray binaries are less than the full potential energy reserve corresponding to (11.1), but this would be expected in the presence of angular momentum and intervening plasma of finite density. Given those factors, as well as the geometry of energy-releasing interactions in relation to the line of sight, observed emissions from binaries containing COs should be softened from those of plausible source conditions. This point would apply to some of the dwarf novae types considered previously that emit (transient) X-rays, and so might be associated with the LMXBs.

LMXB type II transient sources show comparable properties to the dwarf novae, in that, from time to time they produce outbursts lasting typically weeks, with intervals between outburst recurrences ranges from years to decades. The systems' behaviour can be accounted for in a similar general way to those of dwarf novae by allowing the invoked α parameter to take different values between hot and cold phases of the cycle. However, the spacing and intensity of outbursts, as well as their quiescent luminosity, require some additional model features. These concern truncation of the inner disk and inclusion of changes to the rate of mass flow due to irradiation effects on the secondary.

Supersoft sources

A further grouping of LXMBs is that of the *supersoft* X-ray sources (SSXS) that emit X-rays, but only in the range less than 1keV, so with photon energies a couple of orders of magnitude down on those of the HXMBs. The binary character of SSXSs was established following observations with the ROSAT satellite after 1990. Although emitting only relatively low energy X-rays, these sources are highly luminous bolometrically, radiating with total energies of order 10^5 L$_\odot$. Studies of the SSXSs indicate they are, for the most part, consistent with a CV configuration in which the white dwarf continually receives RLOF-transferred material from a companion star at the relatively fast rate of $\sim 10^{-7}$ M$_\odot$ y^{-1} (i.e. Kelvin time-scale mass-transfer). Steady burning of hydrogen in the outer regions of the accreting white dwarf is the posited source of the high luminosity. If continued long enough, this interaction is expected to lead to the Type Ia supernova phenomenon.

LMXBs with black hole components

At present, there are about two dozen X-ray binaries containing COs that are strong black hole (BH) candidates. Most of them (19 sources) have been termed black hole low-mass X-ray binaries (BHLMXBs) because their donor star masses are less than 1 M$_\odot$. More than half of the BHLMXBs have short orbital periods ($P \lesssim 1$ day), and, in a few cases, show a trend of period reduction. These facts point to the binary having experienced common envelope evolution, in which a low-mass companion spirals into the expanding envelope of a massive primary star with dynamically unstable mass transfer. Thereafter, the binary orbit is greatly reduced, the decline of systemic potential energy being transferred to the ejected envelope. If the binary survives the primary's subsequent supernova explosion, whose remnant is the putative BH, it evolves to become a LMXB.

The study of BHLMXBs is of importance in understanding the astrophysics of very strong gravitational fields and binary evolution with common envelope phases. But it is not obvious that low-mass donor stars can really eject the massive envelope of the BH progenitors during the common envelope phase. Calculations for reasonable initial mass distributions and evolving stars predicted a BHLXMB formation rate that was found to be typically two orders of magnitude lower than that derived from observations.

The difference may be accounted for by adopting a higher ejection efficiency. Alternatively, it was proposed that BHLMXBs have evolved from intermediate-mass X-ray binaries containing Ap or Bp donor stars giving rise to high magnetic braking effects, or that a circumbinary disk could significantly affect the mass transfer rate. Recently, a 'failed supernova' mechanism was invoked, where the BH mass can derive from the He or CO core mass of the progenitor within a common envelope scenario, but without the enormous explosion of a regular supernova.

In the theory for forming BHLMXBs, possible angular-momentum-loss mechanisms have included gravitational radiation and magnetic braking, as well as systemic mass loss. The orbital-period derivatives measured for certain BHLMXBs thus provide useful hints on their evolution, or that of their progenitors. Recent measures of BHLMXB period declines are in the range $\dot{P} \approx -0.5$ to -20 msec y^{-1}. Such orbital decays could be interpreted in terms of circumbinary disks, fed from diverted mass loss during the outbursts of BHLMXBs. At the present time there are only a few known examples of well-measured $\dot{P}$ values, but more data will surely help clarify details of the angular momentum evolution of BHLMXB systems.

11.3.4   End Products

Close double neutron stars are believed to be the later evolutionary products of wide BeXRBs, having orbital periods greater than about a year. Neutron-star containing HMXBs with shorter orbital periods should eventually merge. The results of such a process have been associated with Thorne-Żytkow objects.* Although this has been discussed for over 40 years, with the kind of interaction considered not uncommon, particularly with in-spiralling binaries or crowded stellar environments, no such object has been identified with certainty so far.

Close double black holes and BH-NS systems that formed through binary evolution, are proposed as later evolutionary products of the short-period WRXBs. For the formation of the latter systems three models have been proposed: (i) through *common envelope evolution* from wide black hole HMXBs (BHHMXBs). This is basically the same model as that for the formation of NS-NS systems, scaled up to higher masses; (ii) by *in-spiral* due to stable Roche-lobe overflow from BHHMXBs, having typical supergiant HMXB orbital periods upwards from a few days. Only systems with orbital periods less than about 10 days would be likely able to terminate as double black hole systems with orbital periods short enough to merge within a Hubble time; (iii) by *homogeneous chemical evolution* of massive close binaries with orbital periods $< 2 - 3$ days and mass ratios $> 0.7 - 0.8$.

Homogeneous evolution has received more attention in recent stellar modelling. In massive, fast-rotating stars mixing processes induced by rotation can mix centrally fused helium throughout the envelope. Such process may be effective in close binary systems. As these stars will remain like Main Sequence stars for long intervals of time, their evolution is appreciably different from classical binary evolution models. This scenario can lead to tight Wolf-Rayet binaries and, ultimately, to BHHMXBs.

11.4   Gravitational Wave Astronomy

September 2015 witnessed the era-initiating first detection of gravitational waves (GW), from the merger of two black holes, with the advanced Laser Interferometer Gravitational-wave Observatory (LIGO), leading to a Nobel prize in Physics award in 2017. An illustration from the main discovery paper of B. P. Abbott et al. (2016) is shown in Fig 11.12. A subsequent GW detection of the collision of two neutron stars in coincidence with electromagnetic observations inaugurated the field of *multimessenger* experimental and theoretical astrophysics. New detections with a three-detector network have enabled further fact-finding on the nature of GWs, as well as putting better limits on the sky location of GW transients. Multimessenger astronomy confirmed that NS mergers are a source of short gamma ray bursters GRBs.

Ideas about gravitational waves crystallized around the turn of the twentieth century, with the introduction of important new formulations in fundamental physics, culminating in Einstein's general theory of relativity, which presents a gravitational field as a curvature of space-time due to the presence of mass. If this mass is in a state of accelerated motion with respect to an observer, it is natural to expect changes to the corresponding space-time continuum experienced by the observer, even if these are very small in normal situations. Gravitational waves are identified with the propagation of this variation.

---

*K. Thorne and A. Żytkow proposed such stars formed when a colliding neutron star replaces or merges with the core of an originally more typical star. The result may superficially resemble a red giant or supergiant, though the very high surface temperature of the neutron star results in abnormal subsequent evolution.

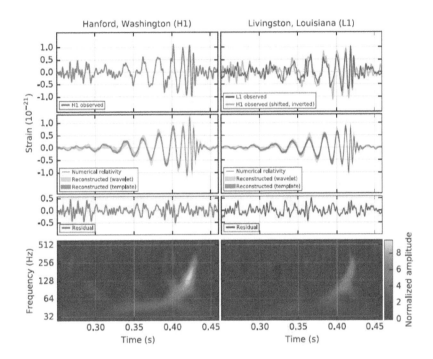

Figure 11.13 Recording of the gravitational wave event GW 150914 as published by B. P. Abbott et al., *Phys. Rev. Lett.*, 116 061102, 2016. Presented under the terms of the Creative Commons Attribution 3.0; https://doi.org/10.1103/PhysRevLett.116.061102. The top panels show the strain measured at the LIGO Hanford and Livingston L1 detectors. The second row presents the strain data filtered into the 35-350 Hz signal band. Suitably parametrized trends, using both numerical relativity calculations and an empirical representation with sine-Gaussian wavelets, are indicated by shaded backgrounds behind the smoothed observational curves. Probabalistic analysis of the match supports the astrophysical modelling with a very high likelihood. Residuals are given in the third panels. The bottom row shows the events in terms of the signal frequency variation with time.

Appropriate exposition of this subject is well outside our present scope; however certain parallels can be made with the more familiar electromagnetic waves, for example, both travel in vacuo at the speed of light. The counterpart to the electric displacement in Maxwell's wave equation, for gravitational waves, is the metric perturbation $h^{\mu\nu}$, also known as the strain tensor. This accounts for changes in the proper separations of particles in the plane perpendicular to the wave direction. The frequency of the wave from a close binary system is twice that of the orbital motion and there are two polarization states. These describe the scale of the metric perturbation parallel to a given reference direction (plus, or h^{+}), and at $45°$ to that direction (cross, or $h^{\times}$), recalling the relationship of the Q and U parameters in the polarimetric double ellipse to the astrometric ellipse of a close binary orbit. GW emission maximises in the direction of the orbital angular momentum vector with a quadrupole harmonic leading term.

Quantitative application of GW theory was greatly stimulated by observations of the Hulse-Taylor binary pulsar in 1974, noted in Chapter 3, and, aside from more basic physical contexts, CO binaries provide interesting sources of gravitational waves that are potentially detectable with facilities like LIGO and its European confrere organization, Virgo.

A key factor is the rate of in-spiralling due to the loss of orbital potential energy associated with the emission of gravitational waves from the binary. In a comparable way to the derivation of (3.89), this can be shown to be given in the post-Newtonian formulation by

$$\frac{dr}{dt} = -\frac{64}{5}\left(\frac{G^3\mu M^2}{c^5 r^3}\right)\phi(e) \; . \tag{11.27}$$

Eqn (11.27) can be integrated directly to yield the result

$$\alpha\tau = r_0^4 \; , \tag{11.28}$$

where $\alpha \sim 0.13 \times 10^{28} M^3$, M being the mass of the system in units of the solar mass, and c the velocity of light. $\phi(e)$ is a function of the eccentricity, usually given as $\phi = (1 + (73/24)e^2 + (37/96)e^4)/(1 - e^2)^{7/2}$. τ is the time scale to the final coalescence of the binary, starting from a given initial separation r_0. This equation determines the kind of observational regime in which effects will be detected. Note the sensitivity of the merger time to M and also the speeding up effect of short τ going with a low initial separation. By setting trial values of τ in (11.28) we can judge that binaries whose separations are on the order of tens of km may coalesce after an appreciable final speed-up lasting a few seconds. In order to be distinct and have such separations, the binaries in question, having masses $\gtrsim 1 M_\odot$, must contain highly compressed stars, i.e. neutron stars or black holes. The period decline of CO binaries with wider separations than a few tens of km, including systems like the Hulse-Taylor binary or perhaps close white dwarf pairs, can be regarded as in a different frequency range for observable effects where GWs are not yet directly observed.

The *power* emitted by the gravitational waves rises with the declining separation as $(G\mu M/2r^2)\dot{r}$, and it is in the final few tenths of a second that sufficient power is generated to allow detection by such instruments as LIGO. The effect is termed a *chirp*, and an illustration is shown in Fig 11.14. Chirp waveforms like the one shown are essentially determined by 4 parameters, including the fiducial quantities of the initial frequency and wave phase, together with the 'chirp mass' M_c. This is related to the total mass M as $M_c = M \times q^{3/5}/(1 + q)^{6/5}$, so typically around 0.4 of the total mass. If M were somehow known independently, then matching the chirp pattern could allow the mass ratio to be determined from this approach. In fact, other information than that available from this simple model of the in-spiral is required to fix the masses. Interestingly, the distance to the source D is the remaining parameter characterizing the chirp.

From a wider perspective, binary black hole (BBH) data should involve other physical parameters such as the spin angular momenta as well as the mass ratio. If the spin and

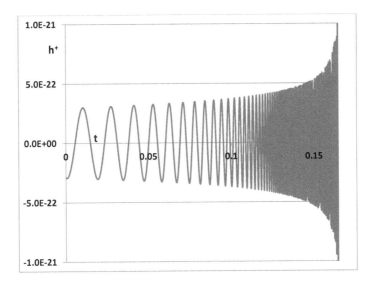

Figure 11.14 Schematic form of a GW 'chirp', associated with the motion towards coalescence of a CO binary system (after J. Wheeler, 2013). The strain h^+ is plotted against time t. The units in this graph, following the form of the underlying function, are scaled to align with those applying to extragalactic black hole mergers such as that of GW150914. The numerical quadrature loses accuracy with the advancing frequency.

orbital angular momenta are not aligned precessional effects occur and the orbital plane will change its orientation. This shows up as modulations in the amplitude and phase of observed GWs, and higher order harmonics appear in the wave description. Analytic solutions of the fully generalized two-body problem in GTR are still under development, but the advent of fast electronic devices with powerful numerical capability has allowed inroads into data interpretation, particularly in the later stages of the interaction. This calls for an ability to relate numerical calculations to the kind of physical parametrization that applies to the in-spiralling.

Apart from the in-spiralling, where post-Newtonian modelling may be applicable, BBH orbital evolution extends to the merger and 'ringdown' phases. The development of post-Newtonian treatment of the orbit shrinkage makes reference to the perturbation parameter $\epsilon = v^2/c^2$, where v is the orbital velocity. Naturally, one may suppose the perturbation approach to lose validity and give way to numerical methods when the orbital velocity $v \to c$. Various separate procedures for numerical calculations of the GW emission from merging black holes have been available over the last decade with a wide range of starting parameters, though difficulties remain for more extreme mass ratios and spin values.

The ringdown phase that follows merger restores the opportunity for analytical treatment on the basis of oscillatory mode formulation and perturbation theory. Numerical relativity still has an important role, however, in providing GW amplitudes to complete the modelling procedure. Recent work in this field has arranged itself into whether the data analysis concentrates in the time or frequency domain, or the extent to which interpolation across a framework of prior numerical relativity waveform models is used. The trends from modelling results in the $(\log(f) : \log(\tilde{h}))$ plane,* with changes of direction at the

---

*$\tilde{h}$ is the Fourier transform of the GW strain.

frequencies of the minimum energy circular orbit, innermost stable circular orbit and ring-down domains, point to the increased information content coming from the complete sequence of the GW emission data.

A catalogue of 11 GW transients associated with CO mergers observed by Advanced LIGO and Virgo detectors, and providing modelling parameters, was published in 2019. Binary black holes (BBH) in this catalogue have total masses between $\sim 19\ M_\odot$ and $\sim 85\ M_\odot$ and range in distance between 320 and 2840 Mpc. No NS-BH mergers were confirmed so far. The catalogued parameters are based on numerically generated models of general relativistic waveforms. In addition to high significance GW detections, the catalogue provides a list of marginal event candidates with an estimated false-alarm rate less than 1 per 30 days.

These early stages of GW-astronomy, briefly sketched in the foregoing, already provided a solution to the problem of the gamma-ray burster as a merging CO phenomenon. They allow tests of GTR and alternative theories of gravity over a new range of extreme conditions as well as permitting further insights into the nature and origins of black holes. Such results are expanding our awareness of frontier areas of physics. In the future, with increasing and improved detective technology, we can expect many different compact binary star systems to be investigated through the medium of gravitational waves. The key distance parameter in BBH coalescence modelling will provide increased precision of the Hubble constant determination, with important implications for fundamental science.

11.5   Bibliography

11.1 The term *Symbiotic Star* was introduced by P. W. Merrill in the *8ème Colloq. Intern. d'Astrophys.*, (Liege) Vol. 20, p436, 1958; and is reserved for those astronomical objects showing a combination of low temperature absorption features together with high excitation emission lines. *White dwarf* was a term introduced by W. J. Luyten after he began a programme of study on them (cf. *Publ. Astron. Soc. Pacific*, Vol. 34, p156, 1922). These stars are also referred to as 'faint blue', depending on the context, for example in A. Sandage & W. J. Luyten, *Astrophys. J.*, Vol. 148, p767, 1967. R. S. Stobie, D. Kilkenny, C. Koen & D. O'Donoghue, *Third Conf. Faint Blue Stars*, eds. A. G. D. Philip, J. Liebert, R. Saffer and D. S. Hayes, L. Davis Press, p497, 1997; noted that many such objects show short period variability. The subject of how such variability is affected by binarity was reviewed by P. Szkody, A. S. Mukadam, B. T. Gänsicke, A. Henden, A. Nitta, E. M. Sion & D. Townsley, *Eighth Pacific Rim Conf. Stellar Astrophys.: (A Tribute to Kam-Ching Leung) Astron. Soc. Conf. Ser.*, Vol. 404, eds. B. Soonthornthum, S. Komonjinda, K. S. Cheng, and K. C. Leung, p229, 2009. D. A. Allen's catalogue of symbiotic stars was published in the *Proc. Astron. Soc. Aust.*, Vol. 5, p369-421, 1984. S. J. Kenyon's review in *The Symbiotic Stars*, CUP, 1986, included information on 133 symbiotic stars and 20 possible examples, as well as tables describing selected observational properties and a spectroscopic summary of a selected sample. Kenyon's work also provides an excellent overview and bibliography of the subject. A catalogue of 188 symbiotic stars and 30 objects suspected of being symbiotic was published by K. Belczyński, J. Mikolajewska, U. Munari, R. J. Ivison, M. Friedjung, in *Astron. Astrophys. Suppl., Ser.*, Vol. 146, p407, 2000.

The heterogeneous quality of systems associated with the symbiotic classification was noted by R. F. Webbink in his contribution to the *Proc. IAU Coll. 103; The Symbiotic Phenomenon*, eds. J. Mikolajewska, M. Friedjung, S. J. Kenyon, & R. Viotti, Kluwer, *Astrophys. Space Sci. Lib.*, Vol. 145, p311, 1988. At least three subgroups of symbiotic stars were identified according to the energy source for their hot components: (i) disk-accreting MS stars, e.g. CI Cyg, see S. J. Kenyon, R. F. Webbink, J. S. Gallagher, J. W. Truran, *Astron. Astrophys.*, Vol. 106, p109, 1982; and J. Mikolajewska, M. Mikolajewski, in *The*

Importance of IUE and Exosat Results on Cataclysmic Variables and Low-Mass X-Ray Binaries, ESA Sci. Tech. Publ., Noordwijk, No. 236, p201, 1985; (ii) wind-fed, shell-burning white dwarfs with extended envelopes, e.g. AG Peg, see J. S. Gallagher, A. V. Holm, C. M. Anderson, R. F. Webbink, *Astrophys. J.*, Vol. 229, p994, 1979; and C. D. Keyes, & M. J. Plavec, *IAU Symp.*, Reidel, No. 88, p535, 1980; (iii) disk-accreting neutron stars, e.g. V2ll6 Oph = GX 1+4, see E. P. Cutler, B. R. Dennis, J. F. Dolan, *Astrophys. J.*, Vol. 300, p551, 1986. Further subdivisions have been proposed, (cf. P. A. Whitelock, *Publ. Astron. Soc. Pacific*, Vol. 99, p573, 1987). The status of the yellow (D'-type) symbiotic stars within this classification scheme appears still open to discussion; see I. S. Glass, B. L. Webster, *Mon. Not. Royal Astron. Soc.*, Vol. 165, p77, 1973; D. A. Allen *The Nature of Symbiotic Stars*, *Astrophys. Space Sci. Lib.*, Vol. 95, p27, 1982.

W. Schmutz, H. Schild, U. Muerset and H. M. Schmid, using high resolution spectrograms of the eclipsing symbiotic system SY Mus, determined absolute parameters of the components that were published in *Astron. Astrophys.*, Vol. 288, p819, 1994. Comparable work of H. Schild appeared in *Astron. Astrophys.*, Vol. 306, p477, 1996. The potentially important role of Raman scattering in relevant spectroscopy was pointed out by H. Nussbaumer, H. M. Schmid, & M. Vogel, in *Astron. Astrophys.*, Vol. 211, L27, 1989. A survey of related polarization effects by H. M. Schmid & H. Schild was published in *Astron. Astrophys.*, Vol. 281, p145, 1994. Modelling of Raman scattering effects was presented by T. J. Harries & I. D. Howarth in *Astron. Astrophys. Suppl. Ser.*, Vol. 121, p15, 1997.

The relationship of the giant components in symbiotic stars to their surrounding Roche lobes was studied by M. R. Garcia in *Astron. J.*, Vol. 91, p1400, 1986. This topic was pursued by S. J. Kenyon & T. Fernandez-Castro, who suggested that binarity enhances mass loss from giants, cf. *Astron. J.*, Vol. 93, p938, 1987.

A distribution of periods of symbiotic stars (from ∼100 to ∼1400 d) was published by J. Mikolajewska in *Baltic Astron.*, Vol. 16, p1, 2007. though this distribution has defied theoretical attempts at a comprehensive explanation; see, for example, Z. Han, P. P. Eggleton, P. Podsiadlowski, C. A. Tout, *Mon. Not. Royal Astron. Soc.*, Vol. 277, p1443, 1995. The topic was raised by R. F. Webbink in *Highlights Astron.*, Vol. 7 (Reidel) p185, 1986; see also P. Podsiadlowski & R. E. S. Clegg in *Dusty Discs: Proc. 9th RAL Workshop on Astron. Astrophys.*, ed P. M. Gondhalekar, p69, 1992. The possibility of a tidally coupled circumbinary disc influencing the orbit was discussed by H. C. Spruit & R. E. Taam, cf. arXiv:astro-ph/0111420, 2001; and A. Frankowski & A. Jorissen, *Baltic Astron.*, Vol. 16, p104, 2007.

The Chandra ACIS-S detection of a bright soft X-ray transient in the Mira AB interacting symbiotic-like binary and X-ray resolution of the system were reported by M. Karovska, E. Schlegel, W. Hack, J. C. Raymond & B. E. Wood, in the *Astrophys. J.*, Vol. 623, L137, 2005. The idea of wind assisted mass loss in symbiotic binaries was quantitatively developed by S. Mohamed & P. Podsiadlowski, in the *15th European Workshop on White Dwarfs: ASP Conf. Ser.*, Vol. 372, eds. R. Napiwotzki & M. R. Burleigh, p397, 2007. Calculations of wind-assisted mass loss in massive star evolution were presented by S-C. Yoon & M. Cantiello, in *Astrophys. J.*, Vol. 717, L62, 2010.

Within the context of binary systems among post-Asymptotic Giant Branch stars S. A. Bell, D. L. Pollacco & R. W. Hilditch (*Mon. Not. Royal Astron. Soc.*, Vol. 270, p449, 1994) called attention to sdO + dK binary UU Sge that allows confident parametrization, due to its eclipsing behaviour (see also, E. Budding & Z. Kopal, *Astrophys. Space Sci.*, Vol. 73, p83, 1980). Such systems are relatively rare, however (see M. Livio, *Planetary nebulae, Proc. 180th Symp. IAU*, eds. H. J. Habing and G. L. M. Lamers, Kluwer, p74, 1997), though they could make up one of the varieties of symbiotic models as the secondary ascends the giant or asymptotic giant branch, according to A. V. Tutukov and L. R. Yungel'son; *Astrofizika*, Vol. 12, p521, 1976. Other symbiotic star types appear to be the products of earlier phases of binary interaction judging by their short orbital periods. Their progenitors should have passed through a phase of common envelope evolution. Scenarios were sketched by B. Paczyński, *Structure and Evolution of Close Binary Systems: Proc. IAU Symp. 73*, eds. P. Eggleton, S. Mitton, & J. Whelan, p75, 1976; H. Ritter, *Mon. Not. Royal Astron. Soc.*, Vol. 175, p279, 1976; R. F. Webbink, *Astrophys. J.*, Vol. 209, p829, 1976; F. Meyer & E. Meyer-Hofmeister, *Astron. Astrophys.*, Vol. 78, p167, 1979; and others. H. E. Bond in *Planetary Nebulae*, ed. S. Torres-Peimbert, *IAU Symp. 131*, Kluwer p251, 1989; called attention to a class of short period 'pre-cataclysmic' binaries consisting of a hot white dwarf primary and lower Main Sequence secondary. Examples include the systems HW Vir (L. L. Kiss, B. Csák, K. Szatmáry, G. Furész, K. Sziládi, *Astron. Astrophys.*, Vol. 364, p199, 2000); and NY Vir (D. Kilkenny, D. O'Donoghue, C. Koen, A. E. Lynas-Gray, F. van Wyk, *Mon. Not. Royal Astron. Soc.*, Vol. 296, p329, 1998), whose eclipses permit more detailed parametrization.

11.2 Turning to the cataclysmic variables (Section 11.2); among the numerous published reviews the large compendium of information in B. Warner's *Cataclysmic Variable Stars*, CUP, 1995, is often cited as a comprehensive resource. The classic paper of Merle F. Walker in *Astrophys. J.*, Vol. 123, p68, 1956; raised the possibility of binarity underlying various groups of stars showing an explosive character in their light output: a theme that was taken up by R. P. Kraft using the descriptor cataclysmic in *Astrophys. J.*, Vol. 135, p408, 1962. The observations of U Gem taken from the AAVSO International Database shown in Fig 11.2 have been contributed by observers worldwide — a service that is highly appreciated.

A *Catalog and Atlas of Cataclysmic Variables* was published by R. A. Downes, R. F. Webbink, M. M. Shara, H. Ritter, U. Kolb, & H. Duerbeck, in *Publ. Astron. Soc. Pacific*, Vol. 113, p764, 2001, and an updated form has been maintained on internet. The catalogue of H. Ritter & U. Kolb, referred to in the caption for Fig 11.4, originally appeared as *Astron. Astrophys.*, Vol. 404, p302, 2003. Fig 11.4 was produced by B. Kalomeni, L. Nelson, S. Rappaport, M. Molnar, J. Quintin & K. Yakut, in *Astrophys. J.*, Vol. 833, p83, 2016. Figs 11.5 and 11.6 refer to the previously cited paper of S. H. Lubow and F. H. Shu in *Astrophys. J.*, Vol. 198, p383, 1975. The spiraling-in and shedding of a large common envelope scenario for the progenitor systems of CVs was presented by B. Paczyński in *Structure and Evolution of Close Binary Systems; Proc. IAU Symp. 73*, eds. P. Eggleton, S. Mitton & J. Whelan, (Reidel), p7, 1976. The mass distribution of WDs in CVs was reviewed by T. P. G. Wijnen, M. Zorotovic and M. R. Schreiber in Astron. Astrophys., Vol. 577, p142, 2015. The same authors reviewed related issues in the Proceedings of the Golden2015 Workshop, arXiv.1512.03310v1, 2015.

The INTEGRAL/IBIS facility, referred to in connection with the high-energy component of CV outbursts, was described by A. Goldwurm, P. David, L. Foschini, A. Gros, P. Laurent, A. Sauvageon, A. J. Bird, L. Lerusse & N. Produit, in *Astron. Astrophys.* Vol. 411, L223, 2003.

The thin disk model, discussed in Section 11.2.2 descends from treatments of galactic structure like those of B. Lindblad, *Zeit. Astrophys.*, Vol. 15, p124, 1938; A. B. Wyse & R. U. Mayall, *Astrophys. J.*, Vol. 95, p24, 1942; C. F. von Weizsäcker, *Zeit., Astrophys.*, Vol. 22, p319, 1943; re-formulated for the stellar accretion disk context in such sources as N. I. Shakira, *Astron. Zh.*, Vol. 49, p921, 1972; N. I. Shakira & R. A. Sunyaev *Astron. Astrophys.*, Vol. 24, p337, 1973; the latter introducing the 'alpha-disk', which relates the kinematic viscosity to the local pressure as in Eqn (11.11). This approach was discussed by A. P. Lightman, *Astrophys. J.*, Vol. 194, p419, 1974; J. E. Pringle, *Ann. Rev. Astron., Astrophys.*, Vol. 19, p137, 1981; and others. *Dynamics of Astrophysical Disks*, ed. J. A. Sellwood, CUP 2004, collects reviews of the applications of disk models to various astrophysical contexts, including close binary systems.

The eclipse-mapping technique, illustrated in Fig 11.7, was presented by R. Baptista, in *Astrotomography — Indirect Imaging Methods in Observational Astronomy: Lecture Notes Phys.*, Vol. 573, eds. H. M. J. Boffin, D. Steeghs & J. Cuypers (Springer-Verlag), p307, 2001, ©Springer Nature; for an illustration relevant to UX UMa see also http://www.ing.iac.es/PR/SH/SH94/spectra.html. C. Knigge, N. Drake, Nick; Knox S. Long, R. A. Wade, K. Horne, R. Baptista, *Astrophys. J.*, Vol. 499, p429, 1998; applied the Hubble Space Telescope (faint object spectrograph) to UX UMa, confirming the 29 sec oscillations observed by B. Warner & R. E. Nather in *Mon. Not. Royal Astron. Soc.*, Vol. 159, p429, 1972; although with intermittent complications of detail. J. Kube, B. T. Gänsicke & K. Beuermann, *Astron. Astrophys.*, Vol. 356, p490, 2000; used eclipse-mapping to study accretion in UZ For. The idea that a line profile provides a mapping of the stellar shape, mentioned in Section 6.3, is recapitulated in the 'Roche tomography' technique presented by C. A. Watson & V. S. Dhillon, *Mon. Not. Royal Astron. Soc.*, Vol. 326, p67, 2001.

A useful summary of the maximum entropy method (MEM) for data analysis can be found in R. W. Hilditch's *An Introduction to Close Binary Stars*, CUP, 2001, pp304-306 (see also J. Skilling, *Maximum Entropy and Bayesian Methods (Proc. 8th Cambridge Workshop)*, ed. J. Skilling, Kluwer, 1989).

There has been a considerable literature on possible mechanisms for producing the explosive behaviour in the light curves of eruptive CVs. Y. Osaki's arguments on the role of the accretion disk's viscous instability (the DIM) were published in *Publ. Astron. Soc. Japan*, Vol. 26, p429, 1974. The 'S-curve' in Fig 11.8 was discussed by J. Smak in *Publ. Astron. Soc. Pacific*, Vol. 96, p5, 1984; where one can find a useful review of the observational evidence on dwarf novae and relevant accretion disk theory. This includes a general explanation of how instability in the accretion disk can produce the outbursts of dwarf novae. Smak (op. cit.) also ventured that other mechanisms than the basic DIM can have a part to play in certain kinds of explosive phenomena. In particular, super-outbursts in SU UMa variables could be triggered by sudden mass-transfer bursts from the cool companion. Such instabilities were discussed by G. T. Bath, *Mon. Not. Royal Astron. Soc.*, Vol. 171, p311, 1975; J. C. B. Papaloizou & G. T. Bath, *Mon. Not. Royal Astron. Soc.*, Vol. 172, p339, 1975; G. T. Bath & J. E. Pringle, *Mon. Not. Royal Astron. Soc.*, Vol. 194, p967, 1981; G. T. Bath & J. E. Pringle, *Interacting Binary Stars*, eds. J. E. Pringle & R. A. Wade. CUP, p17, 1985, D. R. Skillman, D. Harvey, J. Patterson, T. Vanmunster, *Publ. Astron. Soc. Pacific*, Vol. 109, p114, 1997, and others. Modelling of the eruptions of dwarf-novae and low-mass X-ray binaries was reviewed by J-P. Lasota in arXiv:astro-ph/0102072v1, 2001. Further background on the various explanatory ideas about CVs and their outburst phenomena were given by J. K. Cannizzo in *Accretion Disks in Compact Stellar Systems*, ed. J. Wheeler (World Scientific, Singapore), p6, 1993.

The background on relaxation oscillator behaviour, as considered in Subsection 11.2.3.1, is usually associated with electronics. Similar equations to (11.23-26) are found, for example, in S. Varigonda & T. T. Georgiou, *IEEE Trans. Auto. Control*, Vol. 46, p65, 2001. An interpretation of the general relationship between the SU UMa systems and other dwarf novae was published by J. van Paradijs in *Astron. Astrophys.*, Vol. 125, L16, 1983.

The class of CV identified as polars (and intermediate polars), associated with systems like AM Her, was recognized by a few groups around 1977. The field is comprehensively reviewed by M. Cropper in *Space Sci. Rev.*, Vol. 54, p195, 1990. and surveyed by M. L. Pretorius, C. Knigge & A. D. Schwope, *Mon. Not. Royal Astron. Soc.*, Vol. 432, p570, 2013. An example of the use of eclipse effects to constrain modelling of the polar EP Dra was given by C. M. Bridge, M. Cropper, G. Ramsay, J. H. J. de Bruijne, A. P. Reynolds & M. A. C. Perryman in *Mon. Not. Royal Astron. Soc.*, Vol. 341, p863, 2003. The role of accretion inhomogeneities (*blobs*) in explaining the X-ray light curves of AM Her type objects, was presented by J. M. Hameury & A. R. King, in *Mon. Not. Royal Astron. Soc.*, Vol. 235, p433, 1988. The effects of magnetic braking on CV evolution were considered by D. T. Wickramasinghe, J. Li & K. Wu, 1996. *Publ. Astron. Soc. Australia*, Vol. 13, p81, 1996.

AE Aqr was first detected as a radio source by J. A. Bookbinder & D. Q. Lamb using the Very Large Array (VLA) at 1.4 and 4.9 GHz, as reported in *Astrophys. J.*, Vol. 323, L131, 1987. The flux was associated with synchrotron emission from mildly relativistic electrons, powered by the magnetic coupling of a strongly magnetized white dwarf primary to the secondary star, or an accretion disk.

The illustration shown in Fig 11.10 comes from T. S. Bastian, A. J. Beasley & J. A. Bookbinder, in *Astrophys. J.*, Vol. 461, p1016, 1996. A model for the system was produced by G. A. Wynn, A. R. King & K. Horne in *Mon. Not. Royal Astron. Soc.*, Vol. 286, p436, 1997. Although the emission is relatively intense for a stellar source and strongly variable on a wide range of time-scales, the authors did not confirm the presence of the quasi-periodic oscillations or coherent pulsations at 3.6 cm as observed at optical, UV, and soft X-ray wavelengths, nor was a direct correlation established between flaring activity and orbital phase.

11.3 X-ray binaries (XRBs) (Section 11.3) were reviewed by J. Casares, P. G. Jonker, & G. Israelian, in the *Handbook of Supernovae*, (Springer), p1499, 2017; giving particular regard to their evolutionary relationship to the supernovae of Type IbIc and Type II. The apparent gap in the mass distribution of X-ray binaries between about 2 and 5 $M_\odot$ was noted, together with thoughts on how this might bear on the mechanisms of supernova explosions.

R. H. Fowler's pioneering paper on the physics of dense matter was published in in *Mon. Not. Royal Astron. Soc.*, Vol. 87, p114, 1926. It refers to the statistics of the distribution of particle energies in available quantum states in extreme conditions. E. Fermi and P. Dirac considered this distribution for an accumulation of particles that obey the Pauli exclusion principle of quantum physics, according to which no two particles can occupy the same cell of phase-space. This Fermi-Dirac (F-D) formulation is particularly relevant to the behaviour of electrons in conditions of high temperature and pressure. It causes the matter involved to depart, or 'degenerate', from the Maxwell-Boltzmann description of gases in more frequently encountered conditions. The branch of quantum physics that deals with F-D statistics, applying to fermions, presents also the alternative Bose-Einstein statistics, that concern bosons: particles such as photons that do not follow the Pauli exclusion principle.

The relativistic correction to Fowler's pressure-density relation was derived by S. Chandrasekhar in *Astrophys. J.*, Vol. 74, p81, 1931. This classic paper also finds the upper limit to the the mass of a white dwarf that can remain stable and conform to the physical principles presented.

A very comprehensive review of the formation and evolution of collapsed stars in binary systems including white dwarfs, neutron stars and black holes was given by K. A. Postnov & L. R. Yungel'son in *Living Rev. Relativity*, Vol. 17, No. 3, 2014 (with references to almost 900 publications). The review included relevant observational data, giving special attention to the merger rates of compact binary components that would have particular relevance to the forthcoming science of gravitational wave astronomy. The relationship of compact objects (COs) to supernovae and the cosmological significance of such events were also discussed.

X-ray binaries with high mass (HMXBs) were reviewed by E. P. J. van den Heuvel, *Proc. IAU Symp. 346*, 2019; available as arXiv:1901.06939. The review discussed various evolutionary scenarios involving COs, including conditions that could lead to black hole (BH) formation or alternative end-products. The detection of the strong X-ray source Cygnus X-1 in an early space exploration programme preceded by some years its stellar identification and the proposal by C. T. Bolton that it may contain a BH (cf. *Nature*, Vol. 235, p271, 1972). The system was reviewed by J. Ziołkowski in *Mon. Not. Royal Astron. Soc.*, Vol.

440, L61, 2014. The formation and evolution of double neutron star systems were discussed by T. M. Tauris et al., (14 authors) in *Astrophys. J.*, Vol. 846, p170, 2017.

On the basis of an observed correlation between radio and X-ray fluxes, a unified description of jet-accretion coupling in high energy sources involving black holes was proposed by E. Gallo, R. Fender, T. Maccarone & P. Jonker, in *Prog. Theoret. Phys. Suppl.*, No. 155, p83, 2004. The model is substantiated by reference to jets perpendicular to Cygnus X1's accretion disk carrying away part of the energy of the infalling material into surrounding space, as reported by E. Gallo & R. Fender in *Mem. Soc. Astron. Ital.*, Vol. 76, p600, 2005; (see also Gallo et al., *Nature*, Vol. 436, p819, 2005). The absolute parameters of Vela X1, the prototype HMBX, were given by H. Quaintrell, A. J. Norton, T. D. C. Ash, P. Roche, B. Willems, T. R. Bedding, I. K. Baldry & R. P. Fender in *Astron. Astrophys.*, Vol. 401, p313, 2003. The observed flaring and high variability of Vela X-1, as shown in Fig 11.12, was published by I. Kreykenbohm, J. Wilms , P. Kretschmar, J. M. Torrejón, K. Pottschmidt, M. Hanke, A. Santangelo, C. Ferrigno, & R. Staubert, in *Astron. Astrophys.*, Vol. 492, p511, 2008.

The discovery of what are now called (radio) pulsars was announced by A. Hewish, S. J. Bell, J. D. H. Pilkington, P. F. Scott & R. A. Collins, in *Nature*, Vol. 217, p709, 1968. For a comprehensive review with an emphasis on observational techniques one may consult the *Handbook of Pulsar Astronomy*, by D. R. Lorimer & M. Kramer, CUP, 2004. With more of an emphasis on the physical problems posed by pulsars there is W. Becker's collection of reference papers in *Neutron Stars and Pulsars: Astrophys. Space Sci. Lib.*, (Springer) No. 357, 2009;

The propeller model for the spin behaviour in binary pulsars was published by A. P. Illarionov & R. A. Sunyaev in *Astron. Astrophys.*, Vol. 39, p185, 1975. The slowing down of binary pulsar spin rates was further considered by R. E. Davies & J. E. Pringle, *Mon. Not. Royal Astron. Soc.*, Vol. 196, p209, 1981. X-ray pulsations from neutron stars with misaligned rotational and orbital axes were modelled by M. M. Basko & R. A. Sunyaev in *Astron. Astrophys.*, Vol. 42, p311, 1975. X-ray pulsar accretion was reconsidered more recently by K. Parfrey, A. Spitkovsky & A. M. Beloborodov, in *Mon. Not. Royal Astron. Soc.*, Vol. 469, p3656, 2017; and M. Sugizaki, T. Mihara, M. Nakajima, & K. Makishima, in *Publ. Astron. Soc. Japan*, Vol. 69, p100, 2017. Spin-down and spin-up episodes in the BeXRB GX 304-1 were analysed by K. A. Postnov, A. I. Mironov, A. A. Lutovinov, N. I. Shakura, A. Yu. Kochetkova & S. S. Tsygankov, including possible mechanisms of angular momentum transfer, in *Mon. Not. Royal Astron. Soc.*, Vol. 446, p1013, 2015. Constraints on NS properties from X-ray observations of millisec pulsars were produced by S. Bogdanov, G. B. Rybicki & J. E. Grindlay, *Astrophys. J.*, Vol. 670 p668, 2007. Relevant topics, including details of emission processes, were also discussed by W. Nagel, *Astrophys. J.*, Vol. 251, p288, 1981; P. Meszaros & W. Nagel, *Astrophys. J.*, Vol. 299, p138, 1985; U. Kraus, H-P. Nollert, H. Ruder & H. Riffert, *Astrophys. J.*, Vol. 450, p763, 1995; J. E. Trümper, K. Dennerl, N. D. Kylafis, Ü. Ertan, & A. Zezas, *Astrophys. J.*, Vol. 764, p49, 2013.

Further information on the INTEGRAL programme can be found in C. Winkler et al. (15 authors), *Astron. Astrophys.*, Vol. 411, L349, 2003; and a summary of the programme's achievements was given by C. Winkler, R. Diehl, P. Ubertini & J. Wilms, in *Space Sci. Rev.*, Vol. 161, p149, 2011. The number of HMXBs with supergiant companions (SgXRBs) tripled as a result of this programme (cf. R. Krivonos, S. Tsygankov, A. Lutovinov, M. Revnivtsev, E. Churazov, R. Sunyaev, *Astron. Astrophys.*, Vol. 545, p27, 2012; A. J. Bird et al. (9 authors), *Astrophys. J. Suppl. Ser.*, Vol. 223, p15, 2016). As well, the so-called highly obscured sources came to light (the first was Igr J16318-4848, see T. J-L. Courvoisier, R. Walter, J. Rodriguez L. Bouchet & A. A. Lutovinov, *IAU Circ.* No. 8063, 2003), as well as the Supergiant Fast X-ray Transients (SFXTs); (see V. Sguera et al. 12 authors, *Astron. Astrophys.*, Vol. 444, p221, 2005; I. Negueruela, D. M. Smith, P. Reig, S. Chaty, J. M. Torrejón, in *Proc. The X-ray Universe (ESA SP-604)*, ed. A. Wilson, p165, 2006).

These new INTEGRAL discoveries do not seem to fit standard HMXB classification criteria, showing, for example, extreme flaring activity on time-scales of a few hours. The perplexing SFXT sources were reviewed by L. Sidoli in *Proc. XII Multifrequency Behaviour of High Energy Cosmic Sources Workshop*, arXiv:1710.03943, 2017. L. Ducci, V. Doroshenko, P. Romano, A. Santangelo & M. Sasaki *Astron. Astrophys.*, Vol. 568, p76, 2014, estimated that there should be about 40 SFXTs in the Galaxy, but observations are hindered by various circumstantial factors. The external source IC10X-2 was discussed by S. Laycock in *15 Years of Science with Chandra*, Chandra Science Symposium 2014, p40. The topic was reviewed by R. Walter, A. A. Lutovinov, E. Bozzo & S. S. Tsygankov, *Astron. Astrophys. Rev.*, Vol. 23(2), 2015.

'Clumpiness' in a stellar wind was considered by S. P. Owocki & D. H. Cohen, *Astrophys. J.*, Vol. 648, p565, 2006; L. Oskinova, W-R. Hamann, H. Todt & A. Sander, *Astron. Soc. Pacific Conf. Ser.*, Vol. 465, p172, 2012; E. Bozzo, L. Oskinova, A. Feldmeier & M. Falanga, *Astron. Astrophys.*, Vol. 589, p102, 2016. The idea of 'magnetic gating' was discussed by E. Bozzo, M. Falanga & L. Stella, *Astrophys. J.*, Vol. 683, p1031, 2008; C. D'Angelo, R. Caroline & H. C. Spruit, *Mon. Not. Royal Astron. Soc.*, Vol. 406, p1208, 2010; while the 'settling' of quasi-spherical accretion was modelled by N. Shakura, K. Postnov, A. Kochetkova & L. Hjalmarsdotter, *Mon. Not. Royal Astron. Soc.*, Vol. 420, p216, 2010. SgXRB and SXFT properties were compared and their possible relationship studied by P. Pradhan, B. Paul & E. Bozzo, in a study available as arXiv:1711.10510 (2017).

The BeXRB X Per is identified with the pulsar 4U 0352+30: a recent discussion of its behaviour and modelling was given by R. O. Brown, W. C. G. Ho, M. J. Coe & A. T. Okazaki in *Mon. Not. Royal Astron. Soc.*, Vol. 477, p4810, 2018. A. T. Okazaki, K. Hayasaki, & Y. Moritani accounted for different types of X-ray outburst in BeXRBs, referring to accretion scenarios *Publ. Astron. Soc., Japan*, Vol. 65, p410, 2013. General properties of Be stars were reviewed by J. M. Porter & T. Rivinius in *Publ. Astron. Soc. Pacific*, Vol. 115, p1153, 2003.

HMXBs in the SMC were catalogued by F. Haberl & R. Sturm in *Astron. Astrophys.*, Vol. 586, A81, 2016. Detailed analysis of optical photometry for 51 Small Magellanic Cloud Be + X-ray pulsars with the aim of better understanding their nature was carried out by P. C. Schmidtke, A. P. Cowley & A. Udalski in *Mon. Not. Royal Astron. Soc.*, Vol. 431, p252, 2013. A similar study was made by A. J. Bird, M. J. Coe, V. A. McBride & A. Udalski, *Mon. Not. Royal Astron. Soc.*, Vol. 423, p3663, 2012.

The occurrence of stellar black holes in HMXBs was reviewed by J. Casares, I. Negueruela, M. Ribó, I. Ribas, J. M. Paredes, A. Herrero & S. Simón-Díaz, with particular attention to the case of HD 215227, in *Nature*, Vol. 505, p378, 2014; see also E. P. van den Heuvel, in *Proc. IAU Symp. No. 346*, eds. L. M. Oskinova, E. Bozzo, T. Bulik, D. Gies, (CUP) p1, 2019. The emission of γ rays from HMXBs relating to stellar mass BH candidates is actively pursued by international co-operative organizations like the Major Atmospheric Gamma Imaging Cherenkov telescopes (MAGIC) collaboration and the Cherenkov Telescope Array (CTA) consortium.

The book of P. S. Conti, P. A. Crowther & C. Leitherer, *From Luminous Hot Stars to Starburst Galaxies*, CUP, 2008; provides a detailed introduction to the physics of Wolf-Rayet (WR), O and B type stars and their role in shaping the galactic environment. A ULXB discovered in the Circinus Galaxy (GC X-1) was reported by P. Esposito, G. L. Israel, D. Milisavljevic, M. Mapelli, L. Zampieri, L. Sidoli, G. Fabbiano & G. A. Rodriguez Castillo in *Mon. Not. Royal Astron. Soc.*, Vol. 452, p1112, 2015. The authors showed that its properties are probably consistent with a WR + BH binary. From their results they could estimate an upper limit to the detection rate of stellar BH-BH mergers of ~ 16 yr^{-1}.

S. Carpano, F. Haberl, P. Crowther & A. Pollock, in *Proc. IAU Symp. No. 346* p187 (see above), presented NGC 300 X-1 and IC 10 X-1 as the only two robust extragalactic candidates for WR + BH X-ray binaries. For NGC 300 X-1 the authors were able to piece together phased X-ray photometric and radial velocity curves, thence deriving a clearer physical picture of the system. For a recent discussion of microquasars see R. Soria, W. P. Blair, S. Long, T. D. Russell & F. P. Winkler, in *Astrophys. J.*, Vol. 888, p103, 2020. Other reviews were given by M. Bachetti, *Astron. Nachr.*, Vol. 337, p34, 2016; and P. Kaaret, H. Feng & T. P. Roberts, *Ann. Rev. Astron. Astrophys.*, Vol 55, p303, 2017.

ULXBs were originally thought to be rare instances of intermediate-mass BHs accreting at sub-Eddington rates (see Colbert and Mushotzky 1999; Makishima et al. 2000). Alternative models, involving super-Eddington accretion and beaming have been discussed by J. Poutanen, G. Lipunova, S. Fabrika, A. G. Butkevich & P. Abolmasov, *Mon. Not. Royal Astron. Soc.*, Vol. 377, p1187, 2007; A. R. King, *Mon. Not. Royal Astron. Soc.*, Vol. 393, p41, 2009; and others. In one instance, the CO componenet was shown to be a highly magnetised neutron star (see M. Bachetti et al. — 24 authors, *Nature*, Vol. 514, p202, 2014). Among the many follow-up studies in this challenging field are those of Fürst et al. (15 authors), *Astrophys. J.*, Vol. 831, L14, 2016; G. Israel et al. (10 authors), *The X-ray Universe: Conf. Proc.*, eds. J-U. Ness & S. Migliari, online at https://www.cosmos.esa.int/web/xmm-newton/2017-symposium, p104, 2017; S. Carpano, F. Haberl, C. Maitra, *Astron. Tel.*, No. 11158, 2018. For comparisons with BeXBs, see F. Koliopanos, G. Vasilopoulos, O. Godet, M. Bachetti, N. A. Webb, D. Barret, *Astron. Astrophys.*, Vol. 608, p47, 2017; D. J. Walton et al. (17 authors), *Astrophys. J.*, Vol. 857, L3, 2018; G. Vasilopoulos, F. Haberl, S. Carpano & C. Maitra, *Astron. Astrophys.*, Vol. 620, L12, 2018. The subject remains problematic and challenging to physical understanding.

The subject included under the heading low mass X-ray binaries (LMXBs) goes right back to the dawn of X-ray astronomy, with the discovery of Sco X1 in an Aerobee space exploratory mission by R. Giacconi, H. Gursky, F. R. Paolini, B. B. Rossi, *Phys. Rev. Lett.*, Vol. 9, p439, 1962.

Regarding X-ray binaries of lower mass, background on their properties and formation scenarios were given by V. Kalogera & R. F. Webbink, *Astrophys. J.*, Vol. 458, p301, 1996; and *Astrophys. J.*, Vol. 493, p351, 1998; see also W. H. G. Lewin, J. van Paradijs & R. E. Taam's comprehensive account in *Space Sci. Rev.*, Vol. 62, p223, 1993. Dynamical evidence for a BH in the X-ray transient QZ Vul was presented by J. Casares, P. A. Charles & T. R. Marsh, *Mon. Not. Royal Astron. Soc.*, Vol. 277, L45, 1995. Supersoft X-ray sources were reviewed by P. Kahabka & E. P. J. van den Heuvel in *Ann. Rev. Astron. Astrophys.*, Vol. 35, p69, 1997. Insight into their nature was afforded by analysis of RX J0513-69 by K. A. Southwell, M. Livio, P. A. Charles, D. O'Donoghue, W. J. Sutherland, *Astrophys. J.*, Vol. 470, p1065, 1996; see also M. Kato, I. Hachisu, A. Cassatella, *Astrophys. J.*, Vol. 704, p1676, 2009.

A formation model for BHLMXBs, utilizing angular-momentum-loss mechanisms with gravitational radiation and magnetic braking, was presented by F. Verbunt & C. Zwaan, *Astron. Astrophys.*, Vol. 100, L7, 1981; while S. Rappaport, P. C. Joss & R. F. Webbink, *Astrophys. J.*, Vol. 254, p616, 1982; gave attention to their systemic mass loss. Observed LMXBs with suspected BH components, for which there are a few dozen candidates, were reviewed by R. A. Remillard, & J. E. McClintock, including the topic of BH spin determination, in *Ann. Rev. Astron. Astrophys.*, Vol. 44, p49, 2006. Intermediate mass X-ray binaries were discussed by J. Casares & P. G. Jonker in *Space Sci. Rev.*, Vol. 183, p223, 2014.

The possibility of a common envelope phase in XB progenitors was considered by B. Paczyński in *Comm. Astrophys.*, Vol. 6, p95, 1976. This topic was elaborated on by E. P. J. van den Heuvel, in various contributions to the conference on *Accretion-driven Stellar X-ray Sources*, (CUP), eds. W. H. G. Lewin & E. P. J. van den Heuvel, (e.g. p303), 1983; D. Bhattacharya & E. P. J. van den Heuvel, *Phys. Rep.*, Vol. 203, p 1, 1991; T. M. Tauris & E. P. J. van den Heuvel, *Compact Stellar X-ray Sources*, eds. W. Lewin & M. van der

Klis, (CUP), p629, 2006; and more recently reviewed by X-D. Li, in *New Astron. Rev.*, Vol. 64, p1, 2015.

Points of contention, particularly relating to the incidence of BHLMXBs, were raised by S. F. Portegies Zwart, F. Verbunt & E. Ergma, *Astron. Astrophys.*, Vol. 321, p207, 1997; and Ph. Podsiadlowski, S. Rappaport & Z. Han, *Mon. Not. Royal Astron. Soc.*, Vol. 341, p385, 2003. Explanations on how the differences between observations and theoretical predictions might be resolved have been offered by V. Kalogera, *Astrophys. J.*, Vol. 521 p723, 1999; L. R. Yungel'son & J-P. Lasota, *Astron. Astrophys.*, Vol. 488, p257, 2008; P. D. Kiel & J. R. Hurley *Mon. Not. Royal Astron. Soc.*, Vol. 369, p1152, 2006; K. A. Postnov & L. R. Yungel'son, *Living Rev. Relativity.* Vol. 17, No. 3, 2014.

The case of high magnetic braking was examined by S. Justham, S. Rappaport & Ph. Podsiadlowski, in *Mon. Not. Royal Astron. Soc.*, Vol. 366, p1415, 2006; and the role of a circumbinary disk by W-C. Chen, & X-D. Li, *Mon. Not. Royal Astron. Soc.*, Vol. 373, p305, 2006. The 'failed supernova' mechanism was published by C. Wang, J. Kun & X-D. Li, in *Mon. Not. Royal Astron. Soc.*, Vol. 457, p1015, 2016.

High rates of period decline in BHLMXBs have been reported by J. I. González-Hernández, R. Rebolo & J. Casares, *Mon. Not. Royal Astron. Soc.*, Vol. 438, L21, 2014; and J. I. González-Hernández, L. Suárez-Andrés, R. Rebolo & J. Casares, *Mon. Not. Royal Astron. Soc.*, Vol. 465, L15, 2017. The circumbinary disk + common envelope interpretation for such effects was comprehensively reviewed by N. Ivanova et al. (19 authors), in *Astron. Astrophys. Rev.*, Vol. 21, p59, 2013. The subject was also reviewed (in the context of CVs) by H. C. Spruit & R. E. Taam (2001, op. cit.), (see also C. de Loore, J. P. de Grève, E. P. J. van den Heuvel & J. P. de Cuyper, *Mem. Soc. Astron. Ital.*, Vol. 45, p893, 1974; and E. J. P. van den Heuvel, *22 Saas-Fee Advanced Course Swiss Soc. Astrophys. Astron.: Interacting binaries*, p263, 1994). Feeding of a circumbinary disk via mass loss during the outbursts of BHLMXBs has been modelled by X-T. Xu & X-D. Li, *Astrophys. J.*, Vol. 859, p46, 2018.

Thorne-Żytkow objects were introduced by K. S. Thorne & A. N. Żytkow in *Astrophys. J.*, Vol. 212, p832, 1977. The homogeneous evolution model was discussed by S. E. de Mink, M. Cantiello, N. Langer, O. R. Pols & S-Ch. Yoon in *Binaries — Key to Comprehension of the Universe: Astron. Soc. Pacific Conf. Ser.*, eds. A. Prša & M. Zejda, Vol. 435, p179, 2010.

11.4 GW-astronomy (Section 1.4) is still in its infancy at the time of writing. The September-2014 detection of GWs by the Laser Interferometer Gravitational-wave Observatory (LIGO), attributed to a BH merger, was the remarkable event leading to the Nobel Prize in physics being awarded to R. Weiss, K. Thorne & B. Barish, in 2017; see also B. P. Abbott et al., (LIGO and Virgo consortia), *Phys. Rev. Lett.*, 116:061102. The Nobel lecture delivered by K. Thorne can be seen at https://www.nobelprize.org/prizes/physics/2017/thorne/lecture/. A catalogue of GW transients (GWTC-1) was released by B. P. Abbott et al. (1151 authors) in *Phys. Rev. X9*, 031040, 2019.

The general field was recently reviewed by P. Schmidt in *Front. Astron. Space Sci.*, 7:28. doi: 10.3389/fspas.2020.00028, who provides an extensive list of references. J. Wheeler's lecture notes on GWs, available from http://www.physics.usu.edu/Wheeler/GenRel2013/Notes/GravitationalWaves.pdf, give a helpful backgrounder to the subject. An interesting reference on analytical modelling of the complete in-spiral, merger, ringdown sequence is that of S. T. McWilliams *arXiv* 1810.00040V2, 2019. For examples of the frequency domain transform of the GW in-spiral, merger, ringdown sequence see e.g. C. García-Quirós et al. (8 authors), arXiv:2001.10914v2; 2020.

The subsequent detection of the collision of two neutron stars with both electromagnetic and GW observations inaugurated the era of *multimessenger astrophysics*. D. George & E. A. Huerta, *Phys. Lett. B*, Vol. 778, p64, 2018 have applied 'deep learning' data-handling techniques to enable rapid detection and characterization of GWs. Their approach appears well suited to coincident detection campaigns using GWs and multimessenger observations.

Insights into the source binary population requirements giving rise to the detection of GW events were provided by J. J. Eldridge & E. R. Stanway in *Mon. Not. Royal Astron. Soc.*, Vol. 462, p3302, 2016. Eldridge and Stanway inferred relatively high masses and a low metallicity for the progenitor stars of GW 150914, discussing also the implications of their analysis for multimessenger astrophysics.

Index